Advances in
ORGANOMETALLIC CHEMISTRY

VOLUME 32

Advances in Organometallic Chemistry

EDITED BY

F. G. A. STONE

DEPARTMENT OF CHEMISTRY
BAYLOR UNIVERSITY
WACO, TEXAS

ROBERT WEST

DEPARTMENT OF CHEMISTRY
UNIVERSITY OF WISCONSIN
MADISON, WISCONSIN

VOLUME 32

ACADEMIC PRESS, INC.
Harcourt Brace Jovanovich, Publishers
San Diego   New York   Boston
London   Sydney   Tokyo   Toronto

Academic Press, Inc.
San Diego, California 92101

United Kingdom Edition published by
ACADEMIC PRESS LIMITED
24-28 Oval Road, London NW1 7DX

Library of Congress Catalog Card Number: 64-16030

ISBN 0-12-031132-1 (alk. paper)

PRINTED IN THE UNITED STATES OF AMERICA
91 92 93 94   9 8 7 6 5 4 3 2 1

Contents

Formation and Reactions of Organosulfur and Organoselenium Organometallic Compounds

LORNA LINFORD and HELGARD G. RAUBENHEIMER

Hydroformylation Catalyzed by Ruthenium Complexes

PHILIPPE KALCK, YOLANDE PERES, and JEAN JENCK

X-Ray Structural Analyses of Organomagnesium Compounds

PETER R. MARKIES, OTTO S. AKKERMAN, FRIEDRICH BICKELHAUPT, WILBERTH J.J. SMEETS, and ANTHONY L. SPEK

Recent Advances in the Chemistry of Metal–Carbon Triple Bonds

ANDREAS MAYR and HANS HOFFMEISTER

Chemistry of Cationic Dicyclopentadienyl Group 4 Metal-Alkyl Complexes

RICHARD F. JORDAN

ADVANCES IN ORGANOMETALLIC CHEMISTRY, VOL. 32

Formation and Reactions of Organosulfur and Organoselenium Organometallic Compounds

LORNA LINFORD and HELGARD G. RAUBENHEIMER

Department of Chemistry
Rand Afrikaans University
Johannesburg 2000, Republic of South Africa

I

INTRODUCTION

In organic synthetic reactions the scope and applications of organosulfur and organoselenium chemistry have increased tremendously. As explained

1

recently by Zwanenburg and Klunder, "There are numerous cases, for example in modern organic chemistry, in which sulfur-containing groups serve an important auxiliary function in synthetic sequences. Illustrative examples are the reversal of polarity (Umpolung), enhancement of the acidity of C—H bonds and the transfer of chirality from sulfur to carbon" (1). In contrast with many other fields in chemistry, organosulfur complex chemistry has not yet had the stimulus of industrial applications, and interest in the field stems mainly from the curiosity of chemists. This article systematizes the formation and transformation reactions of free and coordinated organosulfur and -selenium complexes in the coordination sphere of metals and shows that transition metals increase the variety of reactions undergone by organosulfur compounds significantly.

Organosulfur complexes of low oxidation state metals are generally less stable than their phosphorus counterparts, but this relative instability lends the complexes their reactivity; organosulfur ligands are usually the reactive centers of the complexes. They can undergo modification, whereas phosphorus ligands hardly ever take part in a reaction and are mainly employed to stabilize organometallic complexes. Selenium compounds have the disadvantage that they are extremely toxic, which may account for their relative scarcity. They usually undergo the same types of reactions as the sulfur analogs.

Although we do not concentrate on the final products formed, but rather on the reactions that lead to these products, the versatility of sulfur-donor groups cannot be overlooked, and examples in which sulfur ligands function as $1e^-$ to $6e^-$ donors are available. Our main interest is the modification of organosulfur or -selenium ligands, and, therefore, simple Lewis adduct formation or substitution reactions between sulfur or selenium compounds and metal complexes are excluded. Reactions resulting in the formation of thiolate complexes are not discussed, and syntheses of organosulfur or -selenium compounds using organometallic catalysts are not included if the metal-containing organosulfur or -selenium species was not isolated and characterized. The few reactions we encountered which involved tellurium reagents are included although they are not mentioned in headings.

Review articles which have appeared in this field have dealt mainly with specific types of compounds, for example, coordinated thioethers (2), thiocarbonyls (3), dithiocarbenes (4), CS_2 complexes (5), heterometallacycles (6), and iron complexes (7). This article covers comprehensively the reactions of organosulfur and organoselenium compounds mediated by transition metals, and material included in earlier reviews is not dealt with in detail.

II

REARRANGEMENTS OF FREE AND COORDINATED ORGANOSULFUR AND ORGANOSELENIUM REACTANTS IN THE COORDINATION SPHERE OF TRANSITION METALS

In this section we describe reactions in which chemical bond breaking and atom migration in organosulfur and -selenium compounds are effected by the addition of metal complexes without the interference of other ligands. We also discuss internal rearrangements of complexes containing organosulfur ligands which occur spontaneously or are initiated by thermal, photochemical, or electrochemical excitation.

A. Thiones

1. Reaction with Iron Carbonyl Complexes

The thione ligand, $S=C(X)Y$, is modified during formation of an adduct with $[Fe(CO)_5]$ or $[Fe_2(CO)_9]$, and (i) cleavage of a $C-X$ or $C-Y$ bond, (ii) cleavage of the $C=S$ double bond (desulfurization), or (iii) formal cleavage of a π $C-S$ bond (leaving the ligand framework intact) can occur. Mechanisms for the reactions have not been studied in detail. The metal carbonyls are activated by UV irradiation, $[Fe(CO)_5]$, or thermally, $[Fe_2(CO)_9]$, and products are mostly dinuclear, although some trinuclear products have been isolated. The $(CO)_3Fe-Fe(CO)_3$ fragment, which appears in many products, needs six electrons to comply with the $18e^-$ rule and the open triangular fragment, $(CO)_3Fe-Fe(CO)_3-Fe(CO)_3$, eight electrons. Thiones carrying aromatic substituents can also react by hydrogen migration, affording cyclometallated dinuclear iron complexes. The products from these reactions are listed in Table I.

Various subclasses of type 1 complexes are known. The intact ligand functions as a $6e^-$ donor. The type 1e complex 11 transforms, by cleavage

TABLE I

PRODUCT TYPES FROM REACTIONS BETWEEN IRON CARBONYLS AND VARIOUS THIONES, $S=C(X)Y$

Type	X, Y	Carbonyl used	Ref.
1			
(a)	$X = NMe_2$, Me, or Ph	$[Fe_2(CO)_9]$	8
	$Y = NMe_2$		
(b)	$X = Ph$, p-MeC$_6$H$_4$, m-MeC$_6$H$_4$,	$[Fe_2(CO)_9]$	9
	p-MeOC$_6$H$_4$, p-CF$_3$C$_6$H$_4$, or p-BrC$_6$H$_4$		
	$Y = OEt$, OMe, OCHMe$_2$, or		
	O-1-adamantyl		
(c)	$X = O(steroid)$	$[Fe_2(CO)_9]$	10
	$Y = SMe$ or SCH$_2$Ph		
(d)	$X = Pr^i$, cyclohexyl, Ph,	$[Fe_2(CO)_9]$	11
	p-MeOC$_6$H$_4$, p-Me$_2$NC$_6$H$_4$, p-MeC$_6$H$_4$,		
	p-BrC$_6$H$_4$, o-CH$_2$=CHC$_6$H$_4$, PhCH$_2$,		
	naphthyl, or SPh		
	$Y = OMe$		
(e)	$XY = S(CH_2)_2S$	$[Fe(CO)_5]$	12
(f)	$X = O$-adamantylmethyl	$[Fe_2(CO)_9]$	13
	$Y = SMe$		
(g)	$X = Me$ or Ph	$[Fe(CO)_5]$ or	14
	$Y = SMe$ or SEt	$[Fe_2(CO)_9]$	
(h)	$X = Me$ or Et	$[Fe_2(CO)_9]$	15
	$Y = OEt$		
(i)	$X = H$	$[Fe_2(CO)_9]$	10
	$Y = O$-cholestanyl or O-cholesteryl		
(j)	$X = H$	$[Fe_2(CO)_9]$	16
	$Y = SMe$, SEt, or SCH$_2$CH=CH$_2$		
(k)	$X = Ph$, CH$_2$Ph, p-MeOC$_6$H$_4$,	$[Fe_2(CO)_9]$	17
	p-Me$_2$NC$_6$H$_4$, p-BrC$_6$H$_4$, C$_4$H$_3$S,		
	α-naphthyl, Pr^i, or C$_6$H$_{11}$		
	$Y = SMe$		
2			
(a)	$X = OR^1 = O$-adamantylmethyl,	$[Fe_2(CO)_9]$	18
	O-cholestanyl, O-cholesteryl,		
	O-ergosteryl, or O-menthyl		
	$Y = SR^2$ with $R^2 = Me$, CH$_2$Ph, or Pr^i		

TABLE I (*continued*)

Type	X, Y	Carbonyl used	Ref.
3	**(b)** X = SCH(Me)F$_c$ (F$_c$ is ferrocenyl) or SCH$_2$F$_c$ Y = SMe	[Fe$_2$(CO)$_9$]	19
4	XY = S(CH$_2$)$_2$S	[Fe(CO)$_5$]	12
5	X = Ph Y = SMe	[Fe(CO)$_5$] and [Fe$_2$(CO)$_9$]	14
6	X = NMe$_2$ Y = SEt (in thione)	[Fe(CO)$_5$]	20
	X = Y = Ph	[Fe$_2$(CO)$_9$]	21

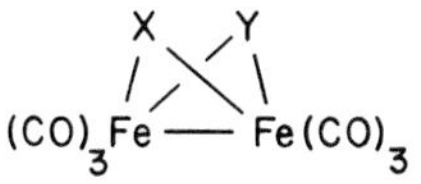

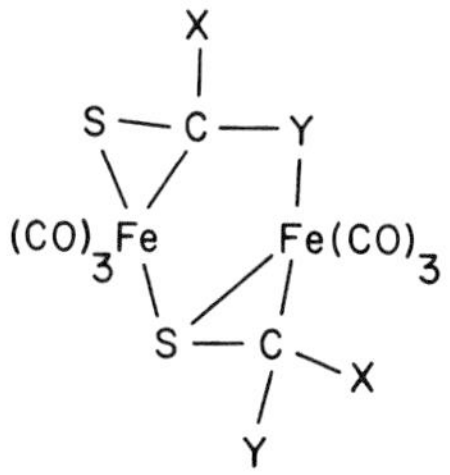

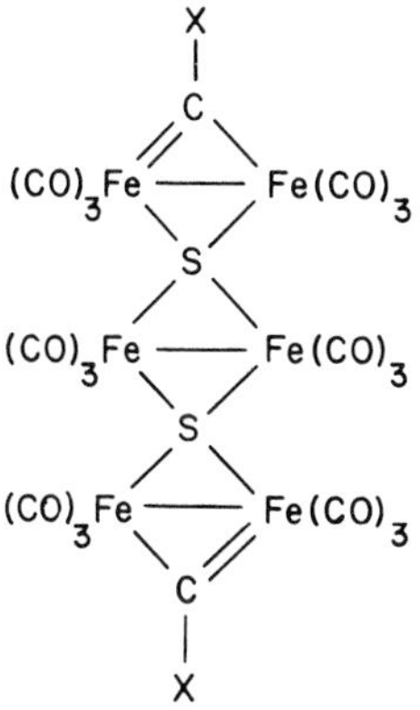

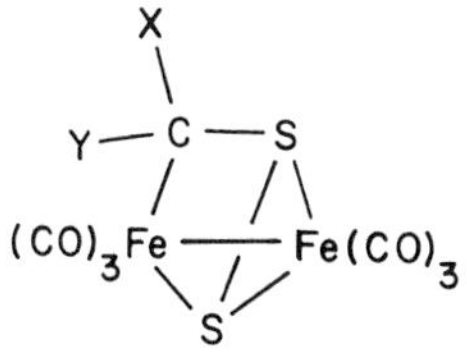

(*continued*)

TABLE I (continued)

Type	X, Y	Carbonyl used	Ref.
7		[Fe$_2$(CO)$_9$]	22
8			
	(a) XY = C(Ph)=CPh	[Fe$_2$(CO)$_9$]	23
	(b) XY = C(O)C(R^{1}R^2)C(R^{3}R^4)O, where R^1 and R^3 are, for example, H or Me and R^2 and R^4, Me or Ph	[Fe(CO)$_5$] and [Fe$_2$(CO)$_9$]	24
	(c) XY = SCH=CHS	[Fe$_2$(CO)$_9$]	25
9			
	XY = SCH$_2$CH$_2$S	[Fe$_2$(CO)$_9$]	12
10 (a)		[Fe$_2$(CO)$_9$]	21
	X = R^{1}C$_6$H$_4$ with R^1 = H, p-MeO, p-Me$_2$N, or p-CF$_3$ Y = R^{2}C$_6$H$_4$ with R^2 = H, p-MeO, p-Me$_2$N, or p-CF$_3$		
(b)		[Fe$_2$(CO)$_9$]	23

and reformation of a CS bond, into complex **12** (43% yield) which is a type **3** complex now containing a terminal dithiocarbene ligand in place of a carbonyl (*26*). Seyferth *et al.* also prepared a complex of type **1j**, with X being H and Y SR [R = Me, Et, or $CH_2C(O)Me$], by deprotonation and subsequent alkylation of $[(CO)_3Fe(SCH_2S)Fe(CO)_3]$ (*15,16,27*). During the preparation of type **1g** complexes, carbonyl insertion into the C—SMe or C—SEt bonds also occurs (*14*).

A σ and a π C—S bond are broken during the formation of type **2** compounds. The resulting fragments function as two $3e^-$ donor ligands. Recently Patin and co-workers found that thermal activation or catalysis by electron transfer can transform a type **1e** compound into a type **2** complex (*13*). This process involves not only rupture of a C—S bond, but also cleavage of an Fe—C σ bond, carbon shift, and formation of a metal–carbene bond with the neighboring iron atom (*28*).

Type **3** complexes are very common, and numerous examples are mentioned in *Gmelin's Handbuch der Anorganischen Chemie* (*29*). The reactions of the thiones $S{=}CR^1R^2$ (R^1 = Ph, R^2 = Ph, p-$MeOC_6H_4$) and $S{=}\overline{CC(Ph)}{=}CPh$, reported independently by Alper and Chan (*21*) and Weiss and co-workers (*23*), involve cleavage of one C—S bond and intermolecular sulfur transfer. The ligands are $6e^-$ donors. The latter reaction also involves an unusual coupling of two cyclopropenethione molecules.

The distribution of electron density in the sulfur-containing ligands is modified to allow the thiones to function as two unsymmetrically opposed $4e^-$ donors in the complex of type **4**. The compound has no metal–metal bond. Behrens and co-workers (*30*) as well as Seyferth *et al.* (*15*) prepared *symmetrical* complexes in which the ligands are bound similarly to the one drawn below the two iron atoms in **4**. In the Behrens compounds, prepared from thioketenes, the α-thio carbon atom is sp^2 hybridized in contrast to the sp^3 hybridization which occurs when the two iron atoms are held together by the O-thioformate groups in Seyferth's compound. The latter group proposed two completely different, competing pathways for the formation of complexes of types **1** and **3**.

A drastic fragmentation, in which both the C—Y single and the C=S double bonds of the thione $S{=}C(X)Y$ are cleaved, occurs during the formation of the highly symmetrical compound of type **5**. As drawn, the bridging carbons on the two butterfly wingtips are $3e^-$ donors. An X-ray structural study indicated significant double bond character in the C—N bonds, and, in another contributing structure, formal positive charges reside on the nitrogen atoms with negative charges on two irons. The fragment Y = SEt is not present in the major product but occurs in a lower yield, unsymmetrical product with the composition $\{(CO)_3Fe\}_5(NEt_2)$-(SEt)S (*20*).

A very unusual intermolecular sulfur transfer from one thione to the sulfur atom of another occurs during the formation of the type **6** complex. According to Alper and Chan, it is likely that a trithiolane or similar species is involved in the formation of this complex (*21*).

In the only known compound of type **7** both the $Fe_2(CO)_6$ fragments are spanned by $3e^-$ donor units of the intact ligand, $\overline{SC_6H_4\text{-}o\text{-}C(S)S}$, in which only an S—S bond has been cleaved. Part A belongs to type **3** complexes in which the α-thio carbon is substituted and thus linked to the other dinuclear iron fragment, B, which is a type **1** complex. One S atom serves as a bridge between the two $Fe_2(CO)_6$ units by acting as a $5e^-$ donor.

Photochemical desulfurization of thiocarbonates occurs with $[Fe(CO)_5]$ to give mononuclear as well as trinuclear (type **8**) carbene complexes (*24*). A similar reaction affords desulfurization of diphenylcyclopropenethione (*22*). The thioketene cluster **13** reacts quantitatively, but in a more complicated way, with a thioketene to form a dinuclear vinylidene complex (**14**).

In addition, the sulfur atom attaches itself to another α-thio carbon to form a 1,1-dithiolato ligand (see type **3**) (*31*).

The second open triangular iron system in Table I (type **9**) is formed from a trithiocarbonate by cleavage of a C=S double bond. A bridging carbene ligand and a sulfur atom α thereto donate four electrons to the metal framework. With the unsaturated derivative of the thione, $S=\overline{CSCH}=CHS$, a corresponding desulfurization with $[Fe_2(CO)_9]$ produces a nonbridging carbene ligand in a carbonyl-substituted derivative of $[Fe_3(CO)_9S_2]$ (compare with type **8**) (*25*).

Aromatic thiones undergo orthometallation with $[Fe_2(CO)_9]$. The ortho proton migrates to the α-thio carbon, and the ligand donates six electrons to the $Fe_2(CO)_6$ system (type **10a**). These Alper compounds can be oxidized (e.g., with Br_2) to give phenyl-substituted thiophenes (*21*). Sulfoxides, like $O=S=C(C_6H_4Me\text{-}p)_2$, carry a positive charge on the sulfur and do not orthometallate, but undergo deoxygenation (*32*). In the group of Weiss, treatment of diphenylcyclopropenethione with $[Fe_2(CO)_9]$ afforded a complex of type **10b** (*23*). This group previously isolated the complex $[C_2H_2C(O)SFe_2(CO)_6]$ with a similar structure (*33*). A similar type of complex, but containing a five-membered dimetalla chelate ring, was prepared by Schrauzer and Kisch (*34*) from a nonaromatic substrate.

2. *Reaction with Other Metal Complexes*

Reactions between thiones and metal carbonyls other than those of iron generally produce much simpler compounds than the complexes mentioned in the previous section. With thiobenzophenones the major reactions, namely, orthometallation, desulfurization, and further reaction of the orthometallated product, lead to compounds of types **A**, (*p*-

A

$RC_6H_4)_2C\!=\!C(C_6H_4R\text{-}p)_2$, **B**, and $(p\text{-}RC_6H_4)_2CH_2$, **C**. With manganese carbonyl, activating substituents, like NR^1_2 (R^1 = Me, Et), are necessary for the formation of type **A** and **C** complexes (*35*). Desulfurization and alkene formation from the resulting carbene, yielding **B**, occurs with moderately electron-donating substituents, as when R is OMe, H, Me, or F. Orthometallation results when these substrates and rhenium carbonyl are irradiated (for initial carbonyl substitution) and then heated (*36*). High yield orthometallation takes place between $[Ru_3(CO)_{12}]$ and 4,4′-dimethoxythiobenzophenone at 70°C to give a symmetrically bonded dimetallic product, **15** (*37*).

Alper accomplished the first sulfur-donor ligand orthometallation of palladium and platinum by reacting a thiobenzophenone, $S\!=\!C(C_6H_4R\text{-}p)_2$ (R = H, Me, OMe, or NMe_2), with sodium tetrachloropalladate or the platinum analog. Refluxing the former product mixture in methanol produces a sulfur–oxygen exchange with formation of aromatic ketones (*38*). Siedle (*39*) also reported a slow orthometallation of the adduct $[Pd(F_6acac)_2]$ with $S\!=\!C(C_6H_4OMe\text{-}p)_2$. Intramolecular orthopalladation of thiopivaloylferrocene **16** to give **17** occurs in satisfactory yield under ambient conditions (*40*).

15

16   17

Reactions of cobalt carbonyl with nonaromatic thiones differ significantly from the aforementioned conversions. S-Alkyl xanthates, for example, react by the rupture of C—R and C=S bonds to afford alkoxy
alkylidyne tricobalt nonacarbonyl clusters, $[RO—CCo_3(CO)_9]$ (41) (vide
infra). On the other hand, dithioesters, $R^1C(S)SR^2$, react with $[Co_2(CO)_8]$
to give alkylidyne tricobalt nonacarbonyl clusters in yields between 49 and
74%. The R groups include methyl, isopropyl, and phenyl (42).

Cyclic trithiocarbonates undergo oxidative addition with the complex
$[Pt(PPh_3)_4]$, resulting, by the cleavage of a C—S bond, in the unusual
heterometallacyclic compounds with structure 18 ($n = 2, 3$) (43).

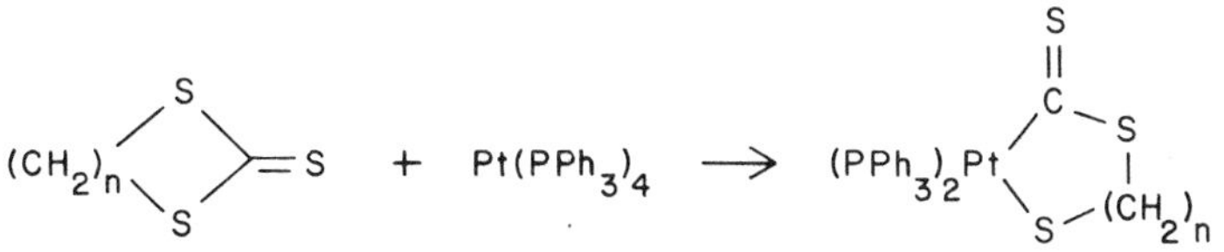

18

Reaction of the metal–metal triple bond in cyclopentadienyl molybdenum or tungsten compounds, $[(p\text{-}R^1C_5H_4)_2M_2(CO)_4]$ $(R^1 = H$ or Me), with thioketones, $S=CR^2R^3$ $(R^2 = R^3 = Ph$, $p\text{-}MeC_6H_4$, $p\text{-}FC_6H_4$, $p\text{-}Me_2NC_6H_4$; $R^2 = OMe$, $R^3 = H$; $R^2 = Bu^t$, $R^3 = $ ferrocenyl; $R^2 = p\text{-}Bu^tC_6H_4$, $R^3 = Bu^t)$ (44), thioesters, $R^2C(S)R^3$ $(R^2 = Ph$, $p\text{-}BrC_6H_4$, or $PhCH=CH$, $R^3 = OMe$ or OEt), thiolactones, $S=CR^2R^3$ $(R^2R^3 = OCH_2CH_2CH_2$ or $OCH_2CH_2CHMe)$ (45), and the dithioester $R^2C(S)R^3$ $(R^2 = Me$, $R^3 = SEt)$ (46) under ambient conditions gives complexes of the general type 19. This type of reaction was also successfully carried out

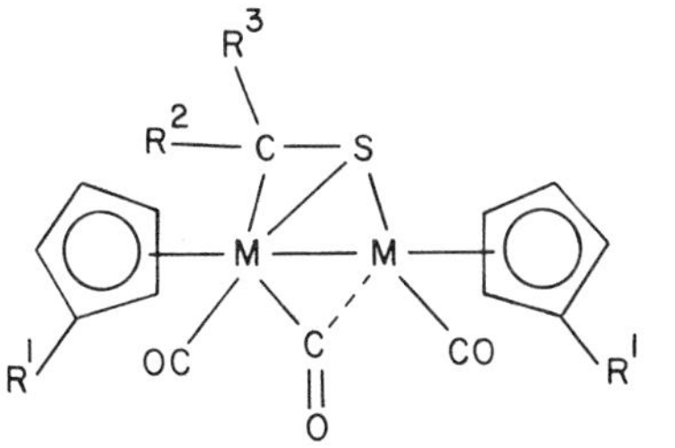

19

using the more complex thiones adamantanethione, d-thiocamphor, d-thiofenchone, $p\text{-}t$-butylthiopivalophenone, and thiopivaloylferrocene (44). In refluxing toluene the dithioesters $R^2C(S)SR^4$ $(R^2 = Me$ or CH_2Ph, $R^4 = Et$ or $p\text{-}MeOC_6H_4CH_2)$ lose their integrity, and complexes 20 $(R^1 = $

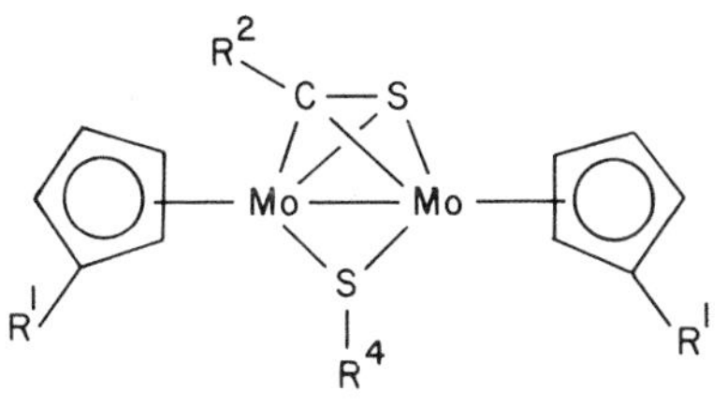

20

H) containing symmetrically bridging thioacyl and thioalkyl functions are obtained with molybdenum as the central metal (46). The dithiolactone $S{=}\overline{C(CH_2)_2CH_2S}$ reacts similarly with the molybdenum complex, but in the product the α carbon is linked to the nonadjacent sulfur atom by an alkyl chain (47).

Thioaroylchlorides, $S{=}CRCl$ (R = Ph, p-FC$_6$H$_4$, p-ClC$_6$H$_4$, p-MeC$_6$H$_4$), are reductively dimerized by [Mn(CO)$_5$]$^-$ to form E and Z isomers of the dithiolene complexes [Mn(CO)$_5$S]$_2$(μ-CR=CR) (see Section IV,C). The unstable Z isomers (21) rearrange to form a metal–metal bonded cis-dithiolene complex, 22 (48).

Vanadocene reacts with thiobenzophenone to give a η^2-CS metal-bonded thiocarbonyl function in 23 (49).

B. Thiolates and Thioethers

The first mononuclear orthometallation involving a sulfur-donor ligand, to produce the compound [MnC$_6$H$_4$CH$_2$SMe(CO)$_3$PPh$_3$], was reported long ago (50,51). Later, Siedle, using NMR studies, showed that the palladium adduct 24 is spontaneously, and with the loss of the protonated organic ligand, converted to an orthometallated product (39).

The aryl group in the same organosulfur ligand as well as in methyl naphthyl sulfide and 2-biphenyl methyl sulfide can be metallated by palla-

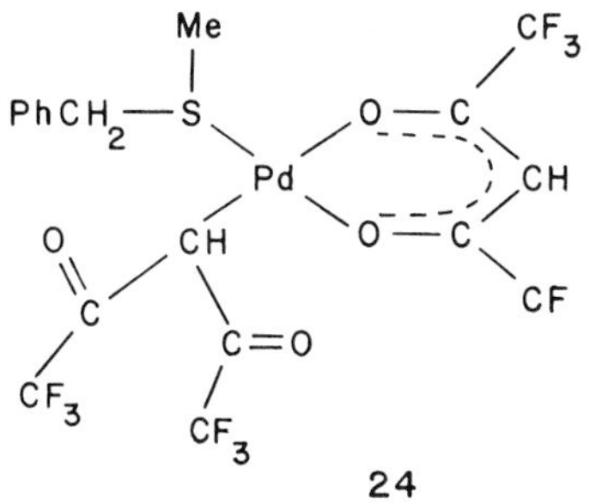

24

dium acetate to give, after subsequent reaction with lithium chloride, dimeric chloride-bridged complexes such as **25**. A similar procedure, using

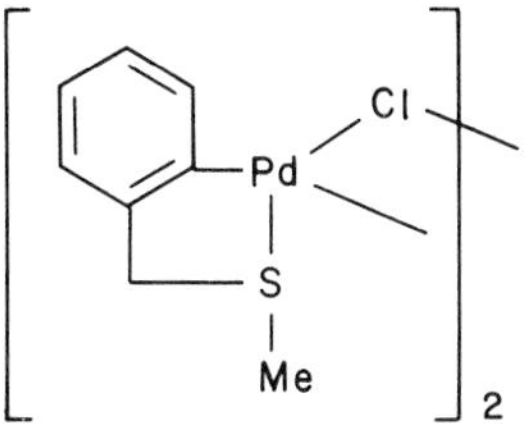

25

2,6-dimethylphenyl methyl sulfide and neopentyl phenyl sulfide, furnishes the cyclopalladated complexes **26a** and **26b** by metallation of the alkyl

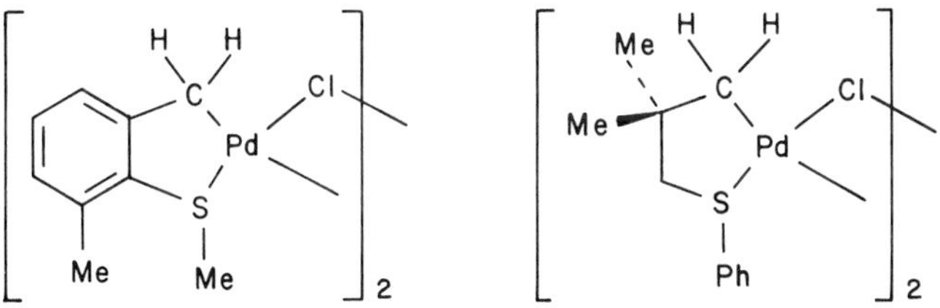

26 a   **26 b**

groups. When using a ligand exchange reaction, 1,3-bis(methylthio-methyl)benzene is metallated at the 2 position of the ring to give a mono-meric compound, **27**, in which both sulfur atoms are coordinated to palladium (*52*).

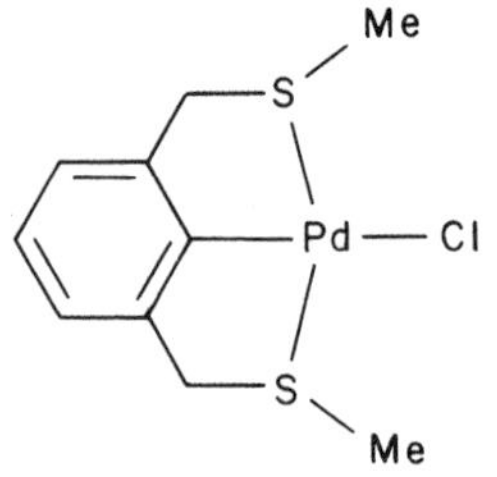

27

Dithioacetals, $CH_2(SR^1)SR^2$, react with $[Fe(CO)_5]$ under UV irradiation to give, after scission of a C—S bond, butterfly complexes $[Fe_2(CO)_6(\mu\text{-}SR^1)(\mu\text{-}CH_2SR^2)C,S]$ $[R^1R^2 = -CH_2SCH_2-$ or $-(CH_2)_3-$; $R^1 = R^2 =$ Me] (28), in which both the SR^1 and CH_2-SR^2 fragments donate three electrons to the $Fe_2(CO)_6$ system (53).

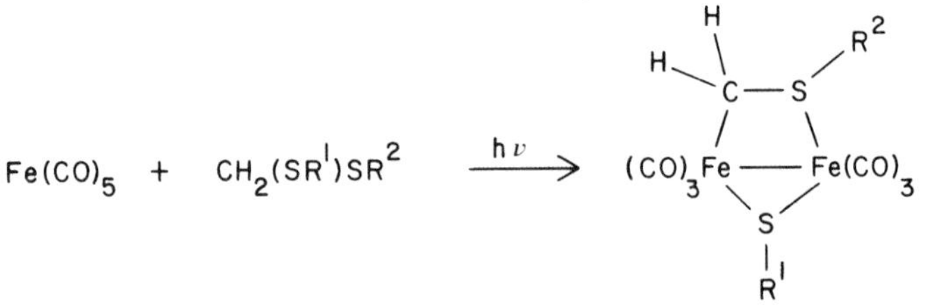

28

Heating (alkylthio)methylzirconocenes, 29, in the presence of PMe_3 effects the intramolecular deprotonation of an agostic hydrogen by a methyl group to form thioaldehyde complexes (30) ($R = Me, Ph, p\text{-}CF_3C_6H_4, p\text{-}ClC_6H_4, p\text{-}MeOC_6H_4,$ or $p\text{-}Me_2NC_6H_4$) (54).

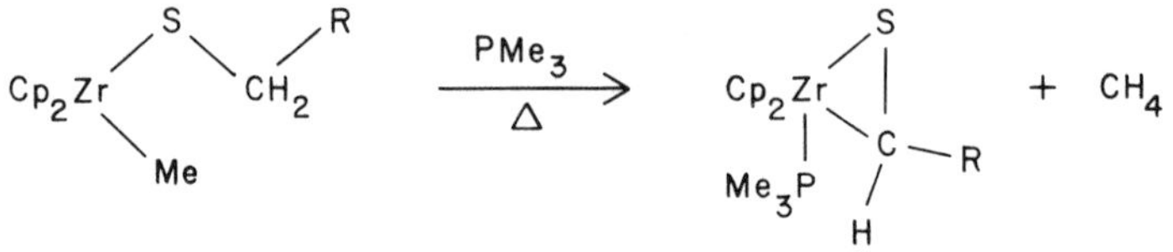

29   30

C. Unsaturated Organosulfur and Organoselenium Compounds

1. Thiophene

Stone and co-workers reported in 1960 (55,56) that thiophene and $[Fe_3(CO)_{12}]$ react to give both a ferrole, 31a (by desulfurization), and a thiaferrole, **31b** (as a minor product). Recently Rauchfuss and co-workers found that methyl substitution of the thiophene improves the yields of both products and that the thiaferroles are precursors to ferroles (57).

Coordinated 2,3-dihydrothiophene (DHT) in a platinum hydride complex (**32**) rearranges slowly, by inserting into the M—H bond, to give **33**. A

similarly bound ligand in the osmium cluster (**35**) results from the insertion of free 2,3-DHT (a proposed intermediate in catalytic hydrodesulfurization) into a metal hydride bond in **34** (58).

2. Vinyl and Allyl Sulfides

Vinyl sulfides, $RSCH=CH_2$ (R = Me, Et, $CH=CH_2$, or Pr^i), react with $[Fe_3(CO)_{12}]$ to give a dinuclear species (**36**) containing $CH=CH_2$ as well as SR fragments (59). This type of oxidative addition was also found in a reaction involving phenyl vinyl sulfide and $[Os_3(CO)_{10}(MeCN_2)]$. The compound formed, $[Os_3(CO)_{10}(\mu\text{-}CH=CH_2)(\mu\text{-SPh})]$, decomposes thermally or photolytically, by the cleavage of a C—H bond at the aryl or vinyl groups, to give a variety of products (60). Adams and Wang reported the

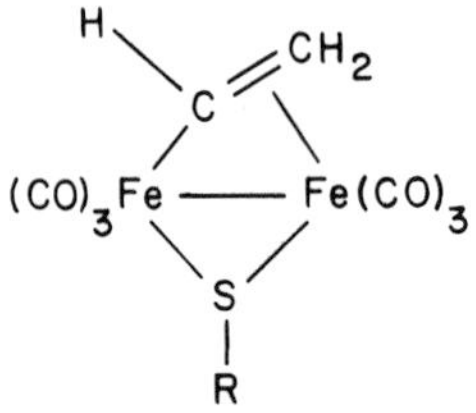

36

decarbonylation and concomitant rearrangement of butterfly osmium carbonyl clusters, **37** (R = Ph or CO_2Me), into rhombohedral ones, **38**, with
quadruply bridging sulfido and alkyne ligands (61).

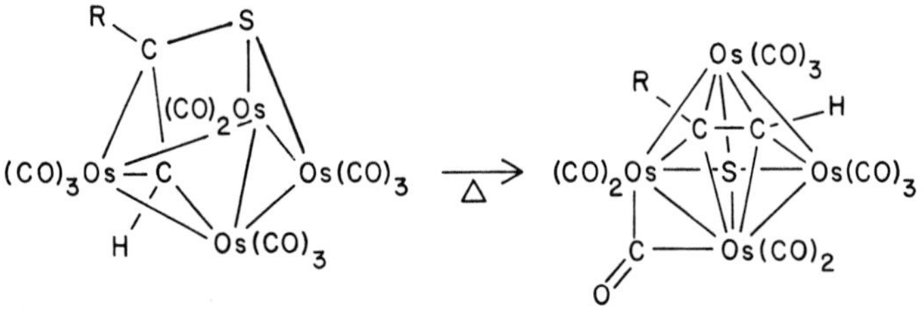

37   **38**

Whereas $[Fe_2(CO)_9]$ reacts with cyclooctatetraenyl and cyclohepta-
trienyl thioethers, by the cleavage of a CS bond, to give S- and SR-contain-
ing carbonyl compounds, $[Ru_3(CO)_{12}]$, with the seven-membered ring
reactant, affords complexes of the types $[Ru_3(CO)_6\{\mu_3\text{-}(\eta^7\text{-}C_7H_7)\}(\mu_3\text{-SR})]$
(**39**) and $[Ru_2(CO)_4\{\mu\text{-}(\eta^7\text{-}C_7H_7)\}(\mu\text{-SR})]$ (R = Me or Bu^t) (**40**) (62) by the
desulfurization of a vinyl and an allyl sulfide.

During the oxidative addition of allylic phenyl sulfides to palladium(0)
complexes to give μ-allyl-μ-sulfide complexes, **41** (R^1–R^5 = H or Me,
R^4 = H, Me, or Ph), reactants having γ substituents (R^4 or R^5) require
more than 20 times as long for completion of the reactions than those with
either no or α or β substituents. The reaction, therefore, probably occurs by
initial electrophilic attack by the palladium on C-γ, expulsion of the SPh
unit, and formation of a metal–metal bond (63).

3. Thioaldehydes and Thioketones

Oxidation of the cobalt atom in an α,β-unsaturated thioaldehyde complex, **42**, by trityltetrafluoroborate (64) or Ag$^+$ ions (65) affords an ionic
dicobalt complex, **43**, containing the organosulfur fragment as a bridging
ligand. A zirconocene thioaldehyde complex, $[ZrCp_2(\eta^2\text{-SCHR})]$, prepared

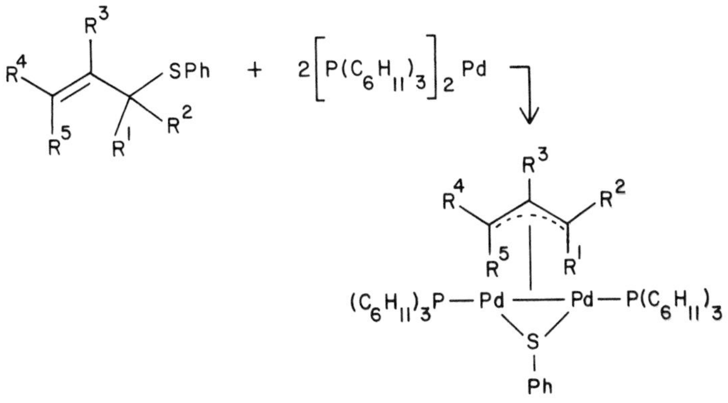

41

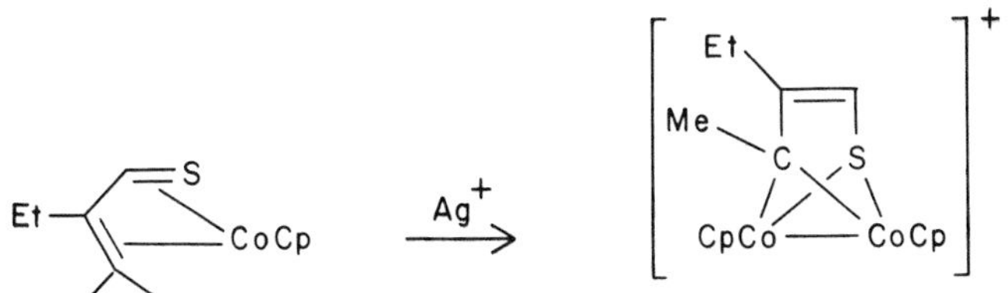

42   **43**

from [ZrMe$_2$Cp$_2$] and HSCH$_2$R (R = Me) in the presence of PMe$_3$, is oxidized by I$_2$ to give a mixture of α and β isomers of 2,4,6-trimethyl-1,3,5-trithiane (66).

The thermolabile adduct between vanodocene and thiobenzophenone, **23** (see Section II,A,2), converts to diphenylfulvene by internal desulfurization of the modified thioketone ligand (49).

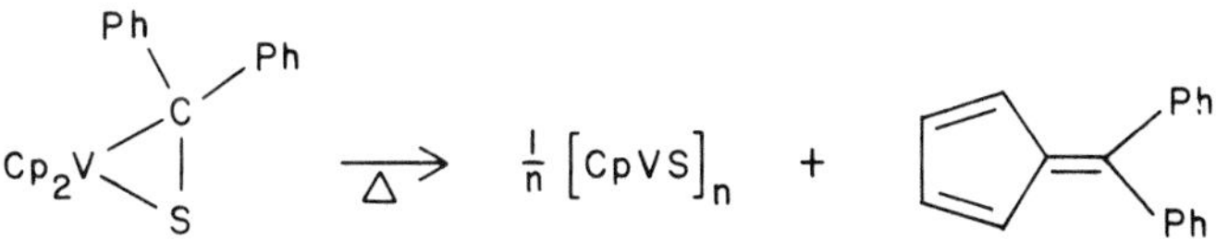

23

4. Organothioformamide

Adams has reviewed the contribution of his group toward rearrangements in triosmium clusters (67). Relevant here are the thermally induced rupture of a C—S bond and formation of a metal–carbon bond in the triply bridged thioformamide cluster, $[Os_3H(CO)_9\{\mu_3\text{-}\eta^2\text{-SCH}=N(p\text{-}FC_6H_4)\}]$ (44), to give via intermediate 45 an open triangular cluster, 46, containing bridging sulfur and formimidoyl ligands.

5. *1,2,3-Thia- and Selenadiazoles*

Schrauzer (*34,68*) and Rees (*69,70*) independently showed long ago that the intermediates in the decomposition of 1,2,3-thiadiazoles or selenadiazoles (47) (R^1 or $R^2 = H$, Ph, Bu^t, or p-Bu^{t}C$_6$H$_4$, E = S or Se), namely, thioketo- or selenoketocarbenes, R^1C—$C(E)R^2$, react with iron carbonyls

to afford complexes of type 48. Later studies showed that there are striking differences in the reactivity of the thia- and selenadiazoles toward [Fe$_2$(CO)$_9$]. Although both yield the above-mentioned type of thioketo- or selenoketocarbene complexes, the sulfur precursors also afford thioketo imine complexes, 49 ($R^1R^2 =$ —(CH$_2$)$_4$—, —(CH$_2$)$_6$—; $R^1 =$ Me, Bu^t,

or p-Bu^{t}C$_6$H$_4$, R^2 = H or Ph) (71), and the selenium derivatives selenoke-
toketene complexes, **50** ($n = 4-6$) (72,73), as major products. Treatment

49   50

of the α-selenoketocyclohexylene diiron complex or the related α-seleno-
ketoketene complex (**50** with $n = 4$) with trimethylamine oxide in alcohol,
ROH (R = Me, Et, Pr^i), affords moderate to good yields of the corresponding diselenide esters (**51**) (74).

51

Cyclooctaselenadiazole in the presence of the CpCo source
[CoCp(PPh$_3$)$_2$] eliminates N$_2$ in refluxing toluene to give a similar type of
ligand bonded to a dimeric CpCo framework (75). In the presence of
[CoCp(PPh$_3$)$_2$], elemental selenium adds to bicyclic selenadiazoles and
thus affords the first diselenolene complexes without electron-withdrawing
substituents (**52–54**) (76).

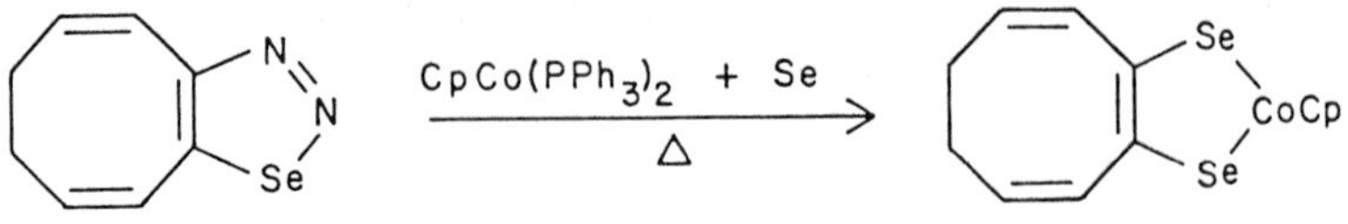

54

6. Fulvenes

The 6,6-dialkyl- and 6,6-diarylfulvenes give simple substitution products, [Mo(CO)$_3$(fulvene)], with [Mo(NCMe)$_3$(CO)$_3$] (77), but 6,6-bis(methylthio)fulvene dimerizes, in different ways, to form **55** and **56** (78).

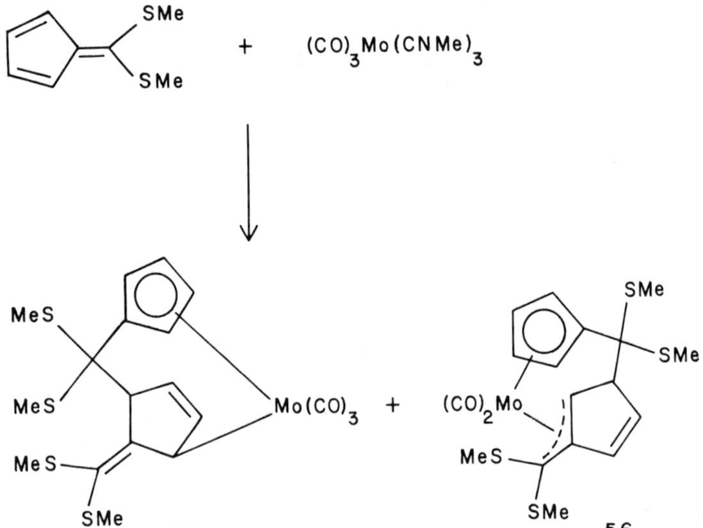

55   **56**

7. Thietes and Dithiolenes

Reaction of thietes with iron carbonyls or cyclopentadienyl cobalt dicarbonyl affords thioacrolein complexes **57** and **58**. On gentle heating or

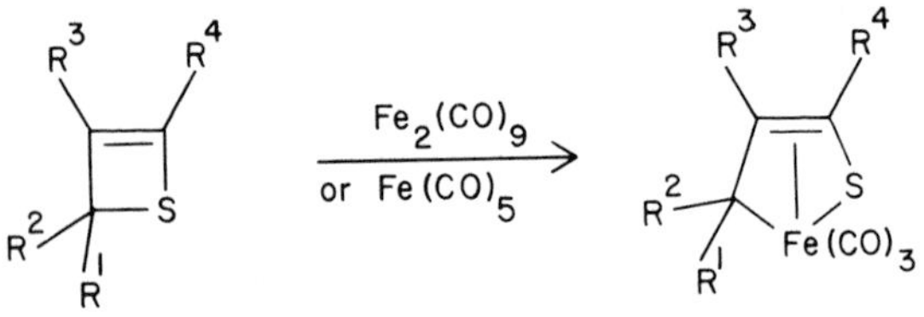

57

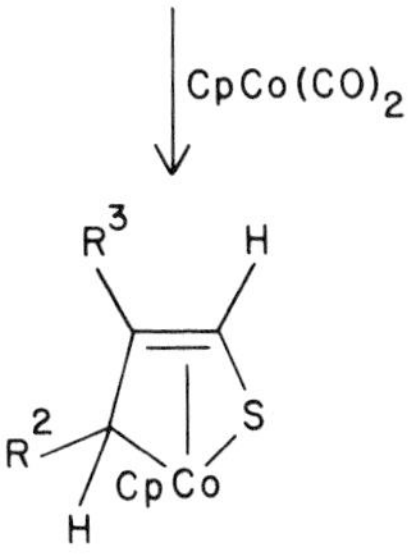

$R^1 = R^2 = R^3 = R^4 = H$

$R^1 = R^4 = H, R^2 = Me, R^3 = Et$

$R^1 = R^4 = H, R^2 = R^3 = -(CH_2)_5-$

58

photolysis the iron tricarbonyl complexes $[\overline{FeCH(R^2)C(R^3)=CHS}(CO)_3]$ (59) ($R^2 = H$ or Me, $R^3 = H$ or Et) lose carbon monoxide to afford dimers, 60, containing a square Fe_2S_2 arrangement (64).

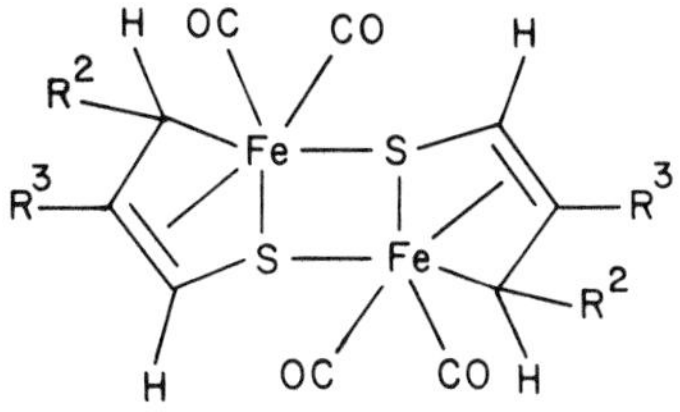

60

D. Thiocarbenes and Selenocarbenes

A systematic study of the synthesis and reactions of various dithiocarbenes has been undertaken by Angelici et al. (4). The ylide complexes obtainable from certain of these carbenes, $[FeCp(CO)_2\{CH(SMe)PR_2H\}]^+$ (61) (R = Ph or c-Hex), undergo pyrolysis at temperatures around 200°C to give cationic phosphine complexes, $[FeCp(CO)_2\{PR_2(CH_2SMe)\}]^+$ (62) (79).

Upon reaction of $[Co(CO)_4]^-$ with $[FeCp(CO)\{C(SMe)_2\}(NCMe)]^+$ (63), a dinuclear species, $[Co(\mu\text{-}CO)(CO)_2\{\mu\text{-}C(SMe)_2\}\{FeCp(CO)\}]$ (64), containing a bridging carbene ligand is formed (80). The nucleophilic addition of the same cobalt complex ion to a complex containing a bridging carbyne ligand (65) affords an open triangular cluster (66) in which the original carbyne carbon is now bridging three transition metal atoms (81).

When the dithiocarbene complex $[RuCl_2CO\{C(SAr)_2\}(PPh_3)_2]$ (67) (Ar = $p\text{-}MeC_6H_4$) (82) is heated, cyclometallation gives compound 68. Ligand ring opening, probably by the formation of an intermediate car-

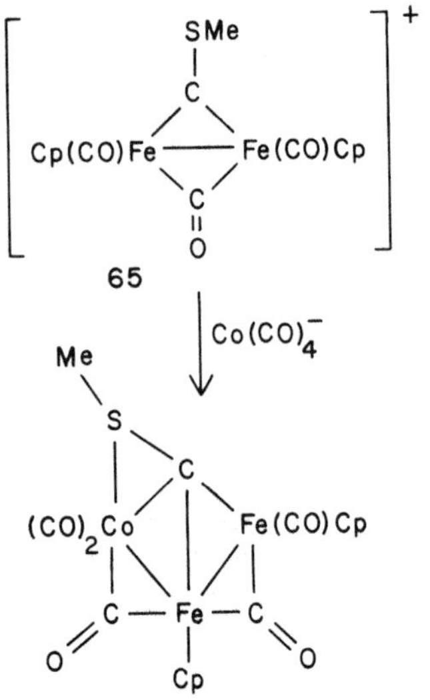

65

66

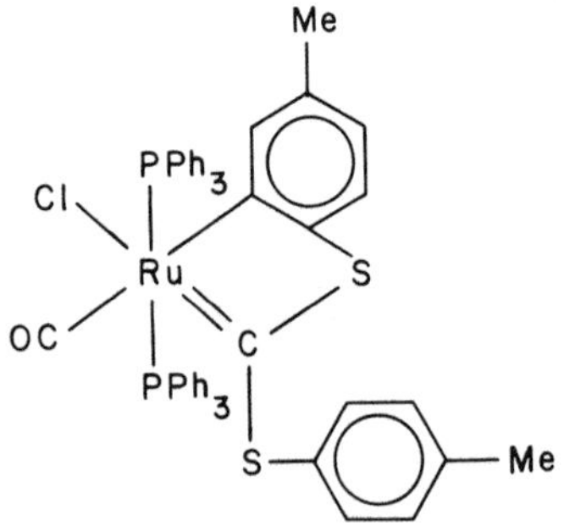

68

byne, occurs on interaction between the dithiocarbene complexes $[W(CO)_3\{\overline{CSC(R^1)}{=}C(R^2)S\}P_2]$ (**69**) [R^1 or $R^2 = CO_2Me$ or CF_3, $P_2 = (R^3{}_2P)_2CH_2$, $R^3 = Me$ or Ph] and $[Co_2(CO)_8]$ (*83*). The original carbene carbon, once again (compare Ref. *81*), bridges all three metal atoms in **70**.

α,β-Unsaturated thio- and selenocarbene complexes of type **71** (M = Cr, W, E = S, Se), which contain an α-phenyl group, spontaneously convert to coordinated indene complexes (**72**) at room temperature by formal insertion of the carbene into a C—H bond of the phenyl group (*84*). The chromium carbene complex is more reactive than the tungsten one and the selenocarbene more than the thiocarbene.

Recently Schubert and co-workers showed that Fischer's (thio)silylcarbene complex, $[W(CO)_5\{C(SEt)SiPh_3\}]$ (**73**), decomposes at room tempera-

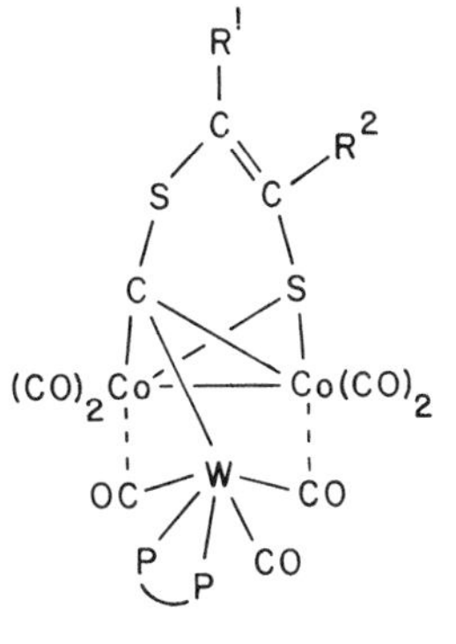

70

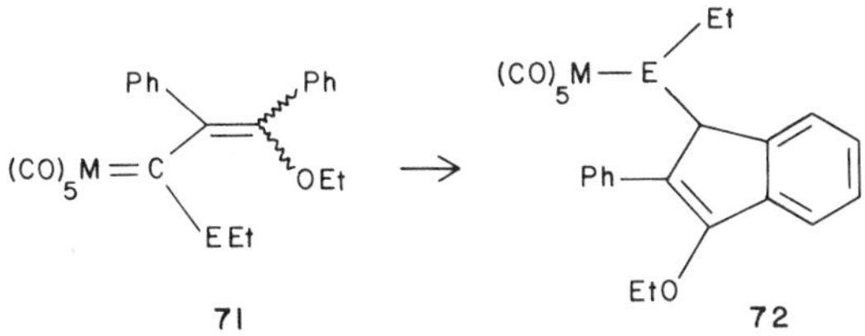

71   72

ture to give a free (**74**) and a coordinated thioketene (**75**)—a conversion made more effective by bubbling CO through the solution. Both products are hydrolyzed on commercial SiO_2 to the corresponding α-silyl(α-thioethyl) carboxylic acids (*85*).

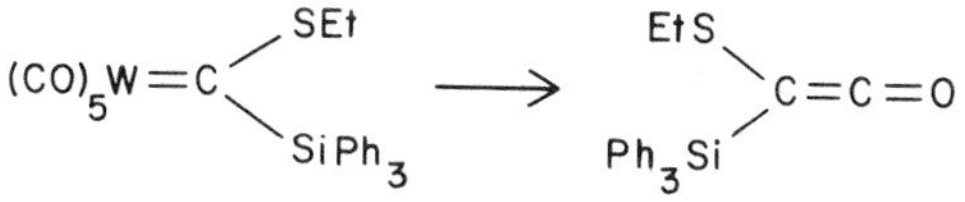

73   74

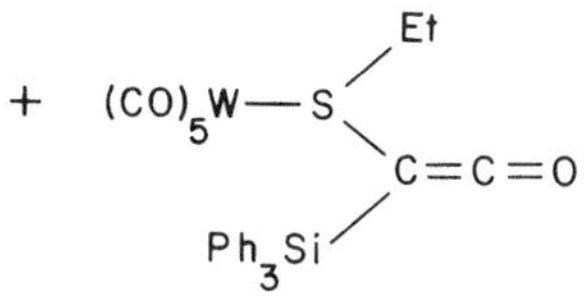

75

E. Metallacycles

Carbonylselenide is ejected from the metallacyclic compound **76** when heated to give a selenoformaldehyde compound, **77** (*86*). An example of a spontaneous decarbonylation (in THF) of an organosulfur chelate is provided by the heterometallacycle **78**, affording **79** (*87*). Muir and co-workers

also report that a comparable tungsten chelate, **80** (R^1 = Me, Et, or Pr^i, R^2 = CF$_3$), rearranges via the heat- and light-sensitive intermediate **81**, which contains a three-membered carbene-donor chelate ring, to yield **82** (*88*). In effect the alkylthio group has migrated across the enone ligand.

Photolysis of the acyl-thioether chelate [FeCp(CO)C(O)C(CF$_3$)=CRS-Me] (**83**) (R = H or CF$_3$) gives a dinuclear iron compound (**84**) without a metal–metal bond (*89*).

The heterometallacyclic alkyne-containing complex **85** (M = Mo or W, $R^1 = Pr^i$ or $p\text{-MeC}_6H_4$, R^1 = Me or Ph) isomerizes into **86** above 20°C (*90*). At 55°C the tungsten compound with R^1 being Pr^i and R^2 Ph rearranges, with incorporation of the alkyne ligand into a metallacycle and extrusion of the thiolate from the chelate ring, to form **87**.

F. Miscellaneous

Mintz and co-workers have studied the decomposition of α-thioalkyl complexes of zirconium. In the α-zirconocenyl thioether **88** (R^1 = Ph,

p-MeC$_6$H$_4$, p-MeOC$_6$H$_4$, or p-ClC$_6$H$_4$, R^2 = Me or Ph), the metal-bonded phenyl group is exchanged for a thioalkyl on a neighboring carbon donor atom by intramolecular nucleophilic attack of the aryl group on the substituted alkyl carbon, and **88** forms (*91*). The mechanism for stilbene elimina-

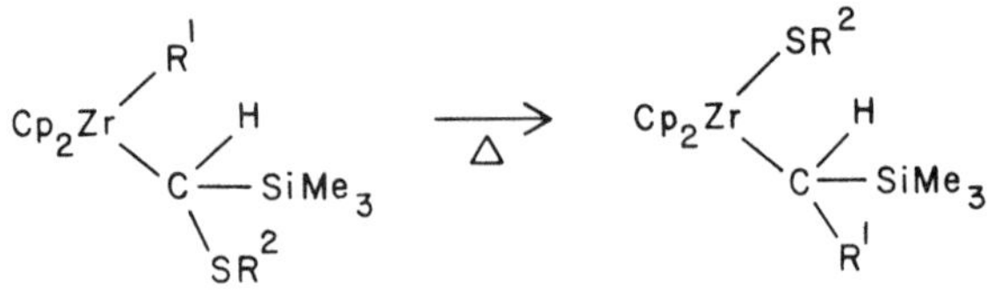

88   89

tion from [ZrCp$_2${CHPh(SPh)}$_2$] (**90**) to give [ZrCp$_2$(SPh)$_2$] is much more complicated (*92*).

The sulfur ylid Me$_2$S(O)CH$_2$ is a very weak base, and, when coordinated to nickel in the complex [Ni(η^2-C$_2$H$_4$)$_2$CH$_2$S(O)Me$_2$] (**91**), spontaneous decomposition to ethane, cyclopropane, and methane occurs (*93*). In another reaction, compound **92** rearranges under UV irradiation to yield, with insertion of iron into a phenyl–carbon bond, a cyclic carbene complex, **93** (*94*).

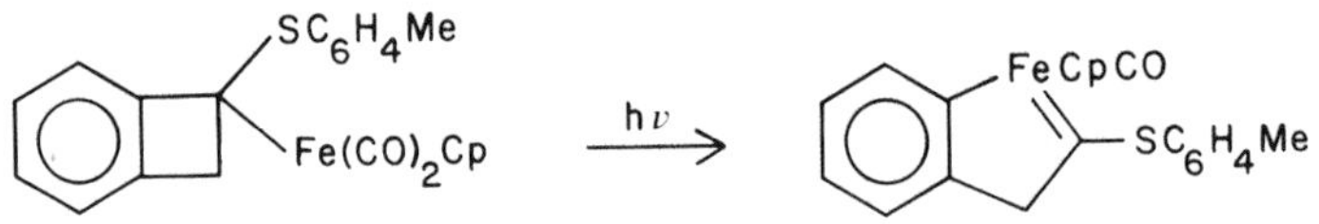

92   93

Roper and Town explain the conversion, with hydrogen, of an osmium-coordinated CS ligand in **94** to SMe in terms of successive hydride transfers from the metal to the thiocarbonyl ligand yielding first a thioformaldehyde complex and then a methanethiolate complex, **95** (*95*). Later this group

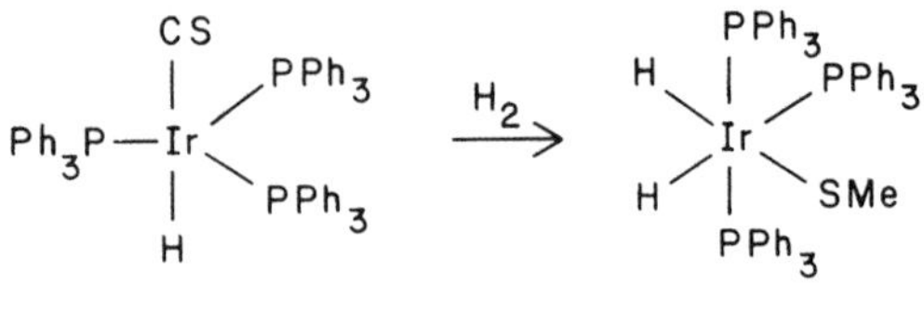

94   95

established that a thioformyl complex, [Os(CO)$_2$(η^2-CH$_2$S)(PPh$_3$)$_2$], which is a logical type of intermediate after the first nucleophilic addition above, reacts rapidly with HCl to give $[OsCl(CO)_2(PPh_3)_2(SMe)]$ (*96*).

Photolysis of $[Fe(Cp)(CO)C(SMe)_3]$ (**96**) produces the compound $[\overline{FeCp(CO)C(SMe)}(SMe)_2]$ (**97**), which is in rapid equilibrium with the form $[FeCp(CO)(SMe)\{C(SMe)_2\}]$ (**98**) (*97*). This isomerization, is rem-

iniscent of the equilibrium postulated by Seebach (*98*), which provides an entry into tungsten dithiocarbene chemistry (see **99**) (*99*).

III

REARRANGEMENTS INDUCED BY DEPROTONATION OF ORGANOSULFUR LIGANDS

We now describe a number of deprotonated S- or Se-containing complexes that rearrange spontaneously. The transformations may involve ligands such as carbonyl or isonitriles.

Deprotonated bridging dithiolate ligands in diiron carbonyl complexes of type **101** (R = H, Ph) rearrange intramolecularly by nucleophilic attack on an iron atom while an electron pair is displaced onto a sulfur atom with formation of an Fe—C bond in **102** (*16*). Neutral derivatives can be formed upon further alkylation, leading to type **1j** complexes (Table I). The bridging ligands $-S(CH_2)_2S-$ and $-SCH_2\text{-}o\text{-}C_6H_4CH_2S-$ rearrange similarly on deprotonation, but $-S(CH_2)_2S-$ undergoes β-elimination to give $[(\mu\text{-}LiS)(\mu\text{-}CH_2=CHS)\{Fe(CO)_3\}_2]$. Dithioformate complexes $[HC(S)SR^1\text{-}\{Fe(CO)_3\}_2]$ (**103**) ($R^1 = Me$, Et, $CH_2=CHCH_2$), also give rearranged anions (**104**) on deprotonation, and they give compounds of the unusual structure type **105** after alkylation with R^2X ($R^2 = Et$ or $CH_2CH=CH_2$, $X = Br$) (*16*).

Cyclic trithiocarbonate complexes of type **1e** [L = CO, P(OMe)$_3$] (Table I) undergo ring opening on deprotonation (*100*). Alkylation of the formed anion affords a new type of carbene complex, **106**, as well as complex **107**,

which belongs to the same class of compounds as the starting material. When heated to 55°C, complex **107** (L = CO) rearranges by methyl migration to give **106**. The type of reaction leading to **107** was observed before when the ylidic compound **108**, formed by deprotonation of an alkylated trithiocarbonate complex, rearranged spontaneously to give the unusual acyclic trithiocarbonate **109** (*101*).

Complex **100** (R = H) represents an organometallic equivalent of 1,3-dithiane. The rearrangement of carbanionic complexes, produced by deprotonation of 1,3-dithiane and other dithioacetals, and thioethers (**110**), containing α-thio protons and coordinated to group 6 metal carbonyls, has been studied in the group of Raubenheimer. A review by Omae (6) includes some of the earlier work (102). Essentially the rearrangement entails a double consecutive internal carbonylation. Neutral heterometallacyclic carbene complexes (**111**, L = CO, phosphine, or phosphite ligands) result

on further alkylation. Addition of α-deprotonated thioacetals or sulfides to $[M(CO)_6]$ (M = Cr or W) also causes cyclization; however, in this case an intermediate carbene is formed, and CO insertion into a metal–carbene bond effects the second carbonylation. The same type of cyclization occurs with $LiCH(SR^1)R^2$ [R^1 = Ph, R^2 = SPh; R^1R^2 = —$(CH_2)_3S$—] and $[Cr(CO)_5CNPh]$, but, instead of a carbonylation in the second step of the transformation, a formal insertion of isocyanide into the metal–carbene bond occurs, to afford a phenylaminocarbene-thioether chelate (103). This mechanism is supported by the isolation of the isocyanide-carbene complex **112**. Its further reaction with ammonia provides another example of a rearrangement caused by insertion of isocyanide, probably via the anionic chelate **113**.

Neutral aminocarbene chelate

Deprotonated isocyanide-thioether complexes, cis-$[M(CNBu^t)(CO)_4$-$\{S(CH_2)_3SCH\}]^-$ (M = Cr or W) (prepared from **114**), also rearrange. In the presence of triethyl phosphine and triethyloxomium tetrafluoroborate, consecutive carbonylation and isocyanide insertion into a metal–carbene bond give trisubstituted aminocarbene-thioether chelates, **115** (*104*).

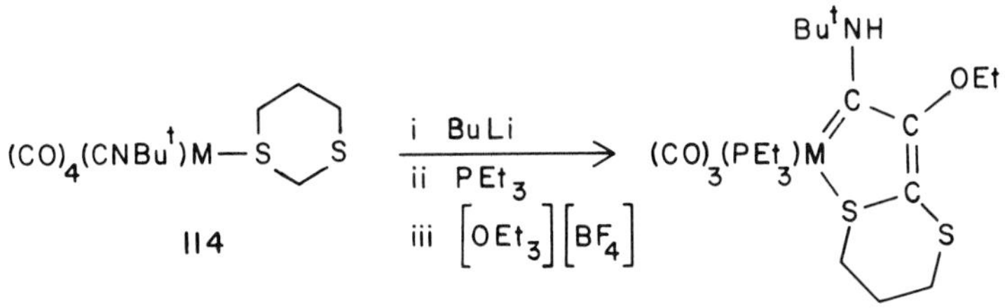

114

115

The anion **117** prepared by deprotonation of the carbene complex [Cr(CO)$_5$\{C(OEt)CH$_2\overline{\text{CHS(CH}_2)_3\text{S}}$\}] (**116**) also rearranges spontaneously by carbonyl insertion to give a neutral *six-membered* chelate (**118**) after alkylation. The analogous tungsten carbene complex affords stereoselectively a vinylcarbene complex (**119**) by opening of the 1,3-dithiane ring (*105*). Reaction of **118** with methylhydrazine produces mainly an NMe- or

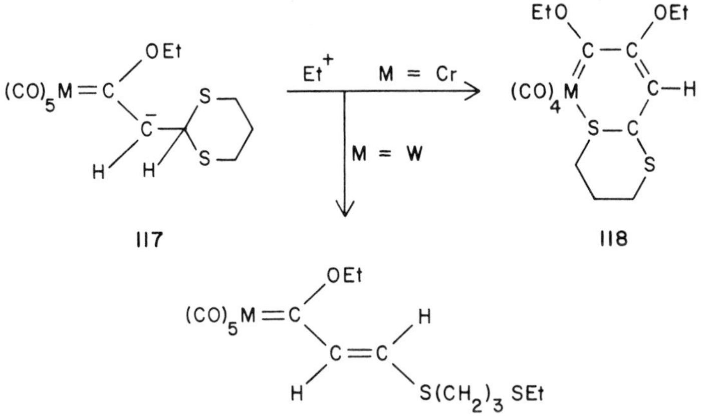

117

118

119

NH-inserted seven-membered chelate product, but a rearrangement product, **120**, also forms in which carbene insertion into a C—S bond effected

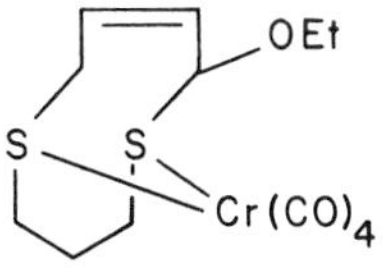

120

coordination of this sulfur atom at the vacated coordination site. Finally, deprotonation of the *C*-methyl group of the carbene complex **121** leads to the neutral methylthiovinyl tungsten complex, **122** (*106*).

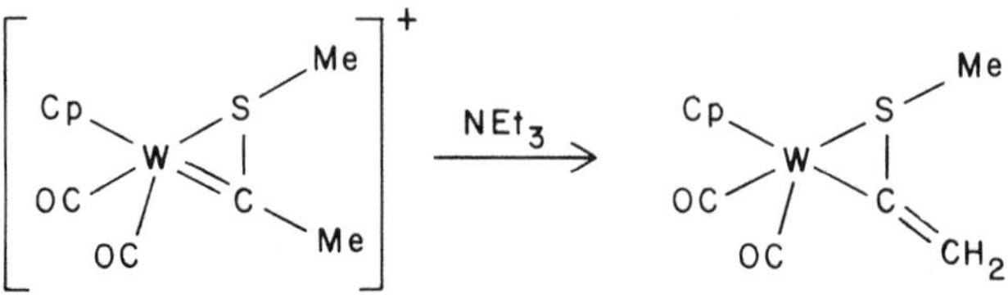

IV

REACTIONS OF ORGANOSULFUR COMPOUNDS WITH ORGANOMETALLIC COMPLEXES

This section focuses on the interactions between organosulfur compounds and coordinated ligands and, linking up with Section II, on the fragmentations of organosulfur compounds by metal carbonyls.

A. *Neutral Organosulfur Compounds*

1. *Thioethers*

Thioethers add reversibly to carbene complexes **123** and **126** to give

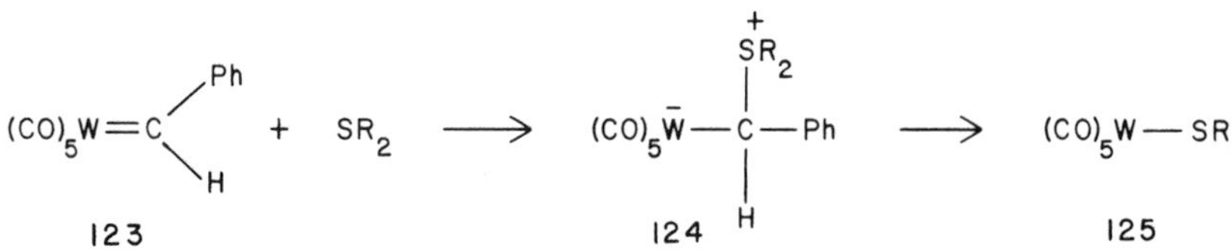

sulfur ylide, **124** [R = Me, Et; $R_2 = -(CH_2)_4-$] or sulfonium salt, **127** [R = Me, CH_2Ph, $CH_2ReCp(NO)(PPh_3)$], complexes. Complex **124** can be used to cyclopropanate olefins. In the absence of other reactants, **124** is converted to a thioether complex, **125**, upon standing at room temperature (*107*). The thioether ligand in **127** (R = Me) is displaced by the nucleophile MeS⁻, and **128**, which is the complex used to produce the dinuclear sulfonium salt form of **127** [R = $CH_2ReCp(NO)(PPh_3)$], forms (*108*).

The cationic ylide complex **130** is formed when SMe_2 adds to the carbenoid ligand in **129** (*109*). The reaction between 2,5-dithia-hex-3-yne

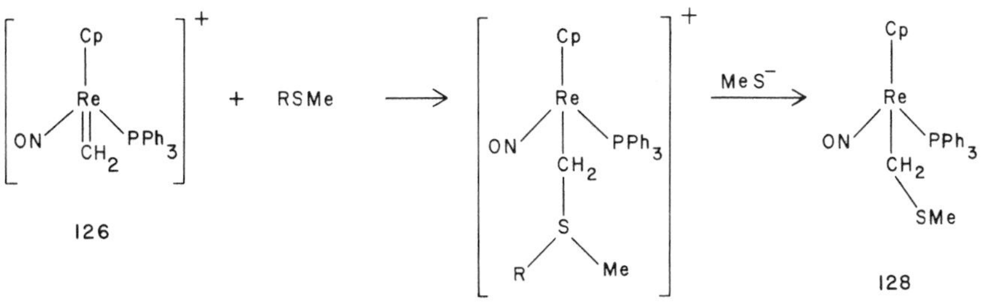

126   + RSMe   ⟶   127   MeS⁻ →   128

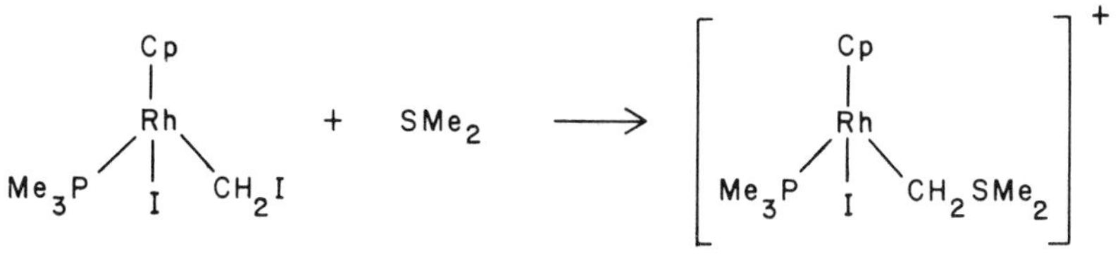

129   + SMe₂   ⟶   130

and [Fe(CO)₅] causes the coupling of two molecules of acetylene and the formation of a ferrole complex (**131**) (*110*).

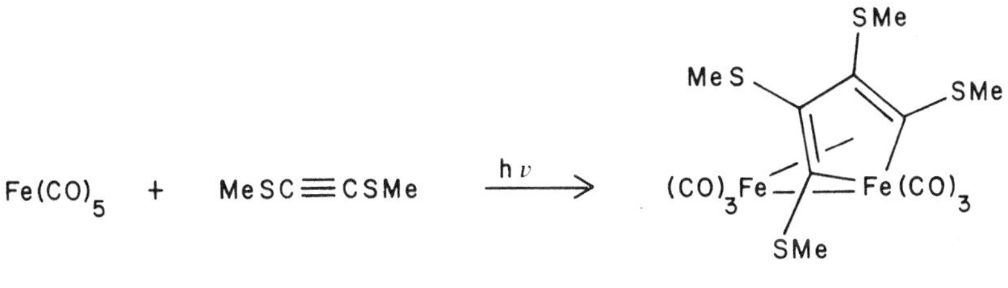

131

Low-valent nickel species catalyze the reaction between Grignard reagents and aryl or alkenyl sulfide in the preparation of alkylarenes, biaryls, and alkenes (see citations in Ref. *111*). Wenkert *et al.* recently isolated compounds **132** and **133** to show that an oxidative addition occurs between diaryl sulfides and tris(tri-*n*-butyl)phosphinonickel (*111*).

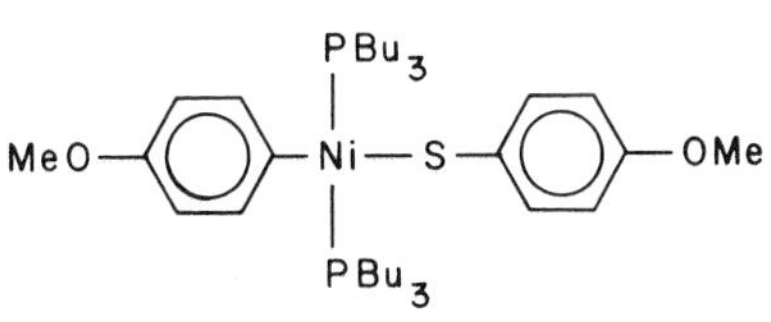

132 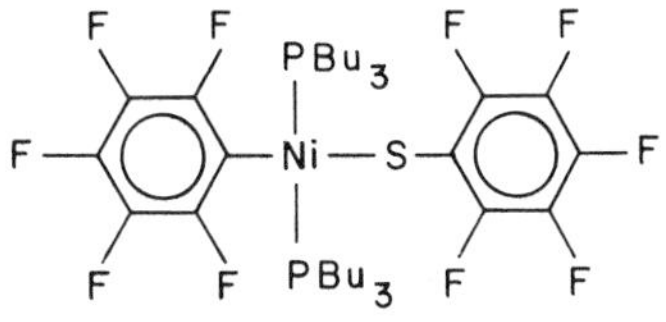   133

2. *Thioketones, Isothiocyanates, and Thioketenes*

The reactions of thiobenzophenone are varied. In the imidoylcobalt-acetone adduct **134**, it displaces the acetone to form complex **135**, which is

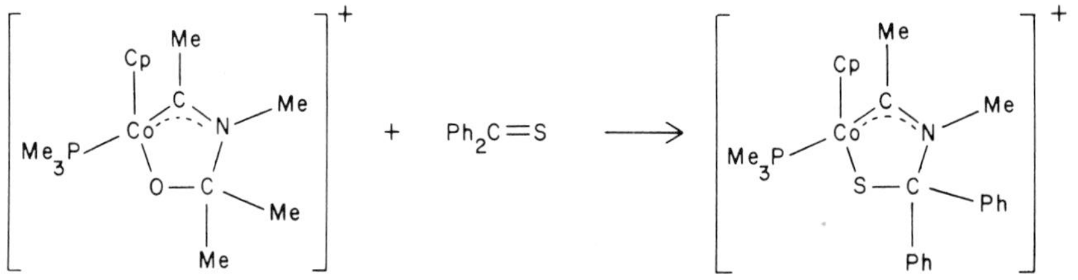

134   135

structurally similar to **134** (*112*). The reaction with $[Os_3H_2(CO)_9(NMe_3)]$ depends on the speed at which thiobenzophenone is added to the reaction mixture. When the reagent is added quickly, compound **136**, in which a

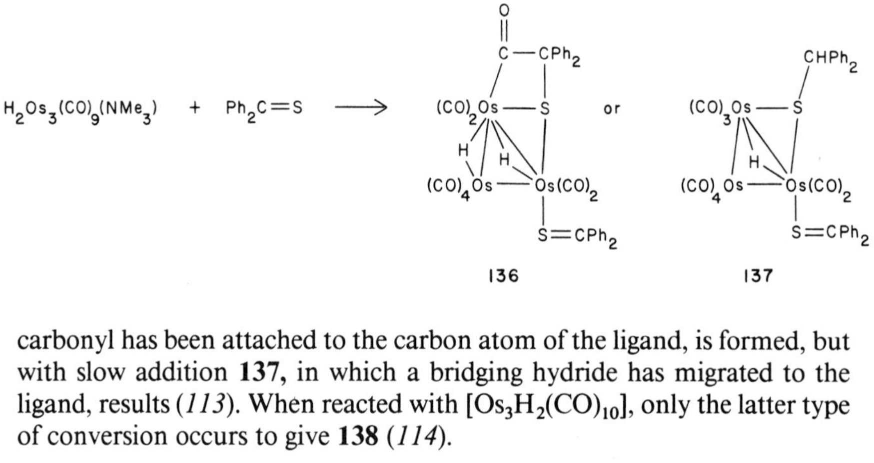

136   137

carbonyl has been attached to the carbon atom of the ligand, is formed, but with slow addition **137**, in which a bridging hydride has migrated to the ligand, results (*113*). When reacted with $[Os_3H_2(CO)_{10}]$, only the latter type of conversion occurs to give **138** (*114*).

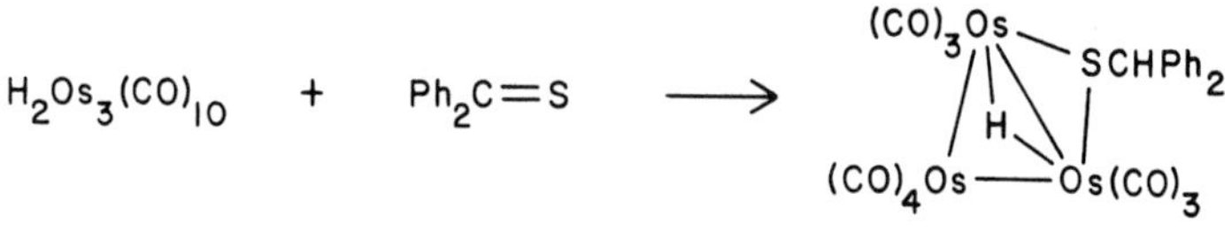

138

Insertion of aromatic and aliphatic thioketones [$S{=}C(p\text{-MeOC}_6H_4)_2$, $S{=}C(p\text{-MeC}_6H_4)_2$, $S{=}C(Ph)(p\text{-MeOC}_6H_4)$, thiocamphor, adamantanethione] into the Zr—H bond of **139** gives the thiolate complexes (**140**) (*115*).

$$Cp_2Zr(H)Cl + R_2C{=}S \rightarrow Cp_2Zr(Cl)SCHR_2$$

139   **140**

Substituted thiobenzophenones react with anionic manganese carbonyl and abstract H^+ from the solvent ($MeOH/H_2O$) to form the dinuclear bridging complexes, **141** (R = H, Me, NMe_2, F, OMe) (*116*). On reacting

$$[Mn(CO)_5]^- + (p\text{-}RC_6H_4)_2C{=}S \rightarrow [Mn(CO)_4\{\mu\text{-}SCH(p\text{-}RC_6H_4)_2\}]_2$$

141

with [W(CO)$_5$I] in the presence of Ag^+ ions, 1,6,6a-λ^4-trithiapentalenes coordinate to tungsten as their thioaldehyde or thioketone valence isomers, **142** [$R^1 = R^4 = H$, $R^2R^3 = -(CH_2)_3-$; $R^1 = R^4 = Me$, $R^2 = R^3 = H$; $R^1 = Bu^t$, $R^2 = R^3 = R^4 = H$] (*117*).

Bis(trifluoromethyl)thioketene, $S{=}C{=}C(CF_3)_2$, dimerizes on reaction with [TiCp$_2$(CO)$_2$] to form a metallacyclic compound [TiCp$_2$SC{=}C(CF$_3$)$_2$}SC{=}C(CF$_3$)$_2$}] (**143**) (*118*).

Carbon–carbon bond formation occurs when the complex [Fe$_2$(CO)$_7${C(Me)CNEt$_2$}] (μC^1, C^2) (**144**), containing a bridging carbene and a terminal aminocarbene, is treated with phenylisothiocyanate, and the complex [Fe$_2$(CO)$_7${C(Me)C(NEt$_2$)C(S)NPh}] (μC^1, C^2, μN) (**145**) forms (*119*).

The heterocumulene compound *p*-tolyl isothiocyanate and the thioketenes, $S{=}C{=}CR_2$ ($R_2 = Me_2(CH_2)_3CMe_2$, Bu^t_2), react with niobium hydride complexes, [NbHCp$_2$(CO)] and [NbH$_3$Cp$_2$], to give insertion products **146** (*120*) and **147** (*121*).

$$[NbHCp_2(CO)] + (p\text{-}totyl)N{=}C{=}S \rightarrow [NbCp_2(CO)SCH{=}N\text{-}p\text{-}tolyl]$$

146

$$[NbH_3Cp_2] + R_2C{=}C{=}S \rightarrow [NbCp_2(SCH{=}CR_2)(\eta^2\text{-}SCCR_2)]$$

147

Reaction of the complex $[Os_3Pt(\mu\text{-}H)_2(CO)_{10}\{P(c\text{-}Hex)_3\}]$ (**148**) with the thioketene 1,1,3,3-tetramethyl-2-thiocarbonylcyclohexane affords, as the major product, $[Os_3Pt(\mu_3\text{-}S)_2(\eta^1\text{-}C=C_6H_6Me_4)(CO)_9[P(c\text{-}Hex)_3\}]$ (**149**), which contains a η^1-vinyl ligand formed by cleavage of the C=S bond of the thioketene (*122*).

The bridging thioformamide ligand in **151** results when RN=C=S

$$[Os_3H_2(CO)_{10}] + RN=C=S \rightarrow [Os_3H(CO)_{10}(\mu\text{-}\eta^1\text{-}SCH=NR)]$$

151

(R = Me, Ph, $p\text{-}FC_6H_4$, $p\text{-}MeC_6H_4$) reacts with $[Os_3H_2(CO)_{10}]$ (*123*). Finally, isothiocyanates insert into a Mn—S bond of **152** (R^1 = Me or Ph, R^2 = Me, Et, or Ph) to form the tricyclic compound **153** (*124*).

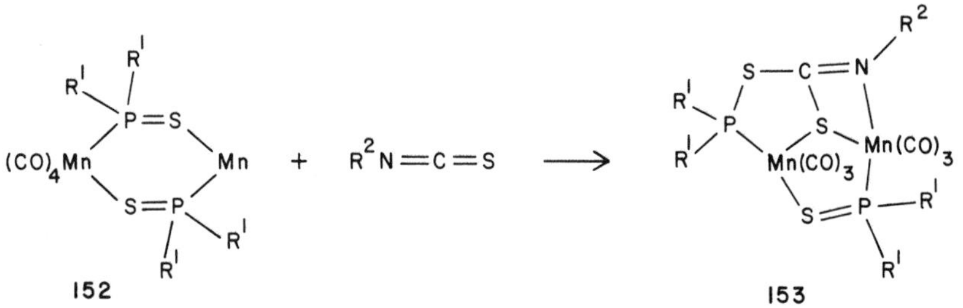

3. *Thiols*

A number of reactions of organothiols with complexes of alkoxycarbenes, isocyanides, or thiocarbonyls, leading to thiocarbene complexes, have been reviewed (*4,125,126*). Further related examples include the conversion of a chlorocarbene to a dithiocarbene by reaction of the ruthenium complex **154** with RSH (R = Me, *p*-tolyl) or $HSCH_2CH_2SH$ to give the dithiocarbene compounds **155** and **156** (*82*).

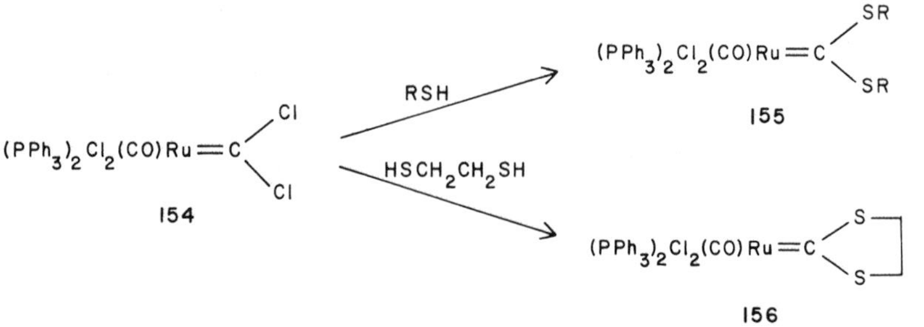

The cationic vinylidene complex **157** adds S—H across the double bond of the vinylidene ligand to give a sulfhydryl (R = H) or a methylthiocarbene (R = Me) complex (**158**) (*127*).

$$[FeCp(CO)(PPh_3)(C=CH_2)]^+ + HSR \xrightarrow{-196° \text{ to } -78°C} [FeCp(CO)(PPh_3)\{C(SR)Me\}]^+$$

157   **158**

When treated with excess EtSH in the presence of base, the trichloromethyl-isocyanide complex **159** leads to the aminothiocarbene complex

$$159 \qquad\qquad 160 \qquad\qquad 161$$

160 and the dithiocarbene complex **161**. Dithiols convert **159** to cyclic dithiocarbene complexes of type **162** (R = H, Me) (*128,129*). In a different

$$159 \qquad\qquad\qquad\qquad\qquad 162$$

type of reaction, propanedithiol links two Os_3 moieties (**164**) when reacting with $[Os_3(CO)_{10}(MeCN)_2]$ (**163**) (*130*).

$$163 \qquad\qquad\qquad\qquad\qquad\qquad 164$$

4. Episulfides

In the reactions of episulfides ring opening usually occurs, as in the reaction between palladium(II) isocyanide complexes of type **165** (L =

$$165 \qquad\qquad\qquad\qquad\qquad\qquad 166$$

PPh$_3$, PMe$_2$Ph, or C$_6$H$_{11}$NC, R = p-MeOC$_6$H$_4$ or C$_6$H$_{11}$) and thiirane, which yields cyclic aminothiocarbene complexes (166). Thiirane does not react with the bulky Bu^{t}NC ligand (131). The thermal fragmentation of thiirane or thiirane S-oxide in the presence of [Fe$_3$(CO)$_{12}$] yields the ethen-edithiolate complex, 3a, by incorporating an additional sulfur atom into

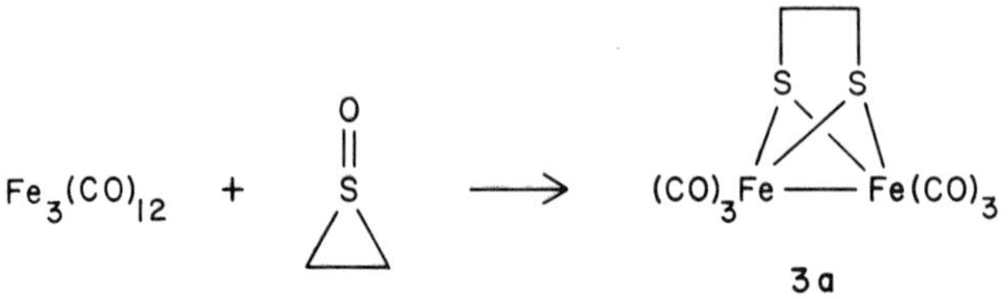

3a

the ligand (132). The same occurs when ethylene sulfide, $\overline{SCH_2CHR}$ (R = H), or propylene sulfide (R = Me) reacts with [MoHCp(CO)$_3$] or [MoCp(CO)$_3$]$_2$ (CpMoL$_n$) to give 168. In this reaction the episulfide is also

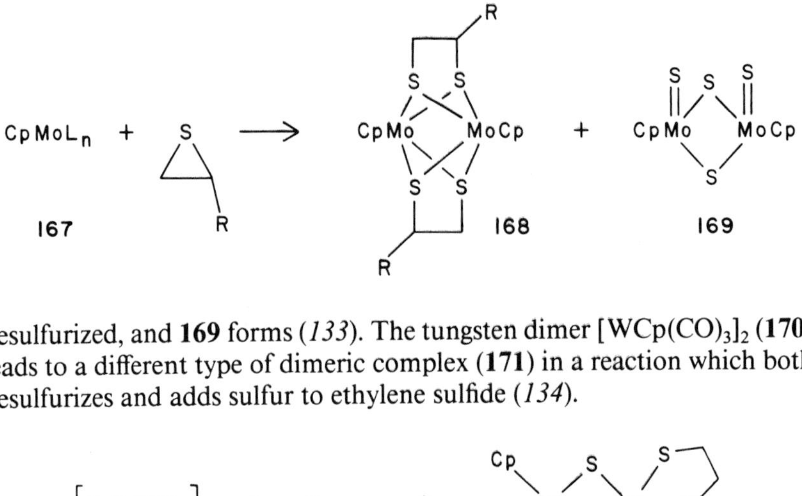

167   168   169

desulfurized, and 169 forms (133). The tungsten dimer [WCp(CO)$_3$]$_2$ (170) leads to a different type of dimeric complex (171) in a reaction which both desulfurizes and adds sulfur to ethylene sulfide (134).

170   171

5. Dithiaheterocycles

Various dithiaheterocyclic compounds fragment unsymmetrically in the presence of iron carbonyls to form bridging dithiolate ligands as in 172 (Table II). In contrast, UV irradiation of dithiane or trithiane in the

TABLE II

DITHIOLATE LIGANDS, S͡S, THAT FORM FROM IRON
CARBONYLS AND DITHIAHETEROCYCLES S͡S

$Fe_x(CO)_y$	S͡S	S͡S (172)	Ref.
$Fe_2(CO)_9$			135
$Fe_2(CO)_9$			136
$Fe_2(CO)_9$			136
$Fe(CO)_5$			12
$Fe_2(CO)_9$	; R = H, Me		136
$Fe(CO)_5$			137
$Fe(CO)_5$			138

$$\text{Fe}_x(\text{CO})_y \quad + \quad \left(\begin{array}{c} S \\ \\ S \end{array}\right) \quad \longrightarrow \quad (\text{CO})_3\text{Fe}\!-\!\!-\!\!-\!\text{Fe}(\text{CO})_3$$

172

presence of [Fe(CO)$_5$] leads to scission of only one C—S bond and the formation of an asymmetric complex, **173** (X = CH$_2$, S), containing a CH$_2$—S bridge (53).

$$(\text{CO})_3\text{Fe}\!-\!\!-\!\!-\!\text{Fe}(\text{CO})_3$$

173

6. *Dithio- and Trithiocarbonates, Xanthates, Dithioesters, and Carbamates*

Rearrangement reactions of the title compounds in the presence of transition metals have been discussed in Section II,A. As mentioned before, dithiocarbonates, R^{1}OC(S)SR^2, and trithiocarbonates, R^{1}SC(S)SR^2, fragment during reaction with [Fe$_2$(CO)$_9$] to give dinuclear products resulting from the formal insertion of Fe into a C—S bond (type 2, Table I). With the dithiocarbonate R^{1}SC(O)SR^2 [R^1 = Me, Et, or CH$_2$Ph, R^2 = Me, Et, or CH$_2$Ph; R^{1}R^2 = —(CH$_2$)$_n$— with n = 2 or 3, —CH=CH—], however, a double fragmentation occurs which leads to the bis{μ-S(alkyl)} hexacarbonyldiiron compounds (**174**) (*139*). The trithiocarbonate *S*-ferrocenylmethyl-*S*-methyldithiocarbonate undergoes a similar double fragmentation to form **174** (R^1 = Me, R^2 = CH$_2$-η^5-C$_5$H$_4$—Fe—η^5-C$_5$H$_5$). These reactions provide a method to synthesize dissymmetrically bridged iron hexacarbonyl dimers. A second product (**175**), consisting of two Fe$_2$(CO)$_6$ units bridged by the two *S*-alkyl ligands and a central sulfur atom, is formed along with **174** in the latter example (*140*).

$$\text{Fe}_2(\text{CO})_9 \quad + \quad \text{R}^1\text{SCSR}^2 \quad \longrightarrow \quad \text{Fe}_2(\text{CO})_6(\mu\!-\!\text{SR}^1)(\mu\!-\!\text{SR}^2) \quad + \quad$$

174   **175**

Several *S*-alkyl xanthates (R^1 = cholestanyl, cholesteryl, α-cholestanyl, adamantanethyl, or menthyl, R^2 = Me, Bu, CH$_2$Ph, or CH$_2$CO$_2$Me) are completely desulfurized (**176**) in reactions with [Co$_2$(CO)$_8$], but in a few cases (R^1 = 2,4,6-trimethylphenyl or α-ferrocenylmethylcarbinyl, R^2 = Me) the RO fragment is eliminated and **177** forms; preference seems to depend on the relative strengths of the C—O and C—S bonds (*41*).

$$Co_2(CO)_8 + R^1OC(S)SR^2 \nearrow \begin{array}{c} [Co_3(CO)_9](\mu_3\text{-COR}^1) \\ \textbf{176} \\ \\ [Co_3(CO)_9](\mu_3\text{-SCMe}) \\ \textbf{177} \end{array}$$

Dithioesters, R^{1}C(S)R^2 (R^1 = Me, Et, Pr^i, Bu, c-Hex, Ph, *p*-MeC$_6$H$_4$, *p*-MeOC$_6$H$_4$, 3,4-Me$_2$C$_6$H$_3$, *p*-Me$_2$NC$_6$H$_4$, 2-thiophenyl, α-naphthyl; R^2 = Me), react with [Co$_2$(CO)$_8$] and are desulfurized to yield alkylidyne tricobalt nonacarbonyl clusters (**178**) (*42*).

$$R^1C(S)SMe + Co_2(CO)_8 \rightarrow [Co_3(CO)_9](\mu_3\text{-CR})$$
$$\textbf{178}$$

The photochemical reaction of *N,N*-diethyl-*S*-ethylcarbamate with [Fe(CO)$_5$] fragments the ligand, and some of the fragments are incorporated into the multinuclear complexes **179** and **5**. As written, the C=NEt$_2$ fragment is formally a bridging cationic carbene ligand (*20*).

179   5

B. Reactions with Anionic Organosulfur and Organoselenium Compounds

1. Reactions of Carbonyl Complexes

The reactions in which lithiated thioacetals undergo double carbonylations with [M(CO)$_6$] (M = Cr, W) to form the chelating carbene complexes

180 upon subsequent alkylation have been dealt with briefly in Section III. A complication arises when R^1R^2 is $CH_2N(Me)CH_2$; two C—S bonds are cleaved and the sulfur atoms are alkylated (141).

Iron pentacarbonyl reacts with α-thiocarbanions to form **181** (R^1 = Ph or p-MeC$_6$H$_4$, R^2 = Ph or CH=CH$_2$) and then with methyl iodide to produce β-keto sulfides, together with some dithioethers and stilbene (R^2 = Ph). The use of an acid chloride (R^3 = CHMe$_2$, CH=CHMe, Ph, C$_4$H$_3$O) as the electrophile leads to the formation of thioesters (142). When the substituted iron carbonyl [Fe(CO)$_4$PPh$_3$] is treated with LiCH(SPh)CH$_2$Bu and [OEt$_3$][BF$_4$], the iron carbene complex [Fe(CO)$_3${C(OEt)CH(SPh)CH$_2$Bu}(PPh$_3$)] (**182**) can be prepared (143).

The action of lithium alkyl- or arylthiolates, RS$^-$ (R = Pr^i, Bu, Bu^t, Ph, p-MeC$_6$H$_4$, p-MeOC$_6$H$_4$, p-ClC$_6$H$_4$, p-FC$_6$H$_4$), on [Co$_3$(CBr)(CO)$_9$] (**183**) in the presence of CO leads to the formation of thioester complexes (**184**) (144).

$$Co_3(CO)_9(\mu_3\text{-}CBr) + RS^- + CO \rightarrow Co_3(CO)_9\{\mu_3\text{-}C(O)SR\}$$

183   **184**

2. *Reactions of Carbene Complexes*

Addition of phenylethynethiolate or -selenate to the carbene complex **185** (M = Cr or W, R = Me or Et, E = S or Se) affords an anionic adduct (**186**) which, unexpectedly, contains a reactive thioketene function. Electrophiles can attack either at the sulfur or at the carbon α thereto to produce coordinated carbenes or thio- and seleno aldehydes and esters (*84,145*).

Reactions of deprotonated thioacetals, LiCHSCH$_2$XCH$_2$S, with the Fischer carbene complexes (**185**, R = Me or Et) lead to new coordinated thiocarbene ligands. The products that form (*146*) suggest that the lithiothioacetals display nucleophilic reactivity at two centers, the negatively charged carbon atom and a sulfur atom, each of which adds to a carbene carbon to form an adduct of type **i** which can undergo two types of ring opening to give the carbene complexes **187** (X = CH$_2$ or S) and **188** (X = S or NMe) (*147*).

Reaction of the α-thiocarbanion of benzothiazole, $\overline{LiC=N(C_6H_4)\dot{S}}$-$o$, with **185** (M = Cr, R = Et) leads, by substitution of the alkoxy carbene ligand, to a Lappert-type carbene complex, $[Cr(CO)_5\{\overline{CN(Et)(C_6H_4)S}$-$o\}]$ (**189**), on alkylation with $[OEt_3][BF_4]$. When the chromium carbene complexes **190** (Ar = Ph or PhC_6H_4) or **185** (M = W, R = Et) are treated with

$$(CO)_5Cr=C\begin{smallmatrix}Ar\\OEt\end{smallmatrix} \quad + \quad HC\begin{smallmatrix}NMe_2\\S\end{smallmatrix} \quad + \quad LDA \quad \longrightarrow \quad (CO)_5Cr-S=C\begin{smallmatrix}Ar\\C-NMe_2\\\parallel\\S\end{smallmatrix}$$

                    190   **191**

$$+ \quad (CO)_5Cr-S=C\begin{smallmatrix}Ar\\C=C\begin{smallmatrix}NHCHMe_2\\Me\end{smallmatrix}\\H\end{smallmatrix}$$

                        192 a

$LiC(S)NR^1_2$ (R^1 = Me or Ph), prepared from $HC(S)NR^1_2$ and lithiumdiisopropylamide (LDA), a complicated series of rearrangements, sometimes incorporating fragments from diisopropylamine, ensues, and three different types of thione complexes (**191–193**) {apart from the substitution complexes, $[M(CO)_5(S=CHNR_2)]$} result (*147*).

$$(CO)_5W=C\begin{smallmatrix}Ph\\OEt\end{smallmatrix} \quad + \quad HC\begin{smallmatrix}NPh_2\\S\end{smallmatrix} \quad + \quad LDA \quad \longrightarrow \quad (CO)_5W-S=C\begin{smallmatrix}Ph\\C=C\begin{smallmatrix}NHCHMe_2\\Me\end{smallmatrix}\\H\end{smallmatrix}$$

                    185   **192 b**

$$+ \quad (CO)_5W-S=C\begin{smallmatrix}NPh_2\\CH_2-NPh_2\end{smallmatrix}$$

                        193

The three-membered ring chelate **194** (R = Me, Et, Pr^i) reacts with thiolates at the carbene center to give the adduct **195** ([W] = $\{HB(pz)_3\}(CO)_2W$, pz = $\overline{NN=CHCH=CH}$). The η^2-$C(SMe)_3$ complex **197** is made from the dithiocarbene complex **196,** which is treated with

$$\left[[HB(pz)_3](CO)_2W \overset{S-Me}{\underset{C-H}{\bigg\langle}} \right]^+ + RS^- \longrightarrow [W] \overset{S-Me}{\underset{\underset{H}{C-SR}}{\bigg\langle}}$$

194   195

MeS$^-$ to attack at the carbene carbon and displace the labile nitrile ligand (*148*).

$$Cp(CO)(MeCN)Fe=C(SMe)_2^+ + MeS^- \longrightarrow$$

196   197

3. Reactions of Carbyne Complexes

Nucleophilic displacement at a terminal methylidyne ligand, with regeneration of a metal–carbon triple bond in **199** (E = S or Se, R = Me, Ph, or *p*-NO$_2$C$_6$H$_4$) to form a new methylidyne ligand, occurs in the reaction of **198** [L = HB(3,5-Me$_2$—C$_3$HN$_2$)$_3$] with thiolate or selenolate anions under phase-transfer conditions (*149*).

$$L(CO)_2Mo \equiv C - Cl + RE^- \rightarrow L(CO)_2Mo \equiv C - ER$$
$$\textbf{198} \qquad\qquad\qquad\qquad\qquad \textbf{199}$$

The reaction of nucleophiles, RE$^-$ (ER = SMe, SPh, CH$_2$Ph, SePh), with the bridging thiocarbyne ligand in **200** leads to bridging carbene complexes (**201**). In the rearrangement observed for two of the products

$$\left[\underset{200}{Cp(CO)Fe \overset{SMe}{\underset{O}{\bigwedge}} Fe(CO)Cp} \right]^+ + RE^- \longrightarrow \underset{201}{Cp(CO)Fe \overset{MeS \quad ER}{\underset{O}{\bigwedge}} Fe(CO)Cp} \longrightarrow \underset{202}{OC \overset{SMe}{\underset{O}{\bigwedge}} ER}$$

200   201   202

(201, ER = SPh, SePh), the added RE^- migrates, and the original substituent at the carbyne carbon is retained (202) (150a). Some of the reactions of carbyne complexes have been reviewed by Kim and Angelici (150b). The mononuclear cationic carbyne complex 203 also adds RSe^- nucleophiles (R = Ph, p-$CF_3C_6H_4$, p-BrC_6H_4, p-FC_6H_4, p-MeC_6H_4, p-$MeOC_6H_4$, 1-naphthyl) at the carbyne carbon to give carbene complexes (204), which rearrange to 205 on heating by migration of the RSe^- group to the metal

and liberation of CO (151). The neutral carbyne complex 206, on the other hand, undergoes double nucleophilic attack on the carbyne carbon and on a carbonyl carbon, and, after hydrolysis, the organic thioester 207 is pro-

duced. The analogous reaction with phenylselenolate does not lead to the corresponding ester but, *inter alia*, to the diselenide complex 208 and the dimeric selenide complex 209 (152).

The formation of the thione complex 211 involves the coupling of two thioformaldehyde units, addition to the carbyne carbon of 210 to form a thiocarbene ligand, sulfur insertion into the metal–carbene bond, and reductive deamination. Instead of adding to the carbyne carbon, the anionic ligand also releases sulfur to the manganese atoms to form **212** (*147*).

$$\left[(MeC_5H_4)(CO)_2Mn \equiv CPh\right]^+ \xrightarrow[\text{ii CH}_3\text{COOH}]{\text{i } \underset{\text{S}}{\overset{\text{S}}{\text{LiCN Me}_2}}} (MeC_5H_4)(CO)_2Mn - S = C \overset{Ph}{\underset{SCH_2 \underset{\overset{\|}{S}}{CNMe_2}}{}}$$

210   211

$$+ \quad \left\{(MeC_5H_4)(CO)_2Mn\right\}_2 (\mu - S)$$

212

4. *Reactions of Acetylene, Vinylidene, and Allene Complexes*

A series of alkenyl complexes (**214, 216,** and **218**) were prepared from the reaction of PhS$^-$ with the alkyne complex **213**, the η^2-allene complex **215**, and the cationic complex **217**. Complex **218**, with the nucleophile linked at the α position, could be converted to an allenacyl complex (**219**) by the addition of CO and a catalytic amount of oxidant, such as [Cp$_2$Fe][BF$_4$] or Ce^{IV}/EtOH. Excess Ce^{IV}/EtOH leads to the formation of an alkenyl ester, (EtO)C(O)C(Me)=C(Me)SPh (*153*).

$$\left[\begin{array}{c} Cp \\ | \\ OC-Fe \\ | \\ (PhO)_3P \end{array} \begin{array}{c} Me \\ C \\ ||| \\ C \\ Me \end{array} \right]^+ + PhS^- \longrightarrow Cp(CO)\left\{P(OPh)_3\right\}Fe - \overset{Me}{\underset{Me}{\diagdown}} \overset{SPh}{\diagup}$$

213   214

$$\left[Cp(CO)(PPh_3)Fe - \overset{CH_2}{\underset{\overset{\|}{CH_2}}{\overset{\|}{C}}} \right]^+ + PhS^- \longrightarrow Cp(CO)(PPh_3)Fe - \overset{H}{\underset{CH_3}{\diagdown}} \overset{SPh}{\diagup}$$

215   216

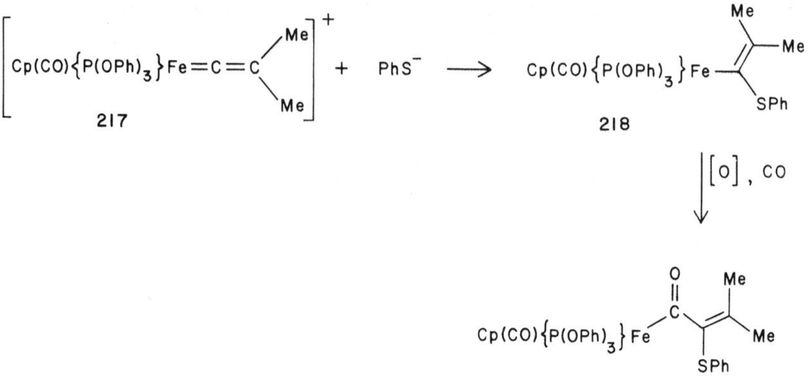

$$\left[Cp(CO)\{P(OPh)_3\}Fe=C=C\begin{array}{c}Me\\Me\end{array} \right]^+ + PhS^- \longrightarrow Cp(CO)\{P(OPh)_3\}Fe-C\begin{array}{c}Me\\Me\\SPh\end{array}$$

217   218

$\Big\downarrow [o], CO$

$$Cp(CO)\{P(OPh)_3\}Fe-\overset{O}{\underset{}{C}}-C\begin{array}{c}Me\\Me\\SPh\end{array}$$

219

C. Reactions of Cationic Organosulfur Compounds

Thioaryl chlorides, $S=C(R)Cl$ ($R = p\text{-}XC_6H_4$, $X = H$, F, Cl, Me), undergo reductive dimerization induced by $[Mn(CO)_5]^-$ to form E and Z dithiolene complexes (220), which can undergo further reaction with acti-

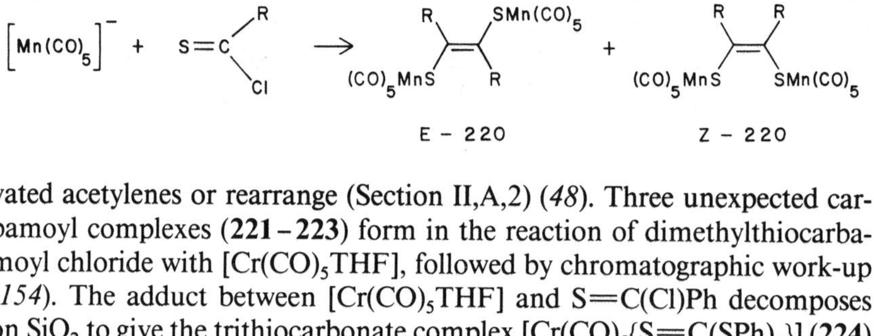

$$\left[Mn(CO)_5 \right]^- + S=C\begin{array}{c}R\\Cl\end{array} \longrightarrow \begin{array}{c}R\quad SMn(CO)_5\\(CO)_5MnS\quad R\end{array} + \begin{array}{c}R\quad R\\(CO)_5MnS\quad SMn(CO)_5\end{array}$$

E − 220   Z − 220

vated acetylenes or rearrange (Section II,A,2) (48). Three unexpected carbamoyl complexes (221–223) form in the reaction of dimethylthiocarbamoyl chloride with [Cr(CO)₅THF], followed by chromatographic work-up (154). The adduct between [Cr(CO)₅THF] and $S=C(Cl)Ph$ decomposes on SiO₂ to give the trithiocarbonate complex [Cr(CO)₅{S=C(SPh)₂}] (224) (155).

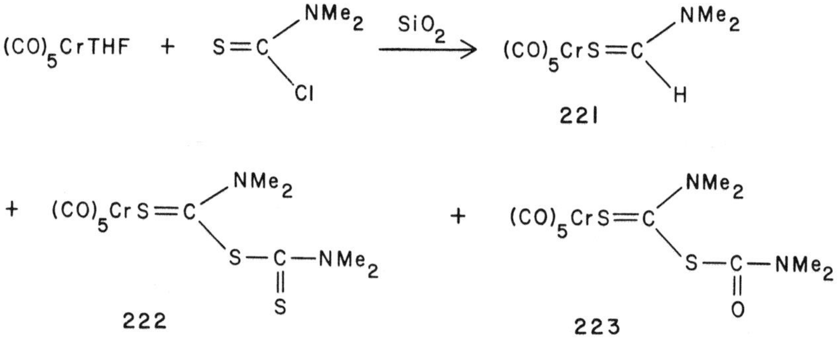

$$(CO)_5CrTHF + S=C\begin{array}{c}NMe_2\\Cl\end{array} \xrightarrow{SiO_2} (CO)_5CrS=C\begin{array}{c}NMe_2\\H\end{array}$$

221

$$+ (CO)_5CrS=C\begin{array}{c}NMe_2\\S-\overset{\|}{\underset{S}{C}}-NMe_2\end{array} \qquad + (CO)_5CrS=C\begin{array}{c}NMe_2\\S-\overset{\|}{\underset{O}{C}}-NMe_2\end{array}$$

222   223

The anionic carbene complexes of type **225** (M = Cr, W) simply add Cl$\overline{\text{CHS}}$(CH$_2$)$_3$S to form **226**, but they rearrange to the vinylcarbene complex **227** with loss of thiophenol when treated with PhCH(SPh)Cl (*105*).

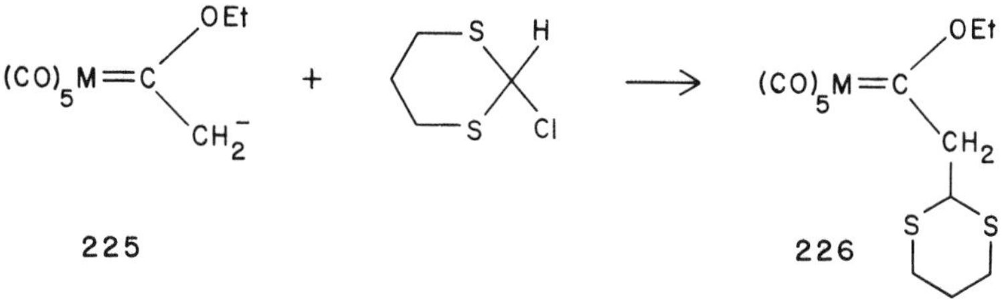

225

226

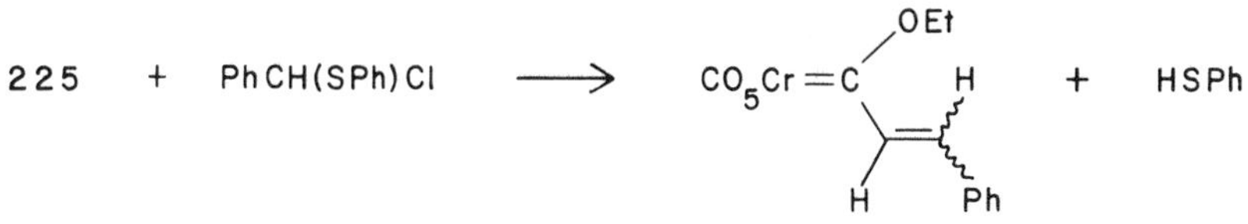

225 + PhCH(SPh)Cl $\longrightarrow$ CO$_5$Cr=C(OEt)(H) + HSPh

227

The anionic adducts **186** abstract R^{2}E^{2+} from the disulfide or diselenide, R^{2}E^{2}E^{2}R^2, which adds to the α carbon of the thioketene function, to form the unsaturated dithioester complexes of type **228** (*84*).

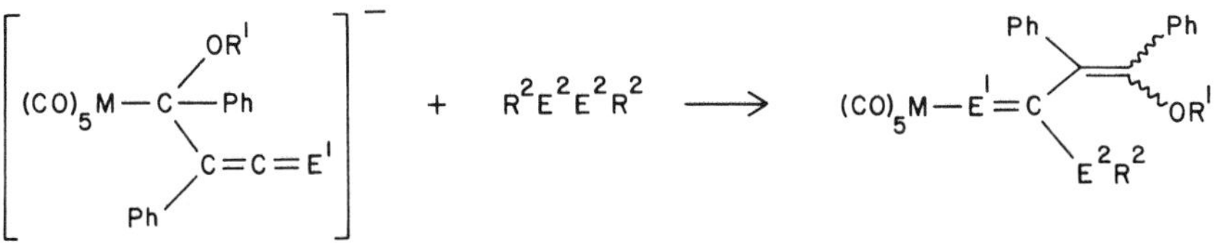

186

228

M	E^1	R^1	E^{2}R^2
W	S	Et	SMe, SEt, SPh, SePh
Cr	S	Me	SEt
W	Se	Et	SePh, SMe, SPh

A stepwise twofold electrophilic attack occurs on the carbyne complex **229** to first give the cationic η^2-carbene complex **230** and then the dicationic dithiatungstabicyclo[1.1.0]butene complex **231** (*156*). A similar η^2-carbene ligand forms in the reaction of the thiocarbyne complex [W(CO)$_2$(CSMe){HB(pz)$_3$}] (**232**) with Me$_2$(MeS)S$^+$ (*157*).

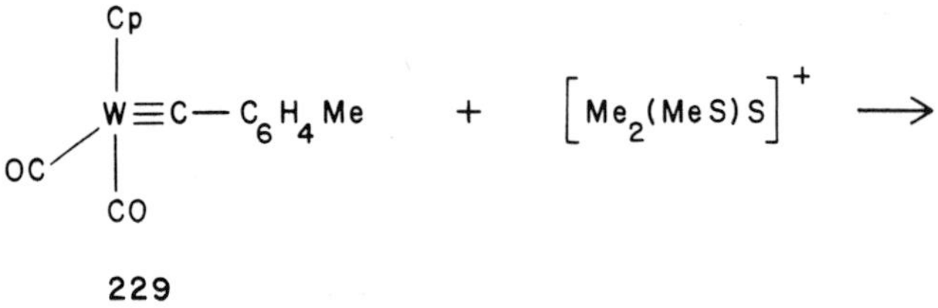

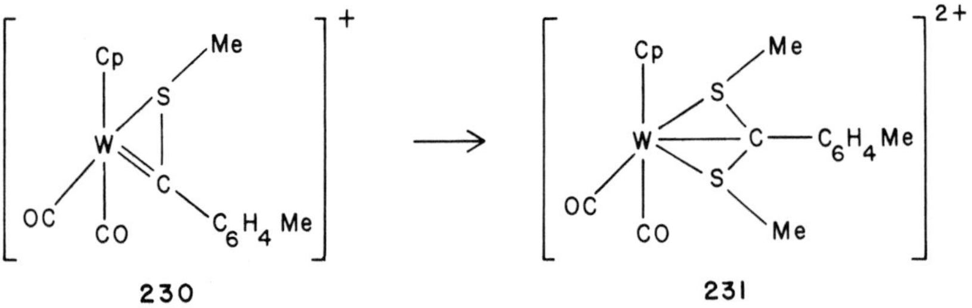

229

230   231

Trichloromethyl thioalkyl compounds of formula $RSCXCl_2$ (**a**, **b**, X = Cl, R = Ph or CH_2Ph) as well as dichloromethyl benzyl thioether react with iron(II) tetraphenylporphyrin (TPP) (**233**), in the presence of a reducing agent (iron powder or $Na_2S_2O_4$), to produce chlorocarbene complexes of type **234** (*158*).

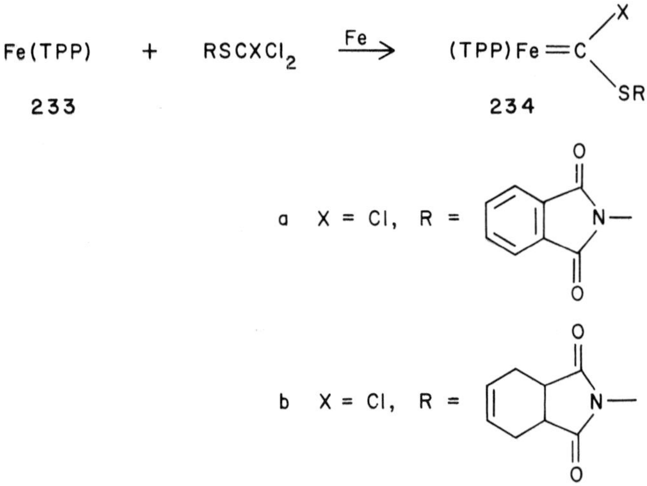

Fe(TPP)   + $RSCXCl_2$ $\xrightarrow{Fe}$ (TPP)Fe=C$\begin{smallmatrix}X\\SR\end{smallmatrix}$

233   234

a   X = Cl, R =

b   X = Cl, R =

V

REACTIONS OF CHALCOGEN ATOM DONORS AND SMALL CHALCOGEN UNITS WITH ORGANOMETALLIC COMPLEXES

Formal E (E = S or Se) insertion, sulfurization of organic carbonyls, and nucleophilic addition to coordinated organosulfur and organoselenium compounds are reviewed in this section. Reactions of H_2S, HE^-, and S^{2-} in which the chalcogen atoms exhibit nucleophilic character, as well as new contributions to CE_2 chemistry, especially reactions with coordinated carbanions and dimerizations, also receive attention.

A. Chalcogen Atom Donors

1. Reactions with Alkyl and Aryl Complexes

The following scheme depicts the sequential insertion of sulfur or selenium into two W—C σ bonds. Each complex can also be synthesized directly from $[WCp(CH_2SiMe_3)_2(NO)]$ (235); the activation barriers increase for each sequential step. The alkylperthio ligand, η^2-S_2R, in 237 is regarded as a $3e^-$ donor, which makes 237 an $18e^-$ complex, whereas 236 is a stable $16e^-$ complex. The reaction with selenium does not lead to the perseleno equivalent of 237, and only 236 and 238 can contain sulfur or selenium (159–161).

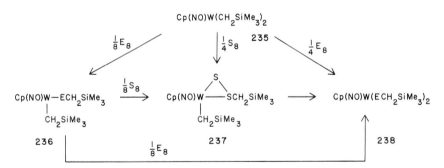

Dichalcogenophenylene complexes (240, M = Ti or Zr, E = S, Se, or Te) can be prepared by combining elemental sulfur, selenium, or tellurium with the diaryl metallocene complexes (239, R^1 = H or Bu^t, R^2 = H, Me, or OMe) (162,163).

The photochemically or thermally induced insertion of sulfur into the

$$(R^1C_5H_4)_2M(C_6H_4R^2) \quad + \quad \tfrac{1}{4}E_8 \quad \longrightarrow \quad (R^1C_5H_4)_2M\begin{matrix} E \\ | \\ E \end{matrix}R^2$$

239   240

Co—C bond of alkyl- and arylcobaloximes (241, R = Et, Pr, pentyl, c-Hex, c-Hex-2-OH, $CH_2CH=CHCH_3$, CH_2Ph, CHMePh, CHMe-p-tolyl) yields organic and organometallic tetrasulfides (242–244) (164).

241   242   243   244

Treatment of [FeCp(CO)$_2$C(O)R] (245, R = Me, Ph) and [M(CO)$_5${C(O)R}] (248, M = Mn, Re; R = Me, Ph) with the sulfurizing agents P_4S_{10} or B_2S_3 converts the acyl to a thioacyl and inserts sulfur into the M—C bond to yield the dithiocarboxylate complexes 246, 247, and 249 (165).

245   246   247

248   249

2. Reactions with Carbene Complexes

Elemental sulfur, selenium, or tellurium reacts with the M=C bond in the electron-rich methylidene complex 250 to give coordinated thio-, seleno-, or telluroformaldehydes (251) (166). The electrophilic methylidene

250   251

complex **252**, however, requires a nucleophile to act as an S atom donor. Addition of $S=PPh_3$ gives a 1 : 1 mixture of the desired thioformaldehyde complex **253** and an ylide complex, $[ReCp(CH_2PPh_3)(NO)(PPh_3)]^+$ (**254**)

252   253

but cyclohexene sulfide provides an S atom without liberating a new nucleophile, with **253** being obtained as the only product (*167*). The selenoformaldehyde complex $[ReCp(\eta^2\text{-}CH_2Se)(NO)(PPh_3)]$, analogous to **253**, is prepared from **252** and triphenylphosphine selenide or potassium selenocyanate (*168*).

The bridging methylidene ligand in **255** also reacts with the sulfur from cyclohexene sulfide to form a thioformaldehyde, which is in a $\sigma + \pi$ bonding mode in **256** and triply bridging in **257**. (Compounds **256** and **257** are interconvertible by the elimination and addition of CO.) One also has access to the thioformyl ligand in **258** by the further decarbonylation of **257** at 125°C with concomitant activation of a C—H bond, resulting in hydride migration (*169*).

Organylisothiocyanates (R^2 = Me, Et, Ph) and -selenocyanates (R^2 = Ph) also act as nucleophilic S and Se atom donors and react with carbene complexes **259** (M = Cr or W, R^1 = H, Me, Br, OMe, CF$_3$, or NMe$_2$), to yield η^1-arylphenylthioketone and -selenoketone pentacarbonyl chromium and tungsten complexes (**260**) (*170,171*). Kinetic results suggest an associative stepwise mechanism with a nucleophilic attack of the sulfur at the carbene carbon in the first reaction step (*172*). It is also possible to isolate thio- and selenobenzaldehyde complexes **262** (M = Cr or W, E = S or Se,

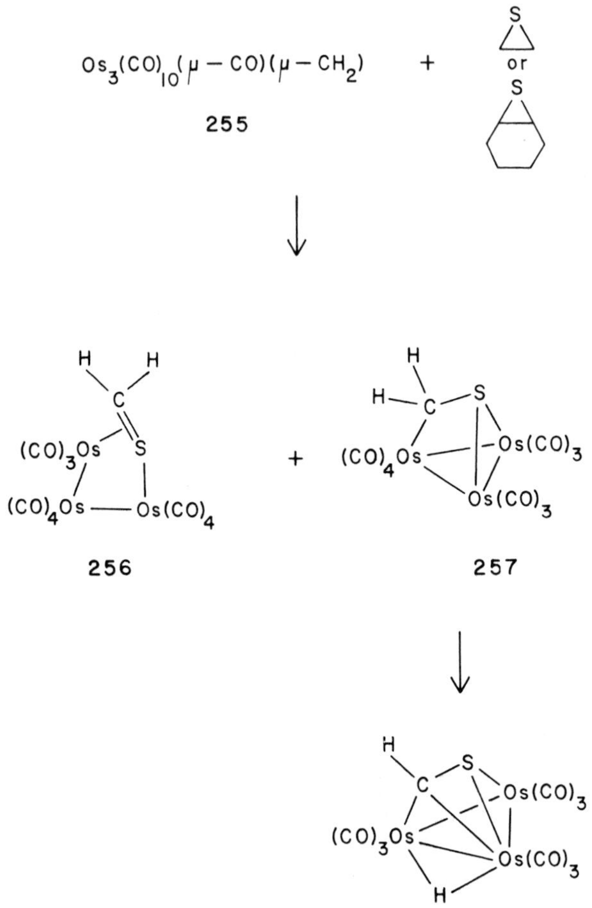

$$Os_3(CO)_{10}(\mu-CO)(\mu-CH_2) \quad + \quad \text{(thiirane or thiabicycloheptane)}$$

255

256   +   **257**

258

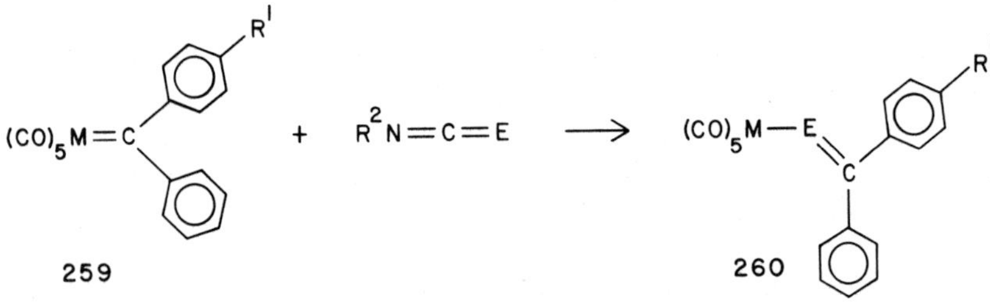

$$(CO)_5M=C \text{(Ar-R')(Ph)} \quad + \quad R^2N=C=E \quad \longrightarrow \quad (CO)_5M-E...C \text{(Ar-R')(Ph)}$$

259   **260**

261   262

R = H, Me, OMe, or CF$_3$) from the reaction of **261** with thiocyanate (or S$_8$) (*173*) and selenocyanate (*174*), respectively. A telluroketone complex, [W(CO)$_5$(Te=CPh$_2$)] (**263**), results from the reaction between [W(CO)$_5$CPh$_2$] and tellurocyanate (*175*). Sulfur, selenium, or tellurium adds to, but does not insert into, the M=C bond in **264** to give **265** (R = H, Me, Ph; E = S, Se, Te) (*176,177*).

264   265

3. *Reactions with Carbyne Complexes*

Reaction of molybdenum and tungsten carbyne complexes **266** with elemental sulfur or selenium transforms the carbyne to the dithiocarboxylate in **267** [M = Mo or W, E = S or Se, L = P(OMe)$_3$ or CO, R =

266   267

CH$_2$Bu^t or *p*-MeC$_6$H$_4$]. The reaction is considered to be charge controlled and to proceed via electrophilic attack by S$_8$ on the carbyne carbon (*178*). An additional complex (**267**, M = W, E = S, L = CO, R = Me) can be obtained from [W≡CMe(Cp)(CO)$_2$] and cyclohexenesulfide (*179*). With the osmium complex **268** (R = *p*-tolyl), however, reaction with sulfur does not proceed beyond the η^2-thioacyl complex (**269**, E = S or Se). Seleno- and telluroacyl complexes result from analogous reactions (*180*).

268   269

A dithiocarbamate ligand acts as the source of sulfur, and its counterion, $[H_2NEt_2]^+$, provides a proton in the reaction of the carbyne complex 270 (R = Me, Ph) with $[H_2NEt_2][S_2CNEt_2]$. The proton in 271 is bonded to the carbon of the former carbyne ligand, which has been converted to a thioformaldehyde ligand (181).

$Cl(CO)_2(py)_2W \equiv CR$ + $\left[H_2NEt_2\right]\left[S_2CNEt_2\right]$ $\longrightarrow$ $(Et_2NCS_2)(CO)W$

270   271

Sulfur reacts with the triply bridging alkylidyne ligand in 272 to convert it to the triply bridging thioacyl ligand in 273 (182). The bridging carbyne ligand in 274 (M = Mo or W, Cp* = Cp or C_5Me_5) reacts with sulfur to form a thioacyl ligand, which transversely bridges the metal atoms in 275.

272   273

In one reaction an additional product (276, Cp* = C_5H_5), with a bridging ligand resulting from the coupling of a CO, the alkylidyne, and S, is also formed (183).

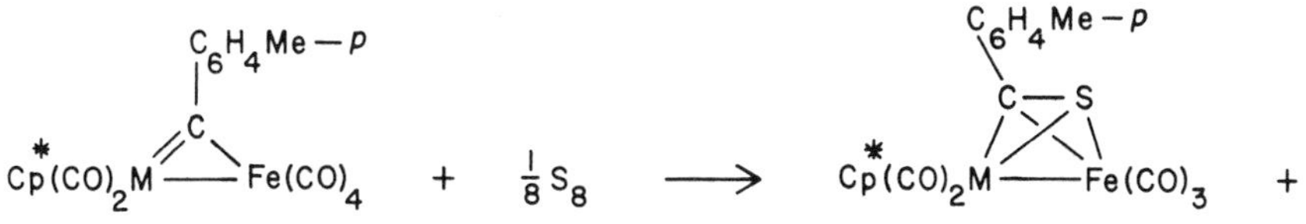

274

275

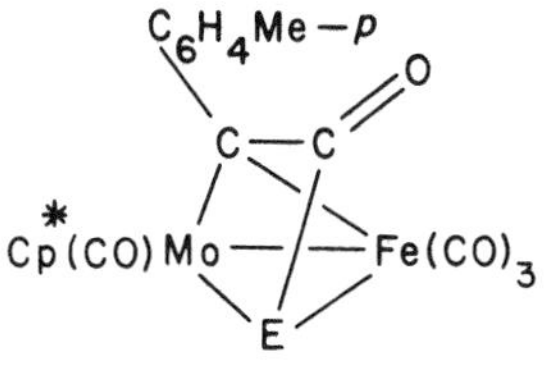

276

4. Miscellaneous

A doubly bridging thiolate ligand in **278** results from the reaction of **277** with elemental sulfur. In effect, the N donor group is replaced by sulfur, and a sulfur atom is inserted into the original Fe—C σ bond (184). The

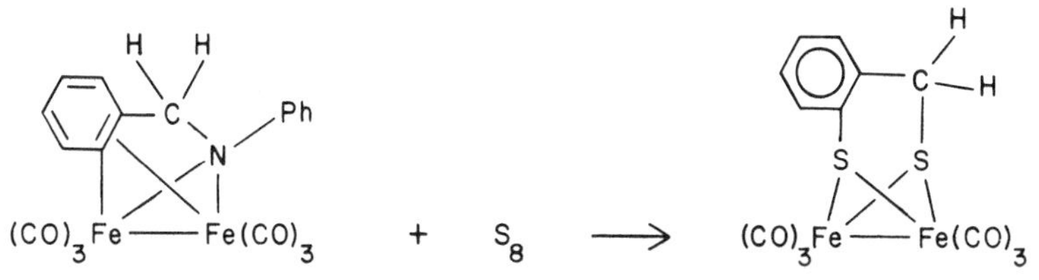

277

278

reaction of allyltrimethylsilane with elemental sulfur in the presence of triiron dodecacarbonyl leads to the addition of sulfur at an olefinic double bond and the formation of **279**. No reaction occurs in the absence of

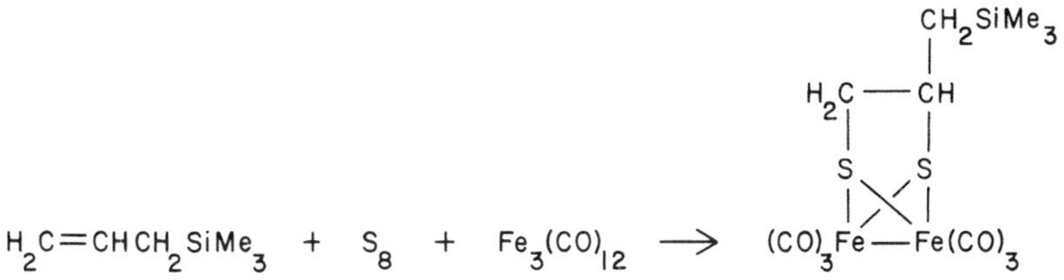

279

$[Fe_3(CO)_{12}]$ (*185,186*). From the reaction between norbornadiene, S_8, and $[Fe_3(CO)_{12}]$ three different complexes (**280–282**) of composition

| 280 | 281 | 282 |

$Fe_2(CO)_6(S_2C_7H_8)$ were isolated (*187*). When cyclohexene reacts with $[Fe_3(CO)_{12}]$ and S_8, sulfur adds to the organic molecule and the metal atoms without affecting the double bond (**283**) (*188*).

283

A heterocyclic carbene complex (**285**) results when cyclohexasulfur is allowed to react with the ylide complex **284**. The cyclization involves the incorporation of two sulfur atoms (*189*).

| 284 | 285 |

The anionic adduct **186**, prepared from $PhC\equiv CE^-$ (E = S or Se) and $[M(CO)_5\{C(OEt)Ph\}]$ (M = Cr or W), adds elemental sulfur or selenium to yield heterocyclic carbene, thione, or selone complexes (**286–291**) after alkylation, treatment with silica gel, or acidification with HCl (*190*). The

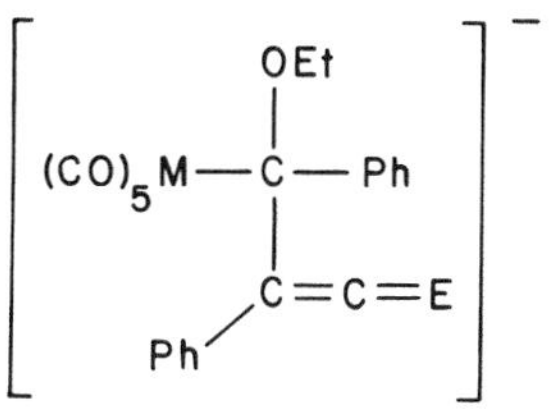

$$\left[\begin{array}{c} \text{OEt} \\ | \\ (CO)_5M - C - Ph \\ | \\ C = C = E \\ Ph \end{array} \right]^{-}$$

186

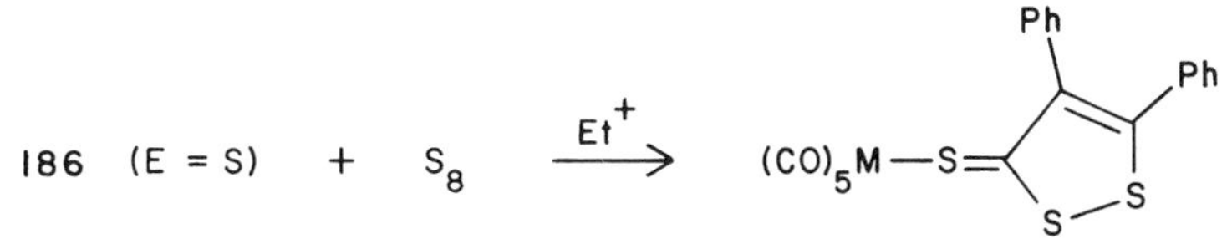

186 (E = S) + S_8 $\xrightarrow{\text{Et}^+}$ (CO)$_5$M—S

286

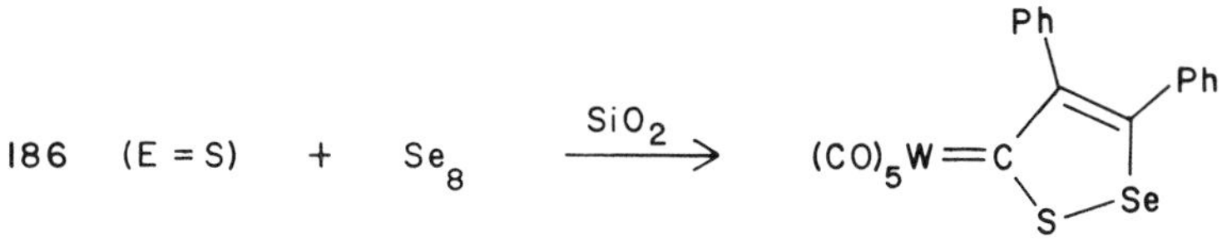

186 (E = S) + Se_8 $\xrightarrow{\text{SiO}_2}$ (CO)$_5$W=C

287

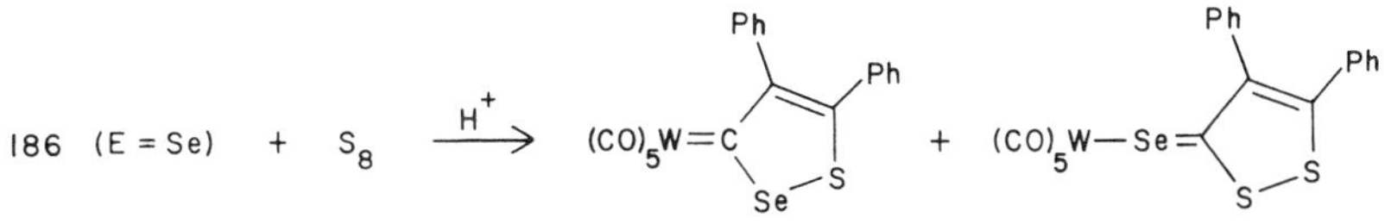

186 (E = Se) + S_8 $\xrightarrow{\text{H}^+}$ (CO)$_5$W=C + (CO)$_5$W—Se

288   **289**

186 (E = Se) + Se_8 $\xrightarrow{\text{H}^+}$ (CO)$_5$W=C + (CO)$_5$W—Se

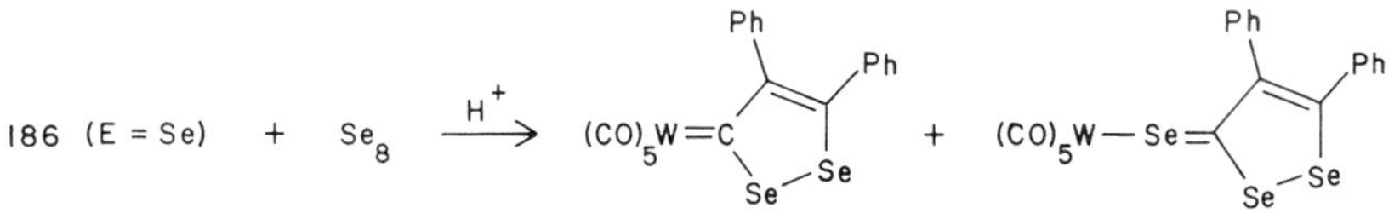

290   **291**

heterocyclic ligand in **292** is released as 4,5-diphenyl-1,2-dithia-4-cyclo-pentene-3-thione (**293**) on treatment of the metallacycle with elemental sulfur (*191*).

292   + S_8 $\longrightarrow$ 293

B. Bisulfide, Biselenide, Sulfide Ions, and Hydrogen Sulfide

The unstable thio-, seleno-, and telluroformaldehydes, CH_2S, CH_2Se, and CH_2Te, can be trapped in the coordination sphere of transition metals

294   +   NaEH $\longrightarrow$ 295

by reacting the carbenoid complex 294 [L = PMe_3, PMe_2Ph, PPr^i_3, CO, or $P(OMe)_3$], with NaSH, NaSeH, or NaTeH (192,193). The chalcogen atom is nucleophilic and can be alkylated (for E = S) or can act as a $2e^-$ donor when substituting THF in [M(CO)$_5$THF] (M = Cr, Mo, W) or [MnCp(CO)$_2$THF] complexes (194). To prepare a Co complex analogous to 295, namely, 297 (E = S or Se), the reaction conditions have to be

296   + CH_2X_2 +   NaEH $\longrightarrow$ 297

changed. Different solvents and a lower temperature are required. This complex can also be alkylated on the chalcogen, but protonation with HBF_4 leads to scission of the Co—CH_2 bond and formation of [CoCp(PMe$_3$)(μ-SMe)]$_2^{2+}$ (298). Complexes containing thio- and seleno-acetaldehyde ligands, such as 299, can be prepared similarly from 296. The related thio- and selenoketone complexes are not obtained, and the four-membered ring chelates (300, E = S, Se) form instead (195).

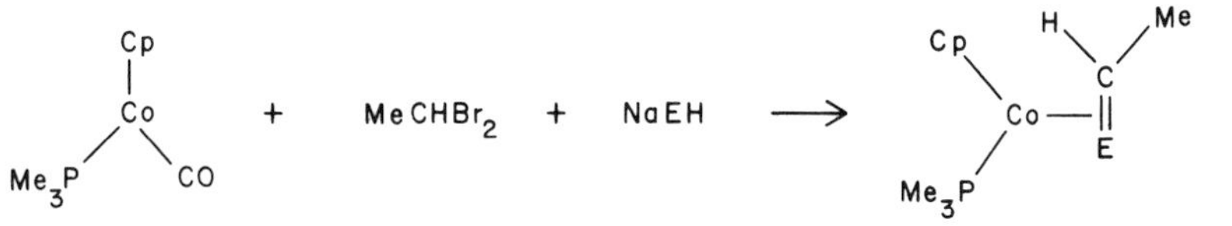

The nucleophilic attack of HS⁻ on the mixed carbonyl isocyanide complex **301** (R = *p*-tolyl) takes place on the carbon of the isocyanide ligand and gives the π-bound *p*-tolylisothiocyanate **302**, whereas MeO⁻ attacks at

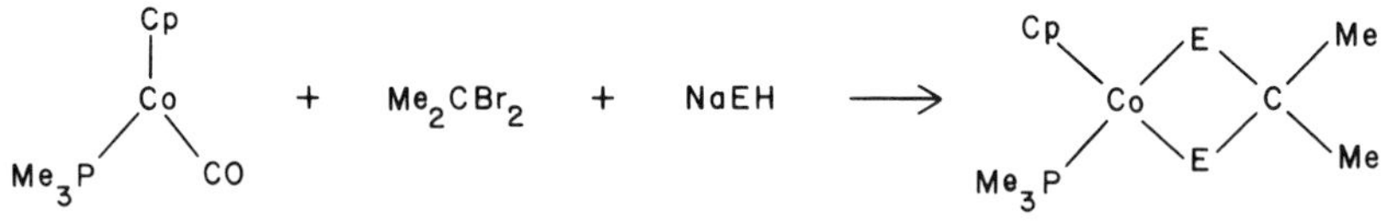

$$[OsCl(CO)_2(CNR)(PPh_3)_2] + HS^- \rightarrow [Os(CO)_2(PPh_3)(\eta^2\text{-SCNR})]$$

301 **302**

the carbonyl (*196*). The proposed carbene complex **303** (R = *p*-tolyl) {prepared from the carbyne complex [Os(CR)Cl(CO)(PPh₃)₂] and Cl₂}, reacts with HE⁻ (E = S, Se, Te) to yield thio-, seleno-, and telluroacyl derivatives

$$[OsCl_2(C(Cl)R)(CO)(PPh_3)_2] + HE^- \rightarrow [OsCl(CO)(PPh_3)_2(\eta^2\text{-ECR})]$$

303 **304**

(**304**) (*180*). The aminochloro carbene in the Ru complex **305** is converted to a dihapto-chalcocarboxamido ligand in **306** by reaction with HE⁻ (E = S, Se, Te) (*82*).

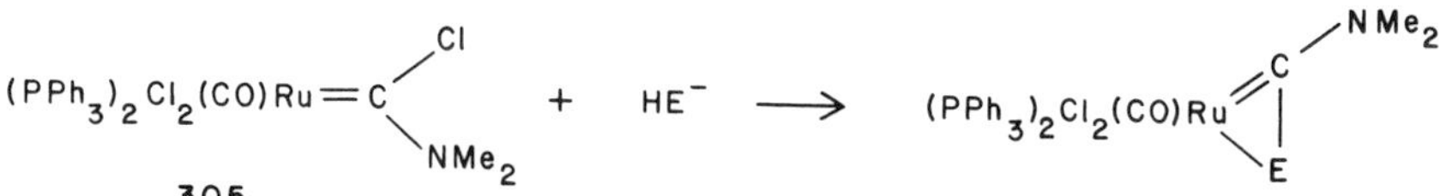

Addition of Na₂S to the methylidene complex **307** gives rise to a trinuclear sulfonium salt (**308**) (*108*). The action of a protic nucleophile, like H₂S, on the prochiral phosphinoketene ligand in **309** (R = Me, Ph) leads to five-membered chiral metallaheterocycles (**310**), which are organometallic derivates of γ-thiolactones (*197*).

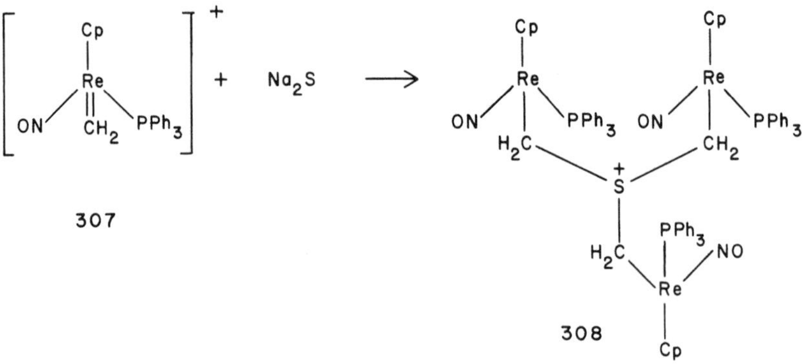

307   308

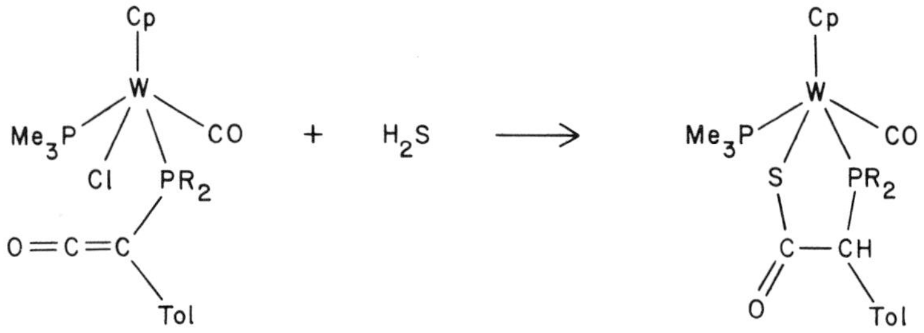

309   310

C. Carbon Disulfide and Carbon Diselenide

1. Insertion Reactions

Of the various bonding modes which dithio complexes, resulting from the insertion of CS_2 into M—X bonds, may adopt (198), types **D** and **E**

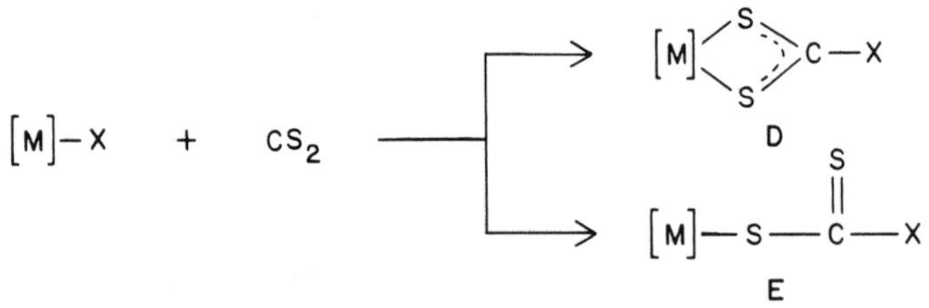

seem to dominate. The insertion of CS_2 into the M—C(sp^3 or sp^2) bonds of metal alkyl or aryl complexes, to form S,S[1]-bonded dithiocarboxylates (type **D**) has been noted for several metal complexes, mainly Os-, Ru-, and Ir-containing ones (198,199). Some type **D** analogs ([M] = FeCp(dppm),

TABLE III

CS$_2$ INSERTION REACTIONS LEADING TO
DITHIOCARBOXYLATE COMPLEXES

	[M]	X	Ref.
311	TiCp$_2$	C(Me)=CHMe	202
	RuCl(CO)	CH=CHPh	203
	RuCl(CO)	CH=CHBu^t	203
	RuCp(PPh$_3$)$_2$	C≡CPh	204
311a	ReCp(CO)(NO)	NHMe	205
	Mn(CO)$_4$(NH$_2$Me)	NHMe	205
	Re(CO)$_4$(NH$_2$Me)	NHMe	205

X = Me, Ph) are in dynamic equilibrium with their unusual η^3-S,C,S^1 isomers (*200*). Gold and copper complexes of both types (X = Cp) are now known (*201*).

Examples of analogous reactions for other M—C(sp^2) and M—C(sp) systems are less abundant, but they have been observed for the Ti and Ru

complexes (**311**, (Table III). The carbamoyl complexes (**311a**, Table III) give rise to type **D** thiocarbamate complexes (**312**, X = NHMe) and resemble organic amines, rather than amides, in their reaction with CS$_2$.

$$[M]—C(O)X + CS_2 \rightarrow 312$$
311a

The formation of the tripod ligand in **314** is suggested to occur via the insertion (type **D**) of CS$_2$ into the Ru—P bond of the Ru(η^2-CH$_2$PMe$_2$) unit of **313**, followed by migration of the hydride ligand to the α-thia carbon (*206*).

A η^3-dithiocarboxylate also (see Ref. *200*) results when CS_2 reacts with the phosphine-π-olefin chelating ligand in **315**. The hydride ligand is transferred to the olefinic system and CS_2 inserts into the metal–carbon bond. The S_2C unit in **316** is formulated as a dithioallylic group which acts as a $3e^-$ donor (*207*).

315   316

In their reaction with CS_2 or CSe_2 the η^2-coordinated acetylene and formaldehyde ligands in **317** and **319** are converted to heterometallacycles [**318** (M = Co or Rh, R = Ph or CO_2Me) and **76**] by a type **E** insertion into an M—C and M—O linkage, respectively. Compound **318** (R = Ph) provides 4,5-diphenyl-1,2-dithio-4-cyclopentene-3-thione on reaction with elemental sulfur, and this reaction sequence is regarded as a possible route to organic heterocycles (*191*); **76**, when heated, gives a selenoformaldehyde complex, $[Os(CO)_2(\eta^2\text{-}CH_2Se)(PPh_3)_2]$ (**77**) (*86*).

317   318

319   76

In the palladium complex [PdMe{OCH(CF$_3$)Ph}(dppe)] (**320**), which contains both a Pd—R and a Pd—OR bond, CS$_2$ inserts into the Pd—OR bond and gives the type **E** complex [PdMe(dppe){SC(S)OCH(CF$_3$)Ph}] (**321**) (*208*). The iron complex [FeCp(CO)$_2${C(O)NHMe}] (**322**), in contrast to the other carbamoyl complexes (Table III), follows the model leading to **E** in its reaction with CS$_2$ and forms [FeCp(CO)$_2${SC(S)NHMe}] (**323**) (*205*).

The usual behavior of CS$_2$ toward M—H bonds is to insert according to model **D** and form a dithioformate ligand (*209*). An insertion into an Fe—H bond {in [FeHCp(dppe)]} has been reported (*210*). The reduction brought about by two molecules of [Os$_3$H$_2$(CO)$_{10}$] proceeds one step further to give a coordinated CH$_2$S$_2$ unit (**324**) (*211*). In the reaction of

$$[Os_3H_2(CO)_{10}] + CS_2 \xrightarrow{\Delta} H_2CS_2[Os_3(\mu\text{-}H)(CO)_{10}]_2$$

324

[Os$_3$(μ-H)$_2$(CO)$_9$(PMe$_2$Ph)], however, the cluster molecule transfers both its hydrides to the CS$_2$, and a C—S bond is cleaved to produce a cluster complex (**325**) containing a bridging thioformaldehyde and an inorganic sulfide (*212*). The other product from this reaction is the bridged dithioformate complex **326**. When the concentration of the starting material in the

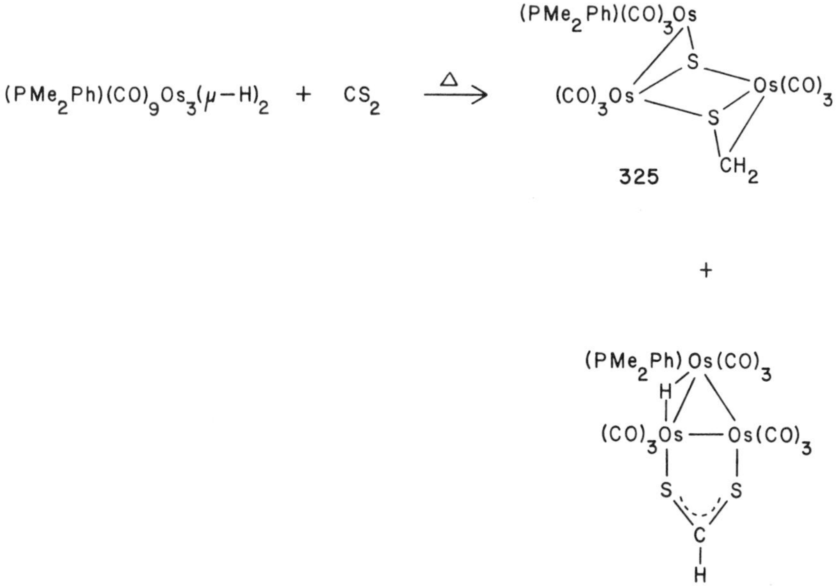

(PMe$_2$Ph)(CO)$_9$Os$_3$(μ—H)$_2$ + CS$_2$ $\xrightarrow{\Delta}$

325

+

326

reaction mixture is high (0.1 M), the phosphine-substituted derivative of **324**, $H_2CS_2[Os_3(\mu\text{-H})(CO)_9(PMe_2Ph)]_2$ (**327**), is obtained as a third major product (*213*).

2. *Addition Reactions*

Unlike the ruthenium alkynyl complex in Table III, the iron complex **328** does not undergo insertion with CS_2, but forms a [2 + 2] cycloaddition product (**329**) containing a 2H-thiete-2-thione (β-dithiolactone) functional group, which can react further with electrophiles, for example, MeI. The CS_2 addition was found to be reversible in solution and in the solid

$$Cp(dppe)Fe - C \equiv CMe \quad + \quad CS_2 \quad \longrightarrow \quad Cp(dppe)Fe - C \cdots C = S$$

          328   **329**

state. The unusual cycloaddition of CS_2 is ascribed to the high electron density on the alkynyl β carbon atom (*214*). This does not, however, account for the difference in the reactions of the Fe and Ru complexes. The vinyl complex **330**, with its highly nucleophilic β carbon, adds CS_2 at that

          330   **331**

carbon and is converted to a thermolabile zwitterionic carbene complex (**331**), which can be alkylated to form $[FeCp\{C(OMe)CH_2C(S)SMe\}(PMe_3)_2]^+$ (**332**) (*215*).

Two molecules of carbon disulfide add stepwise to the carbanionic complexes, generated by consecutive deprotonation of $[Cr(CO)_5\overline{S(CH_2)_3SCH_2}]$ [**110**, $R^1R^2 = (CH_2)_2S$], and a complex (**334**) containing the neutral tridentate ligand $S(CH_2)_3SC\{C(S)SEt\}_2$ results (*216*). The deprotonated amino carbene ligand in **335** is converted to the thiocarbene ligand in **336** on treatment with CS_2 and $[OEt_3][BF_4]$, whereas the anionic carbene complex **337** adds CS_2 to form the vinylcarbene complexes **338** and

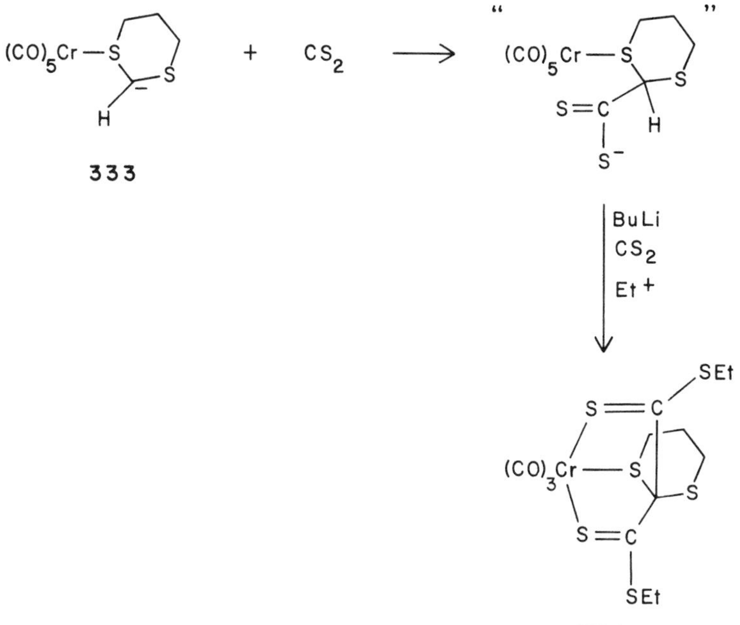

333

334

339. Both **335** and **337** are turned into the triselenocarbonate complex **340** when treated with CSe₂ and [OEt₃][BF₄] (*217*). The anion **341,** derived

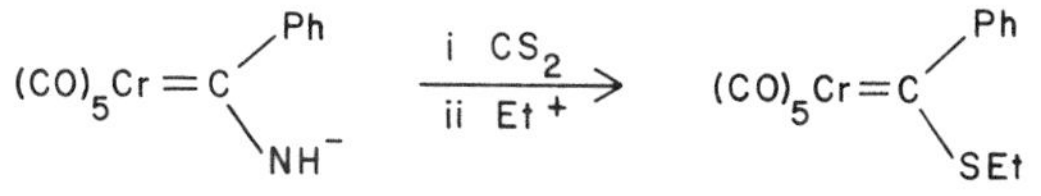

335

336

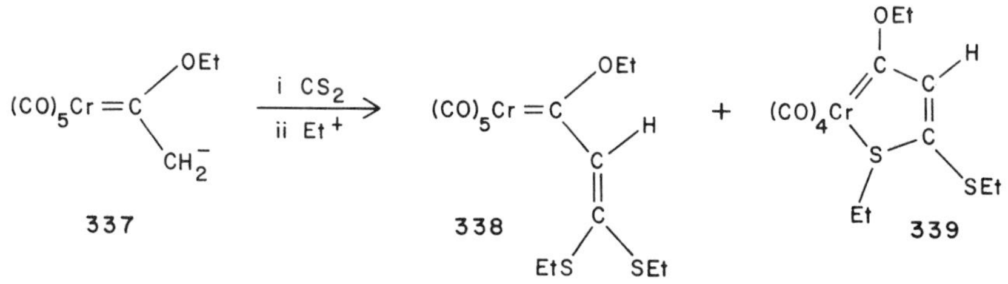

337

338

339

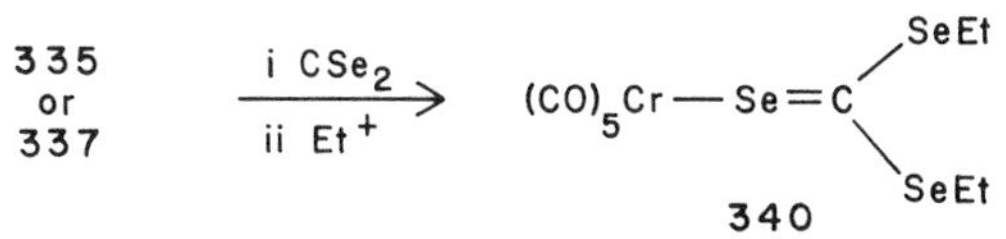

340

from coordinated dimethyl thioformaldehyde, leads to the trithiocarbonate complex **342** on reaction with CS_2, but to **340** and **343** when CSe_2 is added

before alkylation (*154*). When the 1,3-dithiolane-2-thione complex **344** is deprotonated and treated with CS_2, no net addition of CS_2 occurs; however, a study with $^{14}CS_2$ showed that CS_2 was incorporated into the molecule, and a rearrangement was postulated. Alkylation leads to the α,β-unsaturated thioaldehyde complex **345** (*101*).

The acetone ligand of two molecules of the coordinated imidoyl adduct **134** is replaced by CS_2 to form a dimetalla spiroheterocyclic compound (**346**) (*112*). The heterocyclic compounds 1,2-dithiopyrone (**348**, R^1, R^2, R^3, R^4 = Ph or H) and 2-thiopyridone (**349**) can be prepared by the

134   +   CS₂   →   346

347   +   CS₂   →   348

347   +   MeNCS   →   349

addition of CS$_2$ and methylisothiocyanate to the cobaltacyclopentadiene complex **347** (R^1, R^2 = Ph, R^3, R^4 = Ph or H) (*219*).

3. *Dimerization Reactions*

Head to tail coupling of CSe$_2$ (and CSSe or CS$_2$) occurs when the cobalt (**350**), rhodium (**352**, L = PMe$_3$, PMe$_2$Ph) (*220*), and platinum (**354**) (*221*) phosphine complexes as well as the nickel complex **356** (R = Me, Et) (*222*) are treated with CSe$_2$, CSSe, PMe$_2$Ph, and R$_3$PCS$_2$, respectively, to give **351**, **353a** or **353b** ([Rh] = CpLRh), **355**, and **357**.

When the molybdenum diethyldithiocarbamate (dete) complex **358** is treated with CS$_2$ in the presence of PPh$_3$, the phosphine serves to abstract sulfur from CS$_2$. The resultant CS fragments are incorporated into two new bridging ligands of complex **359**. One CS formally inserts onto an M—S

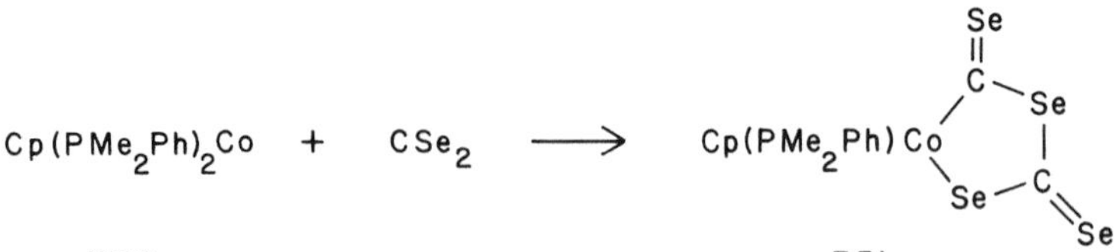

$$Cp(PMe_2Ph)_2Co \quad + \quad CSe_2 \quad \longrightarrow \quad Cp(PMe_2Ph)Co \underset{Se}{\overset{\overset{Se}{\parallel}\;C\;-\;Se}{\bigg|}} C \overset{Se}{\underset{\diagdown}{\diagup}}$$

350   351

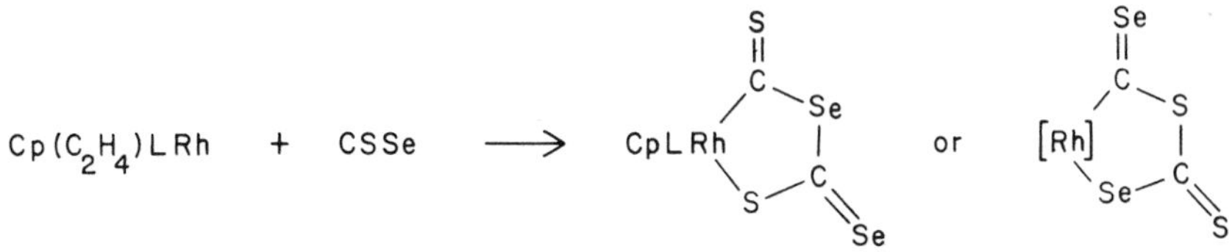

$$Cp(C_2H_4)LRh \quad + \quad CSSe \quad \longrightarrow \quad CpLRh \underset{S}{\overset{\overset{S}{\parallel}\;C\;-\;Se}{\bigg|}} C \overset{}{\underset{Se}{\diagdown}} \quad or \quad [Rh] \underset{Se}{\overset{\overset{Se}{\parallel}\;C\;-\;S}{\bigg|}} C \overset{}{\underset{S}{\diagdown}}$$

352   a   353   b

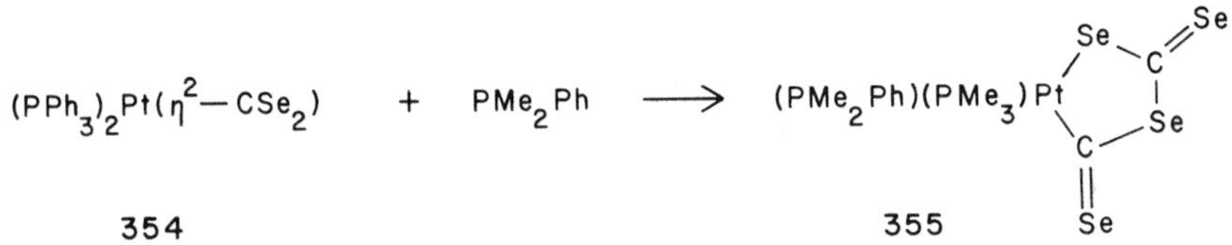

$$(PPh_3)_2Pt(\eta^2-CSe_2) \quad + \quad PMe_2Ph \quad \longrightarrow \quad (PMe_2Ph)(PMe_3)Pt \overset{Se}{\underset{C}{\bigg|}} \overset{C}{\underset{Se}{\diagup\diagdown}} Se$$

354   355   Se

$$Ni(cod)_2 \quad + \quad R_3PCS_2 \quad \longrightarrow$$

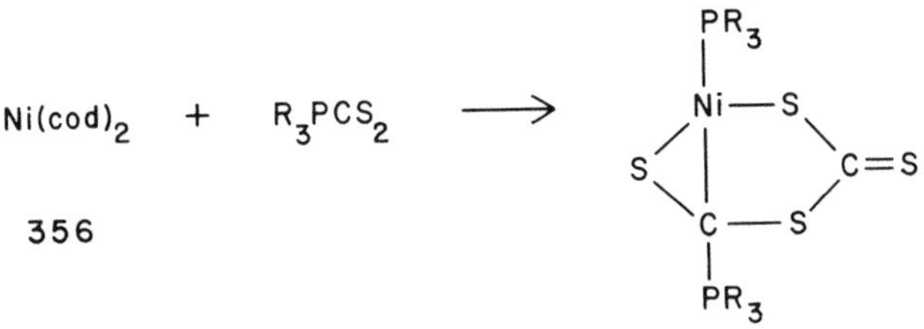

356

357

bond of a dithiocarbamate chelate, and the other couples with CS_2 to form an $-SC(S)SC-$ group (*223*).

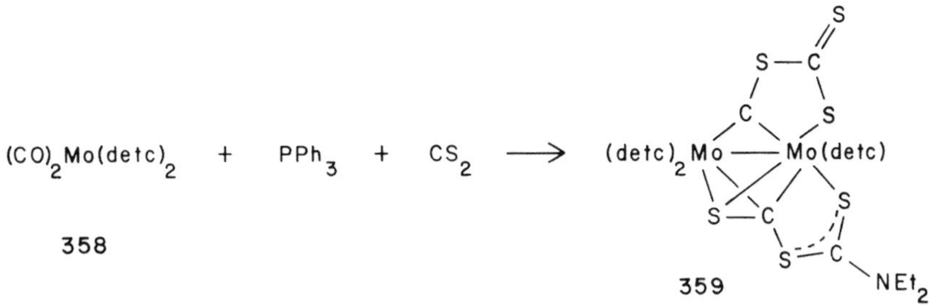

$$(CO)_2Mo(detc)_2 \quad + \quad PPh_3 \quad + \quad CS_2 \quad \longrightarrow \quad (detc)_2Mo \longrightarrow Mo(detc)$$

358   359 NEt_2

The metal-activated head to head dimerization of CS_2 has been reported for Fe (*224*), Me_5C_5—Ni, Cp—Ni (*225*), and Cp—Ti (*226*) carbonyl complexes. The ligands in the Fe and Cp—Ni complexes **360** and **362** are formally regarded as derivatives of ethenetetrathiol, whereas in the Me_5C_5—Ni and Ti complexes **361** and **363** the highly delocalized π-electron systems suggest a major contribution from a bis(dithiolene)-like tetrathio-oxalate structure. The bonding pattern drawn in **361** and **363** does

not represent true bonds (4.5 bonds at C) but rather a superposition of three contributing structures

VI

REACTIONS OF COORDINATED ORGANOSULFUR AND ORGANOSELENIUM COMPOUNDS WITH ELECTROPHILES AND NUCLEOPHILES

The reactions of coordinated CS_2 and of thiocarbonyl complexes have been covered in reviews by Bianchini *et al.* (5) and Broadhurst (3).

A. Reactions with Electrophiles

1. Acetylenes (and PhC≡N—NPh)

Reactions of organosulfur and -selenium complexes with acetylenes (usually activated) either result in formal insertion of the acetylene into an S—C, Se—C, M—C, M—S, or M—Se bond or resemble a cycloaddition reaction to elements of the original complex. These descriptions do not, of course, have any mechanistic implications.

Formal insertion of the acetylene molecule into an S—C, Se—C, or Te—C bond occurs when coordinated thio- and selenoaldehydes, selenoketones, and telluroketone (**364**, M = Cr or W, E = S, Se, or Te, R^1 = H or Ph) react with 1-diethylamino-prop-1-yne (R^2 = Me) or bis(diethylamino)acetylene (R^2 = NEt_2). Regiospecific [2 + 2] cycloaddition, for which a concerted biradicaloid mechanism is proposed (227), followed by stereospecific (Me and Ph cis) electrocyclic ring opening gives the thio-, seleno-, and telluroacrylamide complexes (**365**) (228–230).

$$[(CO)_5M\{E=C(Ph)R^1\}] + R^2C\equiv CNEt_2 \rightarrow E\text{-}[(CO)_5M\{E=C(NEt_2)CR^2=C(Ph)R^1\}]$$

364   **365**

Reaction of the η^2-coordinated dithioformate complex **366** with activated acetylenes (Z = CO_2Me, CO_2Et), however, leads to the insertion of both a CO ligand and the acetylene to afford **367** (231).

366   +   ZC≡CZ   ⟶   **367**

The η^2-vinyl complex **368** (M = Mo, W, R^1 = Et or Pr^i) gives the oligomerization products **369** (M = W, R^1 = Pr^i) and **370** (R^1 = Et or Pr^i) by linking the hexafluorobut-2-yne and η^2-vinyl units on reaction with the alkynes, $R^3C\equiv CR^3$, but both **368** and the bisalkyne compound **371** ($M = W$, $R^1 = p\text{-MeC}_6H_4$) form the insertion product **85** when treated

with $R^2C\equiv CMe$ ($R^2 = Me$ or Ph). The formation of **370** ($M = W$) also involves the unusual transfer of a fluorine atom from a carbon atom to the metal (*90, 232*).

The cyclopalladated benzyl methyl sulfide complex **372** gives a product (**373**) in which the acetylene ($R^1 = CF_3$, CO_2Me, or Ph, $R^2 = CF_3$,

CO_2Me, or Ph) has caused scission of the Pd—C bond. The metallacyclic complexes **373** ($R^1 = R^2 = Ph$) and **375** react with phenylacetylene to

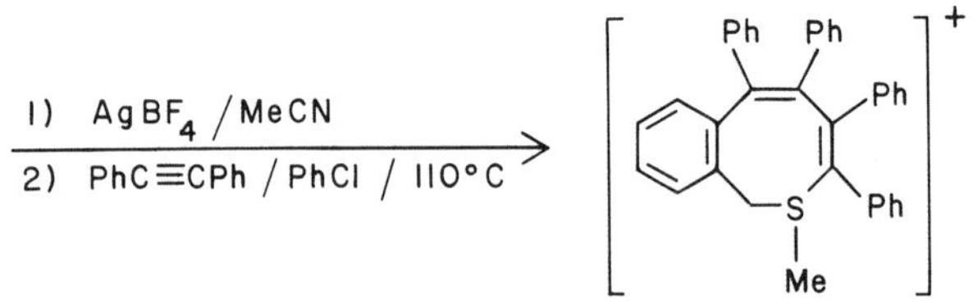

374

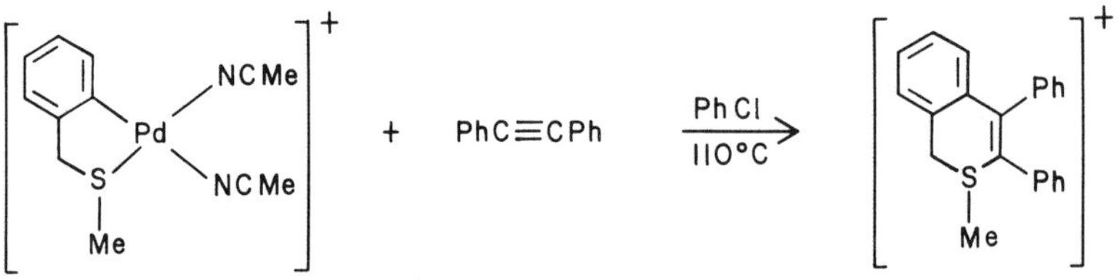

375

376

form intermediates which undergo reductive elimination (with concurrent intramolecular oxidative coupling of the leaving groups) to form the organosulfur compounds **374** and **376**. With the cationic complex **377**, on the other hand, two molecules of acetylene insert between the Pd and the aromatic ring to give **378**. The variation in reactivity of **372**, **375**, and **377**

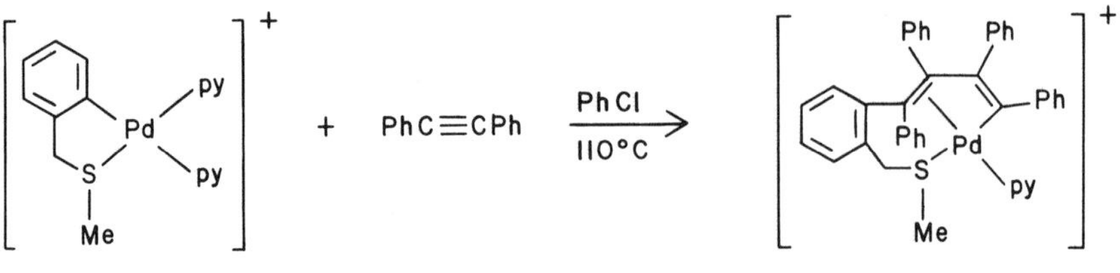

377

378

in the order **375** > **377** > **372** is ascribed to the differences in nucleophilicity of the spectator ligands ($Cl^- > py > MeCN$) (*233*). Formation of C—C bonds also occurs when the unsaturated cluster compound **379** reacts with two molecules of PhC≡CH to yield **380** (*234*).

Various insertion reactions occur when metal thiolate complexes are treated with fluorinated alkynes under different reaction conditions. The manganese thiolate complexes of type **381** ($R = CF_3$, C_6F_5) react with hexafluorobut-2-yne at 20°C in pentane to give an insertion product acting

379   380

as a $3e^-$ donor (382), and the reaction of [FeCp(CO)$_2$SR] (383, R = CF$_3$, C$_6$F$_5$) gives rise to a vinylic ligand in (384), whereas the cobalt thiolate 385 leads to a dimeric product (386), containing a flyover ligand, after the

381   382

383   384

385   386

insertion of two acetylene molecules (235). The alkyl thiolate complexes 387 (M = Fe or Ru, R = H or CF$_3$) form compounds of the well-known type 388 (236). With the dimeric complex [Fe(CO)$_3$SR]$_2$, CF$_3$C≡CCF$_3$ inserts not into the Fe—S bond but, instead, into the Fe—Fe bond (235). [FeCp(CO)SMe] (389), on the other hand, leads to 388 (M = Fe, R = CF$_3$) and 390 (236). Ashby and Enemark described the formation of type 388

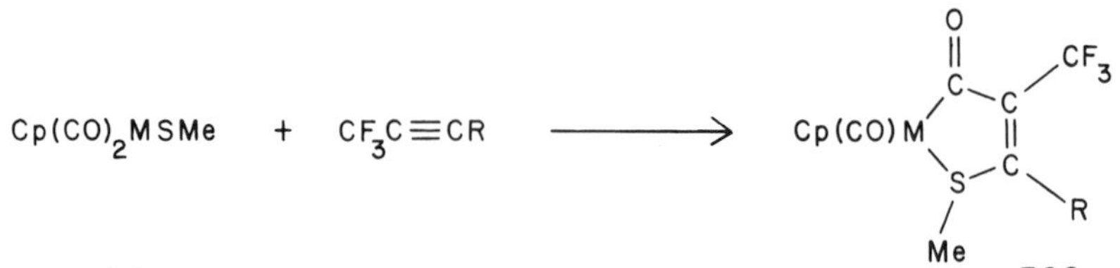

$$Cp(CO)_2MSMe \quad + \quad CF_3C\equiv CR \quad \longrightarrow \quad 388$$

387   **388**

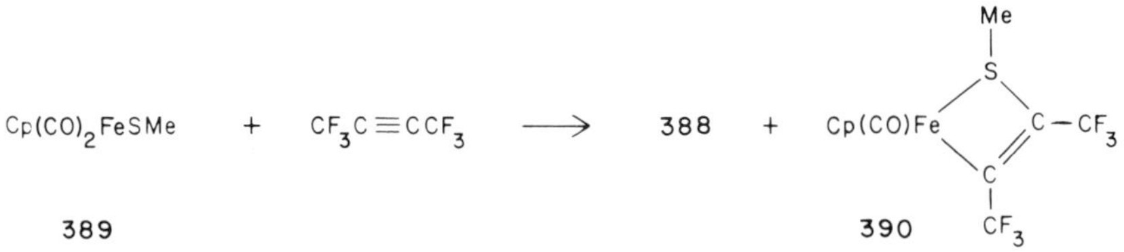

$$Cp(CO)_2FeSMe \quad + \quad CF_3C\equiv CCF_3 \quad \longrightarrow \quad 388 \quad + \quad 390$$

389   **390**

complexes in terms of the $[4\pi + 2\pi]$ 1,3-dipolar cycloadditions of organic chemistry. Accordingly, the S—Fe—C=O fragment possesses four π electrons delocalized over the fragment to make it isolobal with the allyl anion. A correlation diagram showing the similarities between the organometallic and organic reactions is given (*237*).

The highly reactive *E*-dithiolene complexes (*E*-**220**) could be trapped in a [2 + 2] cycloaddition product with activated acetylenes to form **391**

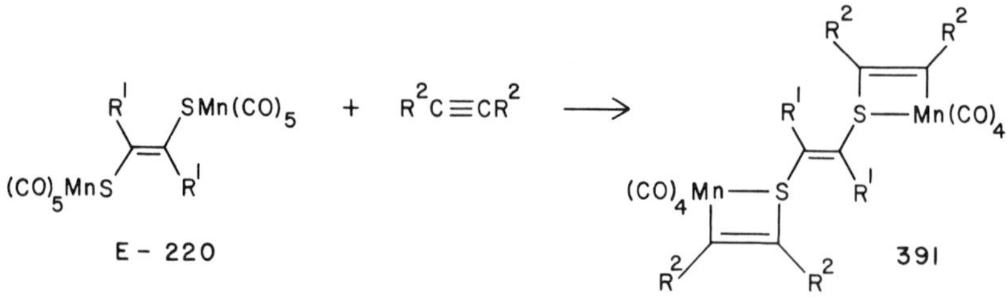

E – 220   391

$(R^1 = Ph, \ p\text{-}FC_6H_4, \ p\text{-}ClC_6H_4, \ p\text{-}FC_6H_4; \ R^2 = CO_2C_6H_{11}, \ CO_2CHMe_2)$, which contains two thiamanganacyclobutene rings (*48*). The anionic thiolate complex **392** reacts with an activated acetylene and an acid chloride to give **393** ($R^1 = Et$, Bu^t, or Ph, $R^2 = H$ or CO_2Me, $R^3 = CO_2Me$ or $C(O)Me$, $R^4 = Me$, Bu^t, or Ph), in which a carbonyl, bridging the vinyl and thiolato ligands, has been incorporated into the organic framework. Protonation instead of acylation of the reaction mixture causes loss of the *tert*-butyl group of thiolate **394** (R^2 and R^3 as for **393**) (*238*).

In the reaction of fluoroalkynes, $RC\equiv CH$ ($R = C_6F_{13}, C_8F_{17}$), with the thiolates [MCp(CO)$_3$SMe] (**395**, M = Mo, W), a cis (40%)–trans (60%) mixture of the vinylic sulfide RCH=CHSMe (**396**) is obtained (*239*).

The addition products, types **G** and **H**, arising from the reaction of activated acetylenes with CS$_2$ complexes **F** were reviewed by Bianchini *et*

al. (*5*). A theoretical study using extended Hückel calculations and qualitative molecular orbital theory to analyze the electronic features of these compounds and their possible interconversion pathways has been done since (*240*). In an extension of this work the Mo complex **397** was treated with HC≡CCN and found to react, essentially, as the unit **ii**, adding three molecules of alkyne to give **398**. When **397** is treated first with an alkyne,

RC≡CR (R = CF$_3$, CO$_2$Me), and then with cyanoethyne, a new type of combination of CS$_2$ with two acetylenes follows, and **400** is formed via **399** in a reaction involving C—C and C—S bond formation (*241*).

The coordinated heterocycle 2,4-diphenyl-1,3,4-thiadiazolethione (**402**)

results when the electron-deficient 1,3-dipole PhC≡N—NPh adds to the CS$_2$ complex **401** [L—L = dmpe, 1,2-C$_6$H$_4$(PPh$_2$)$_2$] (*242*). The metallacyclobutadiene complex **404** is made by the reaction of [Os(CO)(CS)(PPh$_3$)$_3$] (**403**) with PhC≡CPh; the reaction can be seen either as an addition of PhC≡CPh to the Os—CS bond or as an insertion of CS into an Os—C(Ph) bond. Another product from the same reaction (**405**) is readily converted to **404** by the addition of CO (*243*).

The reaction of [Ru(μ-H)$_3$(μ_3-CSEt)] (**406**) with alkynes, RC≡CR (R = Me, Ph), gives two isomeric products, **407** and **408**, in which a C—C

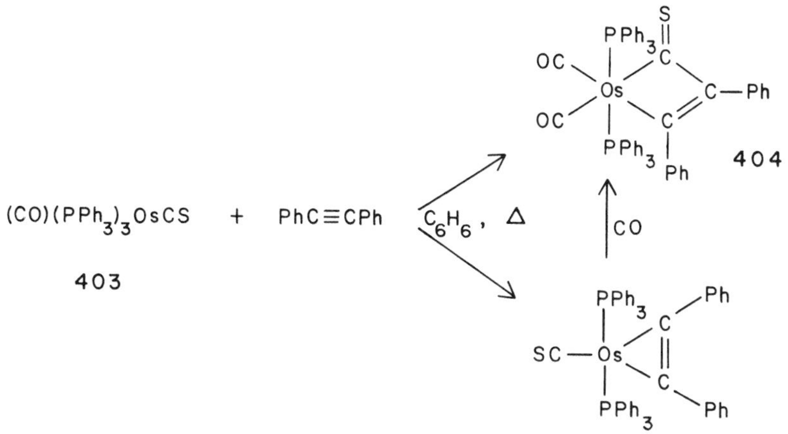

$(CO)(PPh_3)_3OsCS$ + $PhC\equiv CPh$ $\xrightarrow{C_6H_6, \Delta}$

403

404

405

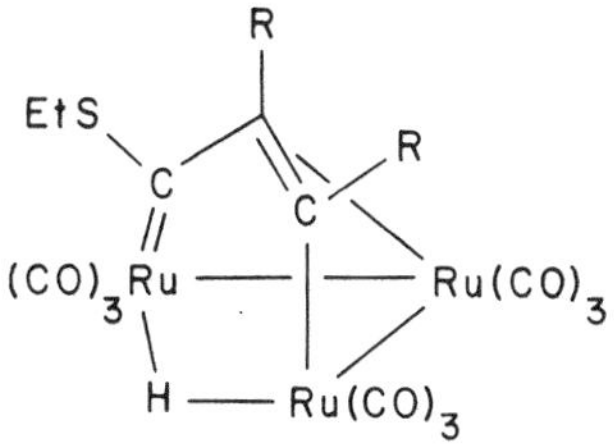

407

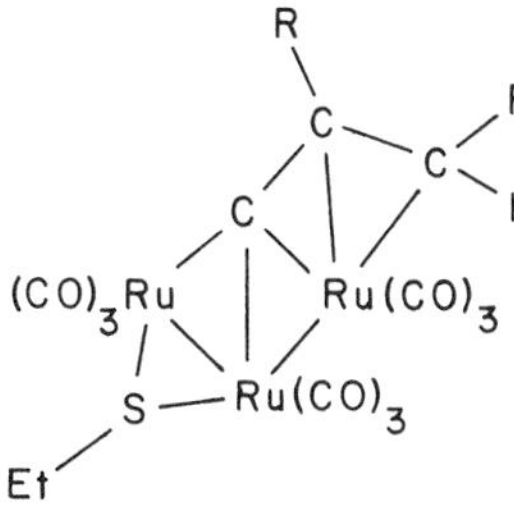

408

bond is formed between the alkyne and the alkylidyne ligand. Compound **407** can be isomerized thermally to **408** (*244*). A series of unsaturated adducts of type **410** were synthesized by the displacement of ethylene from the dimeric molybdenum complexes (**409**, $C_n = CH_2$, CMe_2, $CH=CPh$, $C=S$, R^1, $R^2 = H$, Me, or Ph) (*245*).

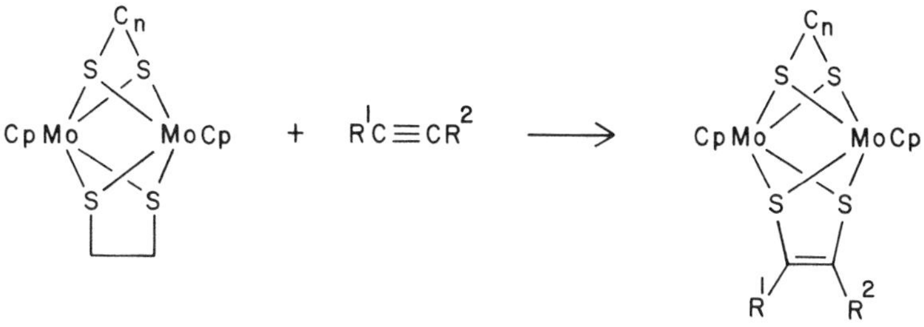

$CpMo$ ... $MoCp$ + $R^1C\equiv CR^2$ $\longrightarrow$ $CpMo$ ... $MoCp$

409

410

2. *Olefins [and $(CF_3)_2C$=N—$C(O)R]$*

The thioaldehyde, selenoaldehyde, selenoketone, and telluroketone complexes $[M(CO)_5\{E$=$C(Ph)R\}]$ (**411**, M = Cr or W, E = S or Se, R = H, Ph, p-BrC_6H_4, p-MeC_6H_4, p-$MeOC_6H_4$, p-$NMe_2C_6H_4$, or p-$CF_3C_6H_4$) react with cyclopentadiene via [4 + 2] cycloaddition to give metal-coordinated thia-, selena-, and telluracyles (**412**). Other dienes investigated

were *trans*-1-methyl-1,3-butadiene, 2-methyl-1,3-butadiene, 2,3-dimethyl-1,3-butadiene, and 1,3-cyclohexadiene (*246–248*).

The reaction between the thiocarbene complex **413** and $(CF_3)_2C$=$NC(O)R$ (R = Ph, p-FC_6H_4) leads to scission of the carbene ligand and transfer of the nucleophilic SMe fragment to the electrophilic center of the acylamine. Protonation, effected either by the rest of the carbene or the solvent, results in a thioether derivative (**414**) (*249*).

3. *Alkylating and Acylating Reagents and Protons*

Alkylation reactions which induce rearrangements or lead to further reactions of the organosulfur ligands are our concern here. Once again, reactions involving thiocarbonyl, dithiocarbene, and CS_2 complexes are not discussed since they have been reviewed (*3–5*).

A class of compounds which have drawn some interest are the sulfonium salt complexes **416**, prepared by alkylation of sulfide derivatives (**415**) (Table IV). These coordinated ylides have found use as stable, stereospecific cyclopropanating reagents yielding compounds of the class **417**.

The dithioester complexes of type **418**, in which the ligand coordinates as an allylic type, $4e^-$ donor, are converted by alkylation to the stabilized, $2e^-$ donor dithiocarbenium complexes (**419**) (*256*). Addition of CS_2, fol-

TABLE IV
Sulfonium Salt Complexes $\{[M]-CHR^1S(Me)R^2\}^+$

[M]	R^1	R^2	Comments	Ref.
NiCpL; L = PPh₃, PPhMe₂, P(OMe)₃	H	Me	Limited use as cyclopropanating reagent	250
FeCp(CO)₂	H	Me	Stable methylene transfer reagent	251,252
MoCp(CO)₃	H	Me	Less stable and less effective than Fe complex	252
FeCp(CO)₂	(CH₂)₄CH=CH₂ (CH₂)₃CH(CH₂)₄C=CH₂	Ph	Intramolecular cyclopropanation, **416** not isolated	253
FeCp(CO)₂	Me	Ph	Ethylidene transfer reagent, **416** not isolated	254
PtClL₂; L = PPh₃, PMePh₂	H	Me	Not tested for cyclopropanation	255

$$[\text{M}]-\text{CHR}^1(\text{SR}^2) \ + \ \text{Me}^+ \ \longrightarrow \ \left[[\text{M}]-\overset{R^1}{\underset{}{\text{CH}}}-\text{S}\overset{R^2}{\underset{\text{Me}}{}}\right]^+ \ \longrightarrow$$

415   416   417

418   419

lowed by alkylation with MeI, transforms an electron-rich alkynyl complex (**328**) to the cationic methyl(dithiocarbomethoxy)vinylidene complex (**420**) via the β-dithiolactone ring in **329** (*214*). A coordinated acetylene ligand (in **422**), not previously prepared, results when the μ²-ketenyl complex **421** is alkylated (*257*). A similar product (**424**) results from [W{C(SMe)C=O}(CO){HC(pz)₃}(PMe₃)] (**423**) (*258*).

Reaction of [Pt(η²-CS₂)dppe] (**425**) with excess MeI results in the formation of a dimeric complex (**426**) held together by a novel bridging CS₂C(SMe)₂ group, formally the condensation product of a CS₂ ligand and a dithiocarbene ligand. This can be achieved by nucleophilic attack of coordinated CS₂ on a carbene ligand and results in the formation of a new carbene. The same product forms when **425** is reacted with the cationic dithiocarbene complex **427** (*259*).

Reaction of the bridging cyanocarbene complexes **428** [Y = SMe or N(Me)C(O)SMe] with $MeSO_3CF_3$ produces the cationic carbene complex **429** [with elimination of MeNCO when Y = N(Me)C(O)SMe] (*260*). An organosulfur compound (**431**) is formed in addition to the neutral alkylated complex **432** on alkylation of the trithiocarbonato complex **430** (*261*).

The anionic adduct **186**, resulting from the reaction between $[M(CO)_5\{C(OR^1)Ph\}]$ (M = Cr, W, R^1 = Me, Et) and Li[EC≡CPh] (E = S or Se), rearranges to the carbene complex **71** on addition of alkylating

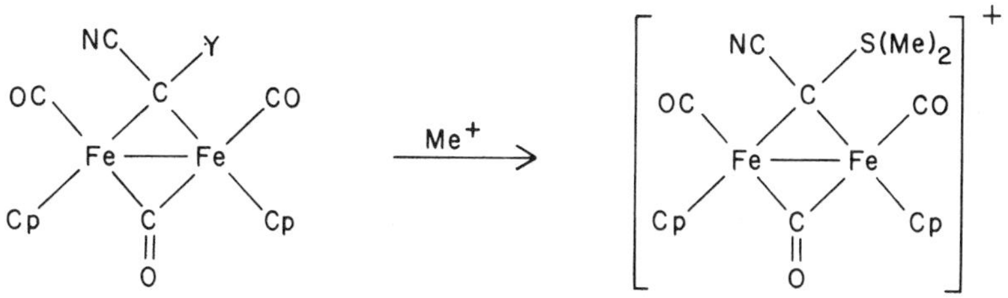

428   **429**

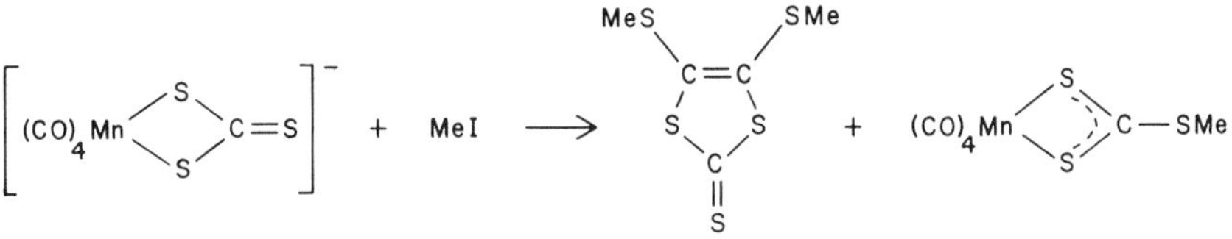

430   **431**   **432**

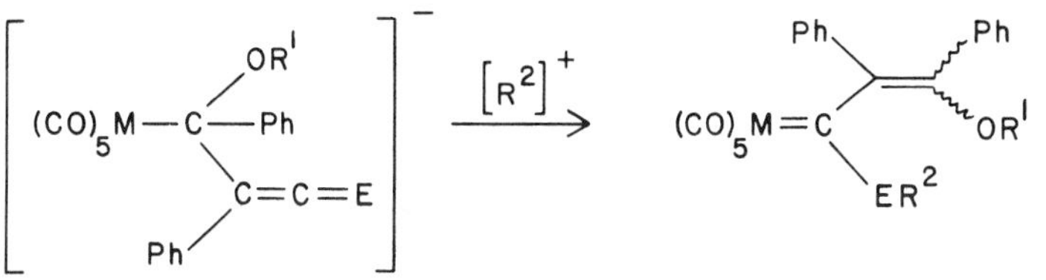

186   **71**

agents ($[R^2]^+$), $[OEt_3][BF_4]$, or MeI (84). Adduct **186** also functions as a nucleophilic synthon for coordinated thio- or selenoacyl anions, $[M(CO)_5(E{=}CR)]^-$ [R = C(Ph)=C(OEt)Ph], when reacting with protons to form thio- or selenoaldehyde complexes of type **433** chemoselectively.

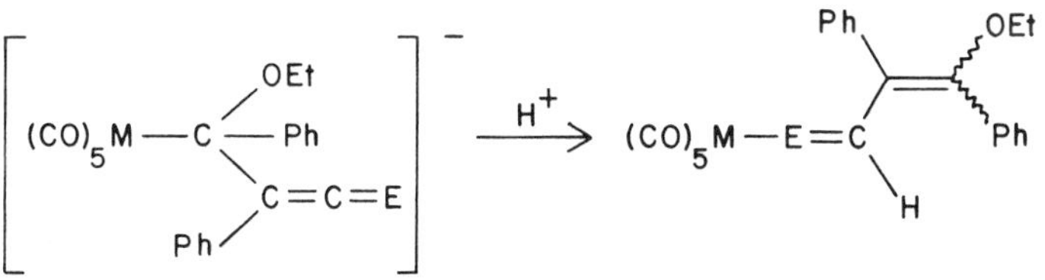

186   **433**

The metallic electrophile $[Fe(Cp)(CO)_2]^+$ reacts similarly to Et^+ with **186** (E = S) to give **434,** whereas $[Au(PPh_3)]^+$, like H^+, affords a tungsten thione complex (435) by bonding to the available nucleophilic carbon atom (262).

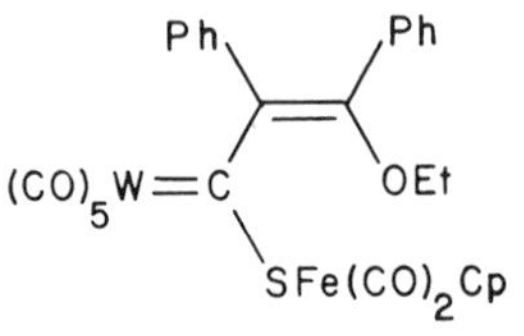

434

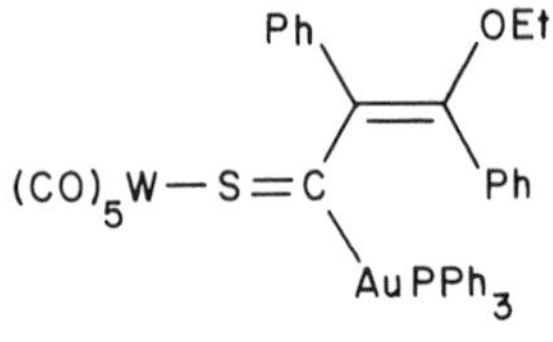

435

Alkylation with MeI causes liberation of the enethiolato ligand from [NbCp$_2$(S=C=CR$_2$)(η^2-SCH=CR$_2$)] (147) as the thioether, CH$_3$SCH=CR$_2$ (436) [R$_2$ = CMe$_2$(CH$_2$)$_3$CMe$_2$]. The remnant is bonded to the metal in the cationic complex [NbCp$_2$(η^2-S=C=CR$_2$)]$^+$ (437) (121). The zirconium sulfide complexes 438 [R$_2$ = (p-MeOC$_6$H$_4$)$_2$, (p-MeC$_6$H$_4$)$_2$, (p-Me$_2$NC$_6$H$_4$)$_2$, (Ph and p-MeOC$_6$H$_4$)], react with, inter alia, acid chlorides and methyl vinyl ketone to produce thioesters (439) and

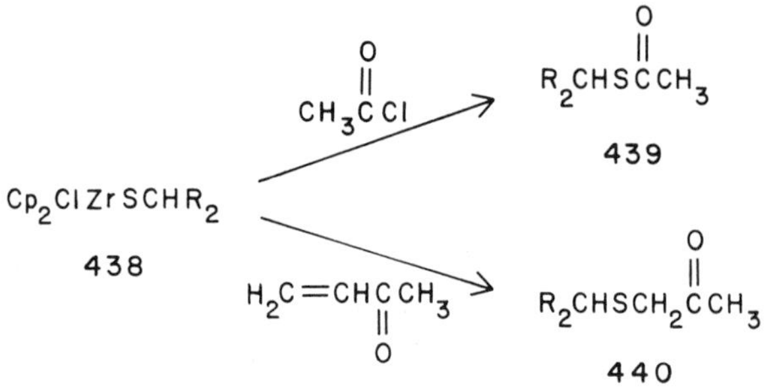

β-keto sulfides (440) (115). Similar products result when the acyltetracarbonylferrate anion 181 is treated with an acid chloride or methyl iodide (Section IV,B,1) (142).

Roper spearheaded the research on coordinated thioaldehydes. In 1977 he established the nucleophilicity of the ligand in an osmium complex (441) which can easily be alkylated at the sulfur atom to give 442. Acidification, however, yields first a methanethiolate complex (443) and eventually liberates methanethiol (263).

The low basicity of the thione sulfur in 444 {[M] = Fp, Re(CO)$_5$} was expected to allow its nucleophilic addition to acetyl chloride or carboxylic acid anhydrides to form S-acyl adducts. These reactions, however, result in desulfurization of the bridging CS_2 unit to form **445** and **446**, whereas alkylation does not have the same result (*264*).

$$FpC(S)S—[M] + MeC(O)Cl \rightarrow ClFpCS + [M]—SC(O)Me$$
$$\textbf{444} \qquad\qquad\qquad\qquad\qquad\qquad \textbf{445}$$

$$FpC(S)S—Fp + (CF_3CO)_2O \rightarrow Fp—SC(O)CF_3$$
$$\textbf{444} \qquad\qquad\qquad\qquad\qquad \textbf{446}$$

A Fischer-type dithiocarbene complex (**447**) which undergoes protonation at the carbene carbon was reported by Le Bozec *et al.* (*265*). The reaction, which produces **448**, proceeds via formation of a *trans*-hydrido-carbene intermediate, isomerization to the cis isomer, and a metal to carbene carbon hydrogen shift.

The carbyne carbon atoms in both [W(CO)$_2${HB(pz)$_3$}(CSMe)] (**449**) (*266*) and [W(CMe)Cp(CO)$_2$] (**450**) (*267*) are nucleophilic and react with protons and the methylthio cation, SMe$^+$, to form cationic carbene complexes. Whereas the first type of carbene complex is typically electrophilic (*257*), the latter one, **451**, is nucleophilic, and treatment with trifluoroacetic acid produces a cationic metallathia cyclopropane complex (**452**) (*268*).

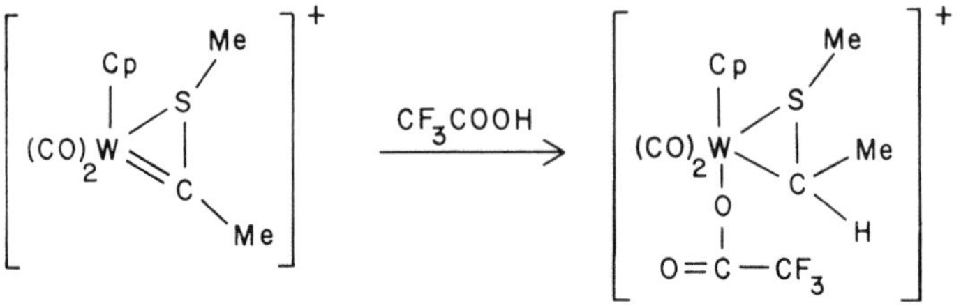

451   452

Methylation of η^2-coordinated thioketene complexes of the type [CoCp(PMe$_3$)(η^2-SC=CR$_2$)] [**453**, R$_2$ = CMe$_2$(CH$_2$)$_3$CMe$_2$, Bu$^t_2$] gives sulfonium ions (*269*), but a proton attacks the double bond to give thioacyl complexes (**454**) (*270*). Certain thio- and selenoaldehyde complexes of

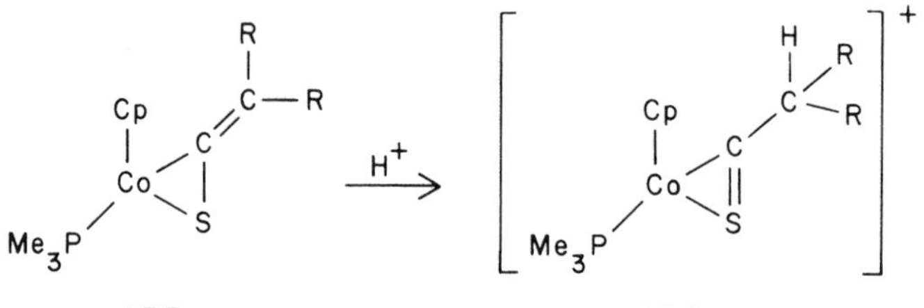

453   454

rhodium (**455**, R = Me or Pr^i) are protonated on the α-thia or α-selena carbon to give dinuclear products (**456**) without metal–metal linkages (*193*).

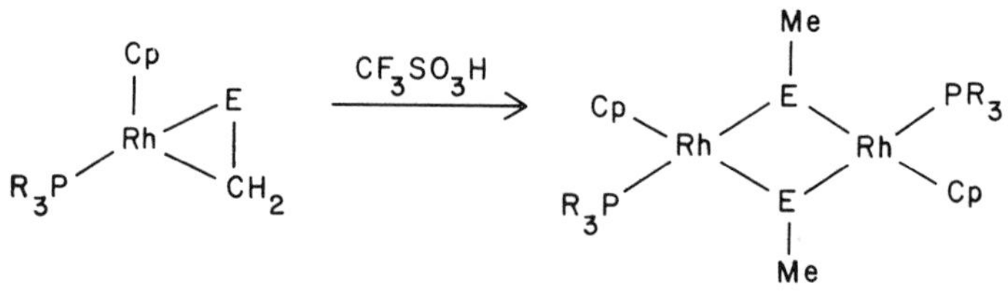

455   456

B. Reactions with Nucleophiles

The reactions of nucleophiles mostly concern thiocarbene and thiocarbyne complexes; some have been reviewed by Angelici *et al.* (*4*).

1. Reactions of Carbene Complexes

Complexes of type **458** can be obtained by the addition of nucleophiles, $R^3[M]$ (MeLi, MeMgBr, C_5H_{11}MgBr, PhMgBr, Me_2CuLi, Bu_2CuLi, Me_2CuLi) to cationic thiocarbene complexes of type **457** ($R^1 = H$ or Me, $R^2 = Me$ or Ph) (*271*).

457 + $R^3[M]$ $\xrightarrow[-78°C]{thf}$ 458

In reactions of the metallacyclic cation $[FeCp(CO)\{C(SMe)SC-(FeL_n)S\}]^+$ {**459**, $[Fe] = FeCpCO$, $L_n = Cp(CO)_2$} with nucleophiles, the

459

results show that the carbene carbon atom bonded to the endocyclic iron atom is the site of addition (*272*), and the reactions are similar to those observed for acyclic dithiocarbene complexes (*4*). Addition of secondary amines, HNR_2 [$R_2 = Me_2$, Et_2, —$(CH_2)_4$—, —$(CH_2)_5$—], yields cyclic aminothiocarbene complexes (**460**), whereas reactions with primary amines, H_2NR (R = Me, Bu, C_6H_{11}), proceed to acyclic isocyanide complexes (**461**), which slowly decompose by elimination of CNR to give **462**. Complex **459** also reacts with monodeprotonated propane-1,3-dithiol to give the fairly stable spirocyclic complex **463**, but the tris (organothio) complexes of type **464** (R = Me, Et), resulting from the addition of mercaptides to **459**, are much less stable and decompose to **462** above − 10°C.

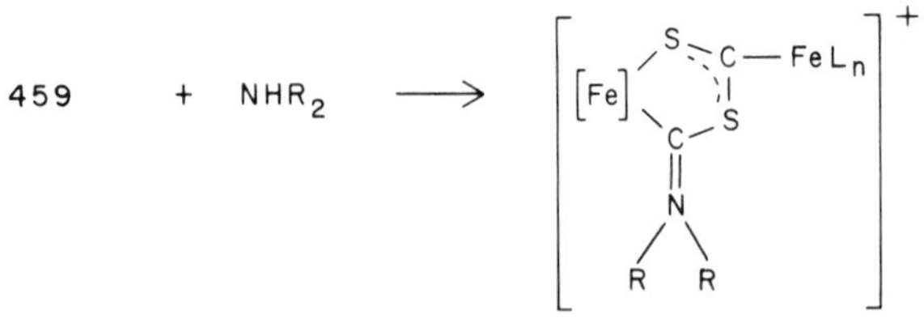

459   + NHR_2 $\longrightarrow$ (structure **460**)

$$\left[[Fe] \overset{S=C-FeL_n}{\underset{C-S}{\Big\langle}} \overset{\displaystyle \|}{\underset{N}{}} \right]^{+}$$

460

459   + NH_2R $\longrightarrow$ (structure **461**) $\longrightarrow$ (structure **462**)

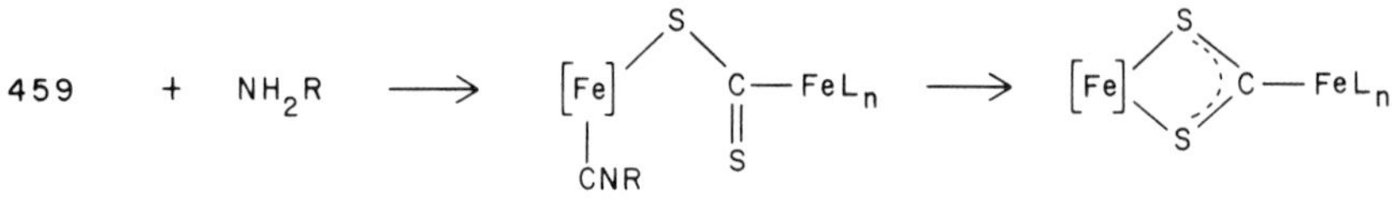

461

462

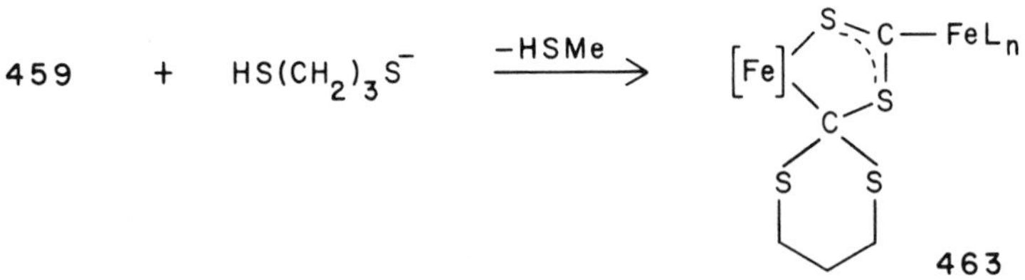

459   + $HS(CH_2)_3S^-$ $\xrightarrow{-HSMe}$ (structure **463**)

463

459   + RS^- $\longrightarrow$ (structure **464**)

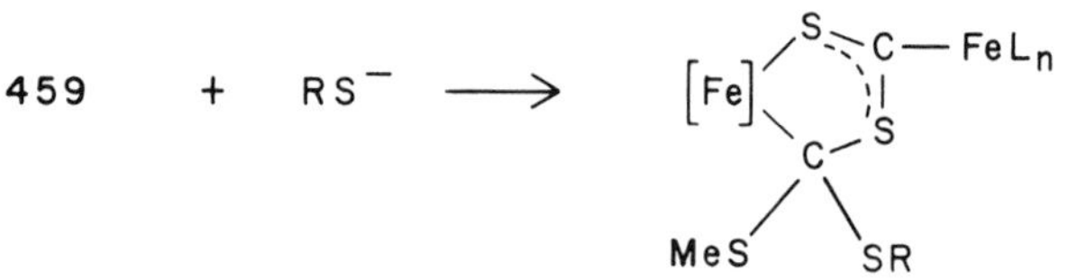

464

Reaction of **459** with cyanide or hydride leads to the nitrile derivative **465** and the hydride adduct **466**, respectively (*272*).

Phosphine derivatives, PR_3 [$R_3 = Ph_2Me$, Ph_3, Ph_2Cl, Cl_3, $(OPh)_3$, $(OCH_2)_3CMe$, HPh_2, H_2Ph, $H(c\text{-Hex})_2$, $H_2(c\text{-Hex})$], add to the carbene

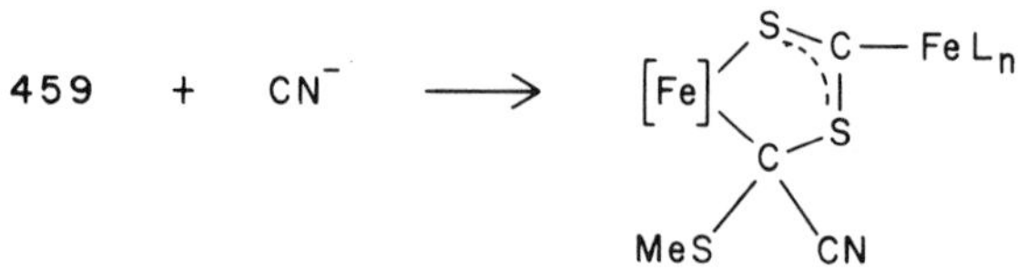

459   + CN^- $\longrightarrow$ (structure **465**)

465

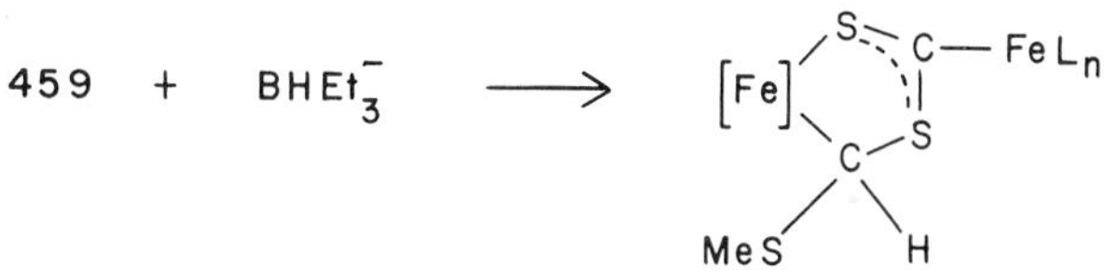

$$459 \quad + \quad BHEt_3^- \quad \longrightarrow \quad [Fe] \begin{array}{c} S = C - FeL_n \\ \\ C - S \end{array}$$

466

carbon of **467** to give the ylide complex **468**, whereas the addition of diazomethane causes rearrangement and forms an S-donor ligand (**469**).

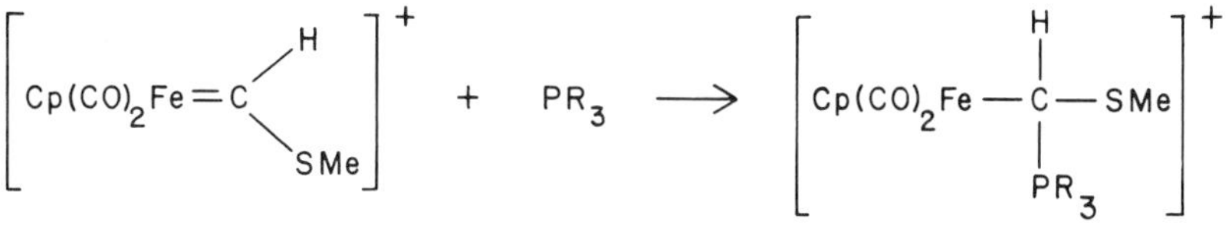

$$\left[Cp(CO)_2 Fe = C \begin{array}{c} H \\ SMe \end{array} \right]^+ \quad + \quad PR_3 \quad \longrightarrow \quad \left[Cp(CO)_2 Fe - \overset{H}{\underset{PR_3}{C}} - SMe \right]^+$$

467   468

$$467 \quad + \quad CH_2N_2 \quad \longrightarrow \quad \left[Cp(CO)_2 Fe - S \begin{array}{c} Me \\ C = CH_2 \\ H \end{array} \right]^+$$

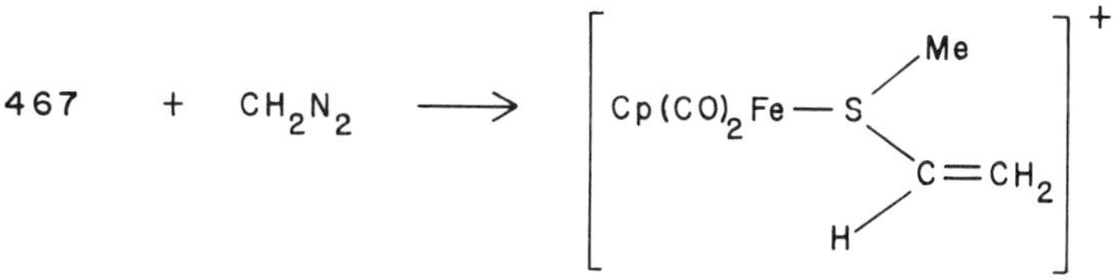

469

No adducts formed with MeCN, THF, SMe_2, or $AsPh_3$ (79). Treatment of the η^2-carbene complex **470** with a selection of bases [B = NaH, $NaBH_4$, NaOMe, NaOPh, (PPN)SH, $NaSCH_2Ph$, LiMe, NEt_3, K_2CO_3, NH_2NH_2, $NHMeNH_2$, NHMeNHMe] leads to the formation of the thiocarbyne complex **471** ([W] = {$HB(pz)_3$}$(CO)_2W$) as well as compound **472**, the

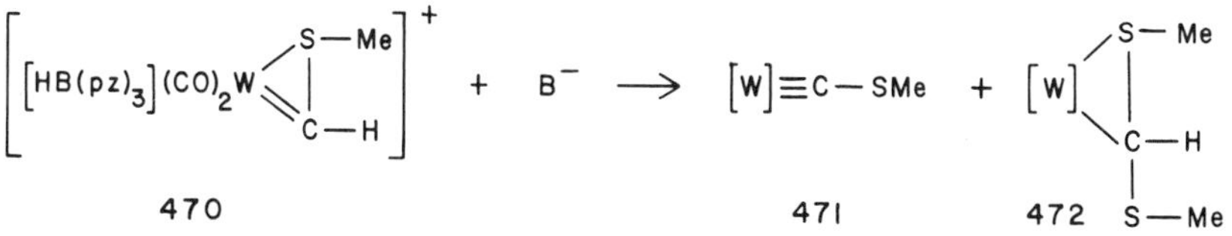

$$\left[[HB(pz)_3](CO)_2 W \begin{array}{c} S - Me \\ C - H \end{array} \right]^+ \quad + \quad B^- \quad \longrightarrow \quad [W] \equiv C - SMe \quad + \quad [W] \begin{array}{c} S - Me \\ C - H \\ S - Me \end{array}$$

470   471   472

formation of which is assumed to involve transfer of MeS^- from **470**. Addition of RS^- (R = Me, Et, Pr^i) to **470** leads to nucleophilic addition at the carbene carbon, producing **473**, a process which can be reversed by the

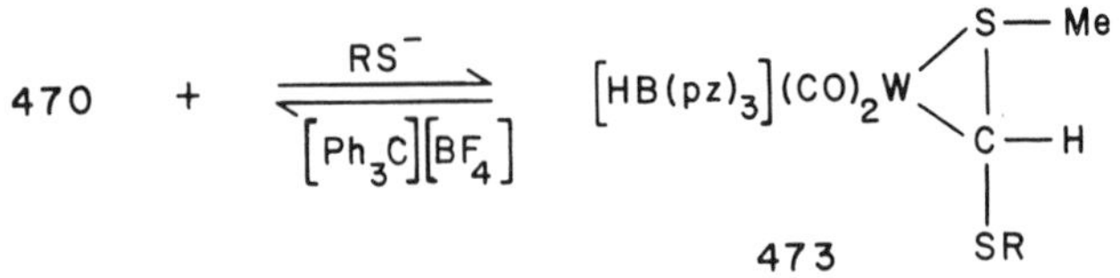

$$470 \;+\; \underset{[Ph_3C][BF_4]}{\overset{RS^-}{\rightleftharpoons}} \qquad [HB(pz)_3](CO)_2W\underset{C-H}{\overset{S-Me}{<\;|\;}} SR$$

473

addition of Ph_3C^+ or H^+. Compound **473** (R = Et) leads to an additional product (**474**) on treatment with $[Ph_3C][BF_4]$ or CF_3SO_3H.

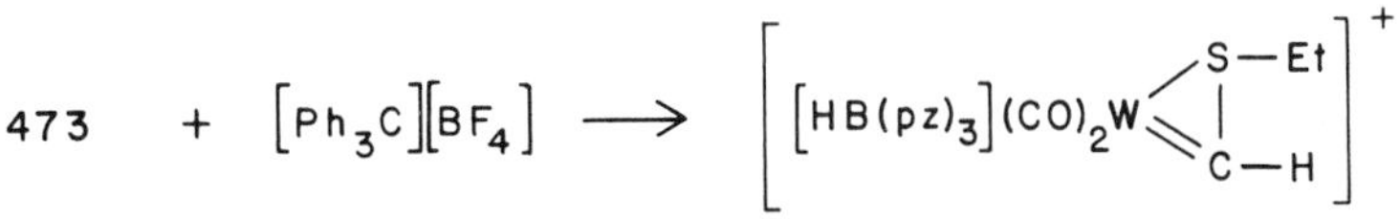

$$473 \;+\; [Ph_3C][BF_4] \longrightarrow \left[[HB(pz)_3](CO)_2W \overset{S-Et}{\underset{C-H}{<\;|\;}} \right]^+$$

474

The η^2-carbene ligand of **470** also reacts with the carbanion $CH(CO_2Me)_2^-$ to form the malonate adduct **475**. LiMe, however, acts as a

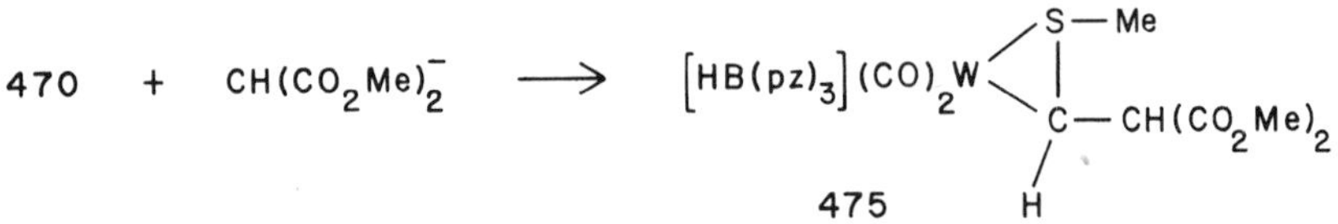

$$470 \;+\; CH(CO_2Me)_2^- \longrightarrow [HB(pz)_3](CO)_2W\overset{S-Me}{\underset{\underset{H}{C-CH(CO_2Me)_2}}{<}}$$

475

base and produces **471** and **472**. In contrast to the aminolysis of η^1-dithiocarbene complexes, which normally leads to aminothiocarbene complexes, the reaction of **470** with secondary amines, HNR_2 (R = Me, Et), produces aminocarbyne complexes of type **476**. When primary amines, H_2NR

$$470 \;+\; HNR_2 \longrightarrow [HB(pz)_3](CO)_2W\equiv CNR_2$$

476

(R = Me, Et, CH_2CH_2OH, $CHMe_2$, Bu^t, p-MeC_6H_4), are used, the resulting aminocarbynes (**477**) are in equilibrium with an isocyanide–hydride tautomer (**478**) (*273*).

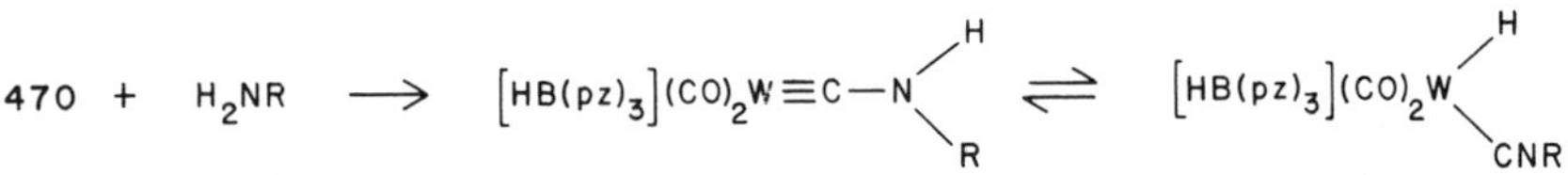

$$470 \;+\; H_2NR \longrightarrow [HB(pz)_3](CO)_2W\equiv C-N\overset{H}{\underset{R}{<}} \; \rightleftharpoons \; [HB(pz)_3](CO)_2W\overset{H}{\underset{CNR}{<}}$$

477   **478**

2. Reactions of Carbyne Complexes

The electron-rich thiocarbyne complex **479** (R = Me, 2,4-dinitrophenyl) is not susceptible to nucleophilic attack at the carbyne carbon, and reaction with PEt$_3$ causes carbonylation to give the η^2-ketenyl complex **480** (257). The same reaction occurs in the complex containing the ligand HC(pz)$_3$ instead of HB(pz)$_3$ (258).

$$\left[HB(pz)_3\right](CO)_2W\equiv CSR \quad + \quad PEt_3 \quad \longrightarrow \quad \left[HB(pz)_3\right](CO)(PEt_3)W\overset{SR}{\underset{C\,\diagdown_O}{\diagup C}}$$

479   480

The bridging carbyne complexes of type **200** (R = Me, CH$_2$Ph, or allyl) add nucleophiles, Nu$^-$ (Nu = H or CH$_2$Ph), to form the bridging carbene complexes (**481**) as mixtures of isomers (150). (The heteroatom nucleophiles are discussed in Section IV,B,3.)

$$\left[\underset{200}{Cp(CO)Fe\overset{\overset{SR}{|}\diagup C\diagdown}{\underset{\diagdown C\diagup}{\underset{\underset{O}{||}}{}}}Fe(CO)Cp}\right]^+ \quad + \quad Nu^- \quad \longrightarrow \quad \underset{481}{Cp(CO)Fe\overset{RS\diagdown\;\diagup Nu}{\overset{C}{\underset{\diagdown C\diagup}{\underset{\underset{O}{||}}{}}}}Fe(CO)Cp}$$

3. Reactions of Thioacyl Compounds

The first reported carbon disulfide complexes of tungsten are of the type [W(CO)$_3$(CS$_2$)(diphosphine)] [**482**, diphosphine = PP = Ph$_2$PC$_2$H$_4$PPh$_2$ = dppe, Ph$_2$PC$_2$H$_4$PMe$_2$, Me$_2$PC$_2$H$_4$PMe$_2$ = dmpe, 1,2-C$_6$H$_4$(PPh$_2$)$_2$ = dppb, and 1,2-C$_6$H$_4$(PMe$_2$)$_2$ = dmpb]. Some of these complexes react with thiiranes (R^1 = H or Me, PP = dppe, dmpe, or dppb) to form heterocyclic chelates (**483, 484**), whereas hydride reagents such as BH$_4$$^-$ or BHEt$_3$$^-$ furnish a coordinated dithioformato ligand (**485**) (274).

Reaction of NaBH$_4$ with the iron complex **486** (L = phosphine ligand) leads to C—C and Fe—Fe bond formation to give compound **487**. Oxidative cleavage of the intermetallic bond produces a complex (**488**) containing a tetrathiooxalate ligand. Sodium amalgam yields **488** directly. When L is one of the more basic, alkyl-containing phosphines, the latter reduction affords the thiocarbonyl complex, [Fe(CO)$_2$(CS)L] (**489**), in relatively low yield (*275*). A thioformyl complex (**490**) is also the minor product when complex **486** accepts a hydride from BH$_4^-$. The major product, **491**, contains a C,S-bonded dithioformato ligand (*231*). When using selenoformaldehyde or telluroformaldehyde complexes of osmium, BH$_4^-$ reacts with their alkylated cations by cleavage of the carbon–heteroatom bond to form [OsMe(CO)$_2$(PPh$_3$)$_2$EMe] (**492**, E = Se or Te) (*86*).

The thioketene dimetallic complex **493** reacts with H⁻ to give an anionic thioacyl complex (**494**). Protonation of **494**, by cleavage of a metal–carbon bond, gives a thioaldehyde complex (**495**) which is stabilized by PPh₃ (*276*).

As mentioned in various sections, nucleophilic addition of phenylethynethiolate or phenylethyneselenolate, PhC≡CE⁻ (E = S or Se), to the carbene complex [W(CO)₅{C(OEt)Ph}] affords thioketene adducts (**186**), which, depending on the electrophile used, function as electrophilic synthons for either coordinated thio- or selenoacyl anions, [W(CO)₅E=CR]⁻, or their structural isomers, anionic thio- or selenocarbene complexes,

[W(CO)₅{C(R)E}]⁻ [R = C(Ph)═C(OEt)Ph]. The electrophilic addition of a variety of electrophiles to these anions has been studied (*84,145,190,262*).

4. *Miscellaneous*

The thiophene complex [Mn(CO)₃thiophene]⁺ (**496**) reacts with hydride ions at the heterocyclic ring to yield **497**. Further protonation results in

decomposition (*277*). Results obtained from another study suggested that the addition of a nucleophile to coordinated thiophenes could be a reasonable first step in the catalytic hydrodesulfurization of thiophene. The enhanced reactivity of the coordinated thiophene in the ruthenium complex **498** (R¹ = H or Me, R² = H or Me) causes scission of an S—C bond to form a butadiene thiolate complex (**499**) on addition of Nu⁻ [Nu⁻ =

MeO⁻, MeS⁻, EtS⁻, PrⁱS⁻, ⁻CH(CO₂Me)₂]. Hydride reacts differently to other nucleophiles and gives, for example, **500** (R¹ = Me, R² = H in **498**) (*278,279*).

Methylene insertion into the Au—S bonds of the heterobimetallic complex **501** results in the formation of the larger ylide ligands in **502** (*280*).

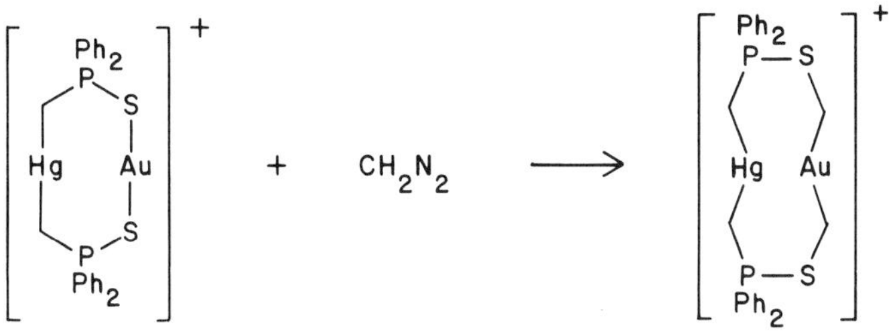

501   502

The five-coordinated complex **503** rearranges on addition of CO to give a dihapto thioacyl compound (**504**) (*243*). The precursor to **503**, [Os(η^2-

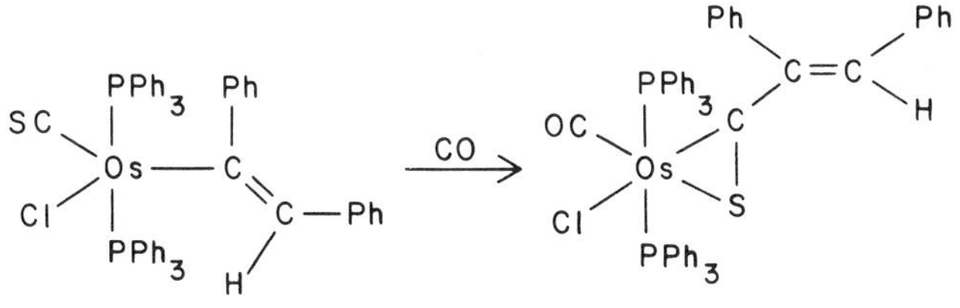

503   504

CPh=CPh)(CS)(PPh₃)₂] (**505**), also reacts with CO to give an organosulfur metallocycle (**506**) in which the sulfur is not a donor atom.

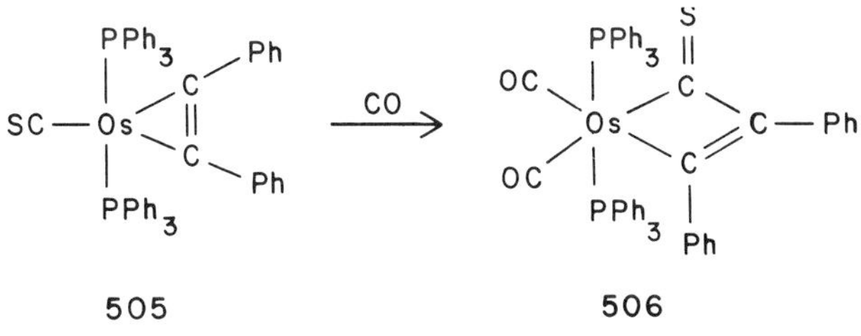

505   506

Functionalization of a thiocarbonyl ligand occurs during the nucleophilic attack on a cyclopentadienyl iron complex by aziridine,

$HN\overline{CH_2CH_2}$, and thiirane, $\overline{SCH_2CH_2}$, to give the cationic aminothiocarbene and dithiocarbene complexes $\{[M]{=}\overline{CSCH_2CH_2NH}\}^+$ (**507**) and $\{[M]{=}\overline{CSCH_2CH_2S}\}^+$ (**508**) $\{[M] = Fe(CO)_2Cp\}$ (*281*).

VII
REACTIONS OF INORGANIC SULFIDE, SELENIDE, AND THIOL COMPLEXES WITH ORGANIC COMPOUNDS

A discussion of reactions of mono- and polychalcogen organometallic complexes as well as conversions of a number of thiol-containing organometallic compounds is presented in this section. A synthetic method developed independently by Herrmann *et al.* (*282*) and Herberhold *et al.* (*283*) enabled the synthesis of thio-, seleno-, and telluroformaldehyde complexes **510** and **511,** by the addition of methylene to metal-bonded chalcogens (**509**) (Table V).

$$[ME] \;+\; CR_2N_2 \;\longrightarrow\; [M]\underset{CR_2}{\overset{E-[M]}{\diagdown}}$$

509   510

The tungsten complex **509j** leads to a dimer containing an M—M bond, **511** (*286*). The selenium and tellurium iron carbonyl compounds (**512**)

$$Cp(CO)_3W\overset{S}{\diagup}\diagdown W(CO)_3Cp \;+\; CH_2N_2 \;\longrightarrow\; Cp(CO)_2W\!-\!\!-\!\!-\!W(CO)_2Cp\;\overset{S-CH_2}{\diagup\diagdown}$$

509 j   511

add PhC≡CH on thermal activation to yield the dimeric butterfly compounds (**513**) (*287*). Reaction of the sulfur-containing osmium cluster **514**

$$(CO)_9Fe_3(\mu_3{-}E)_2 \;+\; PhC{\equiv}CH \;\longrightarrow\; (CO)_3Fe\!-\!\!-\!\!-\!Fe(CO)_3$$

512   513

TABLE V

CHALCOGEN METAL COMPLEXES USED IN
REACTIONS WITH DIAZO COMPOUNDS

Compound	[M—E]	[M]	R	Ref.
509a ⎱ 509b ⎰	[M]——Se——[M]	CpMn(CO)$_2$ (Me$_5$C$_5$)Mn(CO)$_2$	H, Me	282
509c	[M]—Se—Se—[M]	CpMn(CO)$_2$	H	282
509d	[M]—Te([M])—[M]	CpMn(CO)$_2$	H	282
509e	[M]——S——[M]	CpMn(CO)$_2$	H	283
509f	[M]—S—S—[M]	CpMn(CO)$_2$	H	283
509g	[M]—S,S—[M]	(Me$_5$C$_5$)Cr(CO)$_2$	H	284
509h	[M]≡Se≡[M]	CpCr(CO)$_2$	H, Me	284
509i	[M]=Te=[M]	(Me$_5$C$_5$)Mn(CO)$_2$	H, Me	285

with $RC{\equiv}CH$ (R = Ph, CO_2Me) leads to the formation of a carbon–
sulfur bond between the sulfur and the substituted carbon of the acetylene
to form a quadruply bridging thiolato ligand, $SC(R){=}CH$, in **515** (*288*).

$$Os_4(CO)_{12}(u_3{-}S) \quad + \quad RC{\equiv}CH \quad \longrightarrow$$

514   515

In their investigation of $[Fe_2(CO)_6(\mu{-}SH)_2]$ as an inorganic mimic of
mercaptans, RSH, Seyferth and group studied the base-induced reactions
of $[Fe_2(CO)_6(\mu{-}SH)_2]$ (**516**) with activated olefins and acetylenes. Like
mercaptans, **516** undergoes base-induced addition to α,β-unsaturated sys-

tems $[R^1 = H,\ MeC(O)CH,\ MeOC(O)H,\ or\ EtOC(O)CH,\ R^2 = Me,$ $CO_2Me,\ or\ CO_2Et]$ to form $1:2$ adducts (**517**). The $2:1$ addition of ketones

516   517

of the type $R^1_2C{=}CHC(O)R^2$ is prevented by steric hindrance. The two sulfur atoms are in close proximity, and the presence of a tertiary carbon center at one sulfur ligand, after the addition of one molecule of, for example, mesityl oxide, prevents the approach of a second olefin. The SH function, instead, adds to the $C{=}O$ group of the alkanethio function already present and gives the three-carbon-bridged alcohols **518** (R^1, R^2, R^3 = Me or Et), which can be isolated and purified as their trimethylsilyl ethers. Acetylenes, $R^1C{\equiv}CC(O)R^2$ (R^1 = H or Me, R^2 = Me or OMe),

516   + $R^1R^2C{=}CHCR^3$ 518

react with **516** to form $1:1$ adducts (**519**) resulting from two Michael

516   + $R^1C{\equiv}CCR^2$ 519

additions. The product contains a one-carbon bridge between the two sulfur atoms. In the reaction with $MeCO_2C{\equiv}CCO_2Me$, the less strained product **520,** with a two-carbon bridge, is formed (*289,290*). The Michael

516   + $MeOCC \equiv CCOMe$ $\xrightarrow{\text{base}}$

$$MeO_2C-\underset{\underset{S}{|}}{\overset{\overset{H}{|}}{C}}-\underset{\underset{S}{|}}{\overset{\overset{H}{|}}{C}}-CO_2Me$$

$(CO)_3Fe \text{———} Fe(CO)_3$

520

additions leading to bridging sulfur ligands require the presence of activating functional groups like CO_2R, CN, and C(O)R. The substituents on the α,β-unsaturated substrates determine whether a one-, two-, or three-carbon bridge is formed between the two sulfur atoms (*289*).

In contrast with the Fe—SH compound **516,** $(RC_5H_4)_2Ti(SH)_2$ (**521,** R = H, Me) is not nucleophilic unless incorporated into a dimer with molybdenum or tungsten. Complex **522** adds to methyl acrylate and forms the conjugate addition product **523** (*291*).

$Cp_2Ti \text{———} Mo(CO)_4$ (with bridging S—H groups)   + $H_2C=CHCOMe$ $\longrightarrow$ $Cp_2Ti \text{———} Mo(CO)_4$ (with CO_2Me substituents)

522   **523**

Alkynyllithium reagents cleave the S—S bond of $[Fe_2(CO)_6(\mu\text{-}S)_2]$ to give **524,** which can undergo intramolecular addition of sulfur to the α (**525**) or β (**526**) carbon atom of the acetylenic group when treated with electrophiles (Q^+) like H^+, aldehydes, and group 4 organometallic halides. With alkyl halides and acetyl chloride, only S-alkylation or S-acylation is observed (*292,293*).

Acetylenes and ethene insert into the S—S bonds of metal polysulfides [M—S] to yield, with two exceptions (**535** and **540**), dithiolene and 1,2-ethanedithiolato ligands coordinated in a variety of bonding modes (Table VI).

	Q	R
525	H	SiMe$_3$, CO$_2$Me
	SiMe$_3$	CO$_2$Me
	SnMe$_3$	CO$_2$Me
526	H	H, Me, Bu, C$_5$H$_{11}$, Ph, —C(CH$_3$)=CH$_2$, SiMe$_3$
	SiMe$_3$	H, Bu, Ph
	GePh$_3$	Ph
	SnMe$_3$	Me, Bu^t, Ph
	SnPh$_3$	Ph
	PbPh$_3$	Ph
	—CH(OH)Ph	Ph
	—CH(OH)Me	Ph
	—CH(OH)Bu^t	Ph

Activated acetylenes and ethylene insert into the S—S bond of [Fe$_2$(CO)$_6$(μ-S)$_2$] (**527**) under photochemical activation to give dithiolene (**528**) and 1,2-ethanedithiolato (**529**) complexes, respectively (294,295). In [CpFeS$_2$]$_2$ (**530**), both S—S bonds are cleaved, and the dithiolato ligand **531** contains one bridging and one terminally coordinated sulfur (296). The ruthenium analog **532** requires PBu$_3$ to react with acetylenes, R^{1}C≡CR^2, and forms a bridging dithiolene ligand which is folded over toward one metal (**533**) (297). In the reaction of the molybdenum dianion **534**, the acetylenes insert into the Mo—S bonds of the terminal sulfides, rather than into S—S bonds, to form five-membered metalla-2,3-dithiacyclopent-4-ene rings in **535** (298). A series of titanium sulfides and selenides (**536, 538**) are converted to dithiolenes and diselenenes (**537, 539**) at

TABLE VI

REACTIONS OF METAL POLYSULFIDES WITH ACETYLENES AND ETHENE

Compound	[M—S]	Reaction conditions	Product	Ref.
527	$[Fe(CO)_3]_2(\mu-SS-\mu S^1,\mu S^2)$	$ZC{\equiv}CZ(Z = CF_3,$ $CO_2Me, CO_2Et)$; $h\nu$	$[Fe(CO)_3]_2\{\mu-SC(Z){=}C(Z)S-\mu S^1,\mu S^2\}$ **(528)**	*294*
527		$CH_2{=}CH_2$; $h\nu$	$[Fe(CO)_3]_2\{\mu-SCH_2CH_2S-\mu S^1,\mu S^2\}$ **(529)**	*295*
530	$[CpFe(\mu-S)]_2(\mu-SS-S^1,S^2)$	$CF_3C{\equiv}CCF_3$	$[CpFe]_2\{\mu-SC(CF_3)_3{=}C(CF_3)S-\mu S^1,\mu S^2\}$ **(531)**	*296*
532	$[Cp^*Ru(\mu-S)]_2(\mu-SS-S^1,S^2)$ $(Cp^* = C_5Me_4Et)$	$R^1C{\equiv}CR^2$ ($R^1 = Ph,$ $R^2 = H, Ph;$ $R^1, R^2 = H)$; $70°C$; PBu_3	$Cp^*Ru\{\mu-SC(R^1){=}\overline{C(R^2)S-\mu S^1,\mu S^2}\}RuCp^*$ **(533)**	*297*
534	$[(Mo(O)(\mu-S))_2(\mu-SS-S^1,S^2)]^{2-}$	$Z_2C{\equiv}CZ(Z = CO_2Me)$	$[\overline{Mo(O)(\mu-S)\{C(Z){=}C(Z)S\}}]_2^{2-}$ **(535)**	*298*
536	$[(RC_5H_4)_2TiS_x]_n$ $(x = 5, n = 1;$ $x = 3, n = 2;$ $x = 2, n = 2;$ $R = H, Me, Pr^i)$	$ZC{\equiv}CZ$ (Z $= CO_2Me,$ $CO_2Et)$; Δ	$[(RC_5H_4)_2\overline{TiSC(Z){=}C(Z)S}]$ **(537)**	*299* *300* *301* *302*
538	$(RC_5H_4)_2TiSe_5$ $(R = H, Me)$	$ZC{\equiv}CZ$ (Z $= CO_2Me,$ $CF_3)$	$[(RC_5H_4)_2\overline{TiSeC(Z){=}C(Z)Se}]$ **(539)**	*301*
536	$[(RC_5H_4)_2Ti]_2S_4$ $R = H, Me, Pr^i$	$ZC{\equiv}CZ$ (Z $= CO_2Me)$	$[(RC_5H_4)_2\overline{TiSSC(Z){=}CZ}]$ **(540)**	
541	$[CpW(S)(\mu-S)]_2$	$HC{\equiv}CR$ (R $= H, Ph)$	$[CpW(\mu-S)\{SC(R^2){=}C(R^2)S\}]_2$ **(542)**	*134*
543	$[(MeC_5H_4)Mo(S)(\mu-S)]^2$	$H_2C{=}CH_2$	$[(MeC_5H_4)Mo(\mu-SCH_2CH_2S-\mu S^1,\mu S^2)]_2$ **(544a)**	
		$HC{\equiv}CH$	$[(MeC_5H_4)Mo(\mu-SCH{=}CHS-\mu S^1,\mu S^2)]_2$ **(544b)**	*303*
545	$[(MeC_5H_4)V(\mu-S)]_2(\mu-SS-S^1,S^2)$	$CF_3C{\equiv}CCF_3$; $60°C$	$[(MeC_5H_4)V(\mu-S)]_2\{\mu-SC(CF_3)S{=}C(CF_3)S-\mu S^1,\mu S^2\}$ **(546)**	*304*

elevated temperatures (*299–302*) but to the metalla-2,3-dithiacyclopent-4-ene (**540**) at room temperature (*302*). A tungsten dithiolene complex (**542**) has also been prepared (*134*). In contrast to **534**, the Mo sulfide dimer **543** reacts with ethylene and acetylene through its bridging sulfurs to form the known dimers **544a** and **544b** (*303*). Acetylene addition to the vanadium complex **545** is proposed to occur at the bridging sulfur atoms, while the μ-S$_2$ ligand rearranges. (*304*).

The molybdenum dimer **547** can be converted, by alkylation, to a cation (**548**), which can be transformed into an anion (**549**) by two-electron chemical reduction. Treatment of **547** with excess methyllithium leads, after work-up in air, to the bis (μ-methylthiolate) **550**, which can also be prepared by addition of methyllithium to **548**. The addition of vinylmagnesium bromide to **548** yields **551**. The anion reacts with alkynes, activated olefines, and an allene to form **552**, **553**, and **554**, respectively (*305*).

The cationic complex [{Mo(MeC$_5$H$_4$)}$_2$(μ-S)(μ-SH)(μ-SCH$_2$S-μS^1, μS^2)]$^+$ (**555**), reacts with THF to convert it to a 4-(hydroxy)butanethiolate ligand in the cationic product [{Mo(MeC$_5$H$_4$)}$_2$(μ-S){μ-S(CH$_2$)$_4$OH}-(μ-SCH$_2$S-μS^1, μS^2)]$^+$, **556** (*306*).

The reaction products obtained when the hydrosulfide ligand in

TABLE VII

ELECTROPHILIC ADDITION TO THE COMPLEX [W(CO)$_5$SH]$^-$

Electrophile	Product	Type of ligand	Ref.
Me$_2$C=O + H$^+$	[W(CO)$_5$S=CMe$_2$] (558)	Thioketone	307
p-YC$_6$H$_4$CHO + H$^+$ (Y = Me$_2$N, MeO, Me)	[W(CO)$_5$S=CHC$_6$H$_4$Y-p] (559)	Thioaldehyde	307
RN=C=O (R = Me, Ph)	[W(CO)$_5$SC(O)NHR]$^-$ (560)	Thiocarbamate	308
Ph$_2$C=C=O	[W(CO)$_5$SC(O)CHPh$_2$]$^-$ (561)	Thioacetate	308
RN=C=S (R = Me, Ph)	[W(CO)$_4$(S$_2$CNHR)]$^-$ (562)	Chelated dithiocarbamate	308
Ph$_2$C=C=NPh	[W(CO)$_4$N(Ph)C(CHPh$_2$)S]$^-$ (563)	Thioimidate	308
Ph$_2$C=C=NPh + H$^+$	[W(CO)$_5$S=C(NHPh)(CHPh$_2$)] (564)	Thioamide	308
RN=C=NR + H$^+$ (R = Pr^i, C$_6$H$_{11}$)	[W(CO)$_5$S=C(NHR)$_2$] (565)	Thiourea	308

[W(CO)$_5$SH]$^-$ (557) interacts with organic electrophiles (sometimes followed by acidification with CF$_3$CO$_2$H) are summarized in Table VII.

VIII

ADDENDUM

This chapter has been updated to include literature published up to the end of 1990.

1. *Section II, F*

Doyle and Angelici reported the thermal rearrangement of a tungsten complex containing an organosulfur ligand similar to that in 97. The complex, [W(CO)$_2${η^2-C(SMe)(SR)(SMe)}{HB(pz)$_3$}] (566) (R = Me or Et), undergoes irreversible cleavage of a C—S bond and migration of MeS or EtS to give the carbyne complexes, [W(CSR^1){HB(pz)$_3$}(SMe)(SR^2)] (567) (R^1 = R^2 = Me; R^1 = Me or Et, R^2 = Et or Me). When in 566, R = Ph or p-MeC$_6$H$_4$ photolysis is required to effect the conversion to 567 (R^1 = Me, R^2 = Ph or p-MeC$_6$H$_4$) (309).

Irradiation of [Os$_3$(CO)$_{11}${S(CH$_2$)$_2$CH$_2$}] (568) with UV light causes ring opening and the activation of a C—H bond to produce [Os$_3$(CO)$_{10}$(μ-SCH$_2$CH=CH$_2$)(μ-H)] (569) (310), whereas the thermally induced rearrangement of [Os$_3$(CO)$_{10}$(μ-SCH$_2$CMe$_2$CH$_2$)] (570) leads to compounds 571 and 572, and also involves the activation of a C—H bond

571   572

(311). The addition product between $[M_3(CO)_{10}(\mu_3\text{-S})]$ and MeC≡ CNMe$_2$, **573** (M = Fe or Ru), isomerizes upon heating to **574** (312).

573   574

2. Section IV, A, 1

Thiophene (or substituted thiophenes) adds oxidatively to [Rh-(C$_5$Me$_5$)(Ph)(H)(PMe$_3$)] (**575**) and benzene is eliminated to form **576**

576

(313), whereas trimethylene sulfide adds to the M=C double bond of the titanocene methylidene complex, [TiCp$_2$(CH$_2$)(PMe$_3$)] (**577**) to form [TiCp$_2${(CH$_2$)$_4$S}] (**578**) (314).

The hexyne, MeSC≡CSMe, rearranges upon reaction with [RuCp-(Cl)(PMe$_3$)$_2$] (**579**) to give the vinylidene complex, [RuCp{C=C(SMe)$_2$}(PMe$_3$)$_2$]$^+$ (**580**) (315) (compare with **131**).

3. Section IV, A, 3

Although the thiolysis of alkoxycarbenes to form thiocarbenes is not a new reaction, the new method, in which a catalyst is used, minimizes the formation of unwanted side products and dramatically increases the yield. In the presence of NEt_3, the resulting thiocarbene ligands can be condensed with aldehydes, $CH(O)Ph$ and $CH(O)CH=CHPh$, to yield the new thiocarbene complexes $[Cr\{C(SR)CH=CHPh\}(CO)_5]$ (581) (R = $CH_2CH=CH_2$, Et or Ph) and $[Cr\{C(SR)CH=CHCH=CHPh\}(CO)_5]$ (582) (R = Ph) (316).

4. Section IV, B, 4

The reaction of the sulfonium alkyne complex, $[WCp(\eta^2MeSC\equiv CSMe)(\eta^2\text{-}MeSC\equiv CSMe_2)(Cl)]^+$ (583) with RS^- (R = Ph, $p\text{-}MeC_6H_4$) differs from that of the alkyne complex, 213 (which adds RS^- to the alkyne ligand) and SMe_2 is displaced from the sulfonium group to give $[WCp(\eta^2\text{-}MeSC\equiv CSMe)(\eta^2\text{-}MeSC\equiv CSR)]$ (584) (317). Similarly, EtS^- displaces SMe_2 from the sulfonium complex, $[RuCp\{C=C(SMe)(SMe_2)\}-(PMe_3)_2]^{2+}$ (585) to form $[RuCp\{C=C(SMe)(SEt)\}(PMe_3)_2]^+$ (586) (318).

Angelici and co-workers reported the addition of MeS^- to the thiocarbonyl ligand in $[Fe(C_5Me_5)(CO)_2(CS)]^+$ (587) to yield the dithioester complex, $[Fe(C_5Me_5)(CO)_2\{C(S)SMe\}]$ (588) (319).

5. Section IV, C

As in the carbyne complex, 229, MeS^+ [from $Me_2(MeS)S^+$] adds to the metal-bonded carbon atom of the vinylidene ligand in $[RuCp\{C=C(SMe)_2\}(PMe_3)_2]^+$ (580) and $[RuCp\{\eta^2\text{-}C(SMe)=C(SMe)_2\}(PMe_3)_2]^{2+}$ (589) forms (318).

6. Section V, A

The methylidene complex, $[TiCp_2(CH_2)(PMe_3)]$ (577), can also add sulfur (from S_8, $S=PPh_3$ or SCH_2CHR, R = Me or Ph) to produce $[TiCp_2(\eta^2\text{-}CH_2S)(PMe_3)]$ (590), containing a thioformaldehyde ligand similar to those in 251 and 253 (314).

7. Section VI, A, 1

The insertion of acetylenes into carbon-heteroatom double bonds (see 364–365) was extended to include carbene complexes and the reaction between $Et_2NC\equiv CMe$ and $[M\{C(Ph)SR\}(CO)_5](M = Cr$ or W, R =

$CH_2CH{=}CH_2$, c-Hex, Bu^t, Ph or CH_2Ph) to produce E-[M{C-$(NEt_2)C(Me){=}C(Ph)SR$}$(CO)_5$] (591) was reported. An indanone complex, 592, was isolated from the same reaction mixture (320).

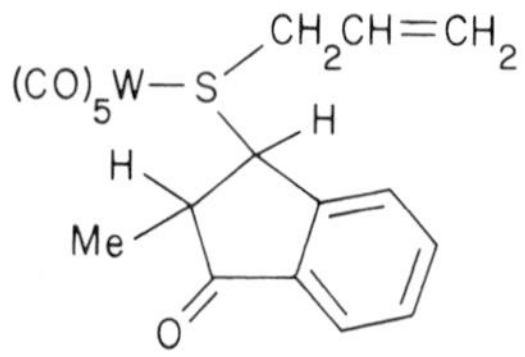

592

The reaction leading to the formation of 367 was extended to include the phosphine ligands, PMe_2Ph and $P(OMe)_3$ (321).

The reaction between [$FeCp(CO)_2SMe$] and $HC{\equiv}CCN$ yields a new complex, [FeCp{C(O)C(CN)${=}$CHSMe] (593), related to the known complex, 388 (322), whereas treatment of [$Fe_3(\mu$-Cl)$(CO)_9(\mu_3$-$SBu^t)$] (594) with

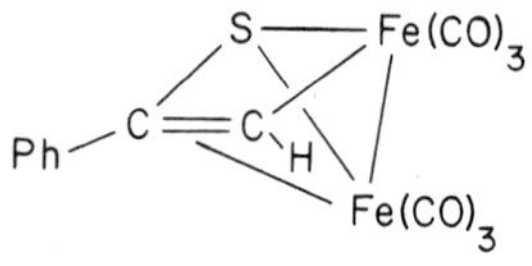

595

$HC{\equiv}CPh$ leads to the dimeric compound, 595. When 595 is prepared in a one-pot reaction between [$Fe_2(CO)_9$] or [$Fe_3(CO)_{12}$], Bu^tSH and $PhC{\equiv}CH$, carbonyl insertion occurs to form the ligand in 596 (which is similar to that in 388) as well as the new ligand in 597 (323). Another

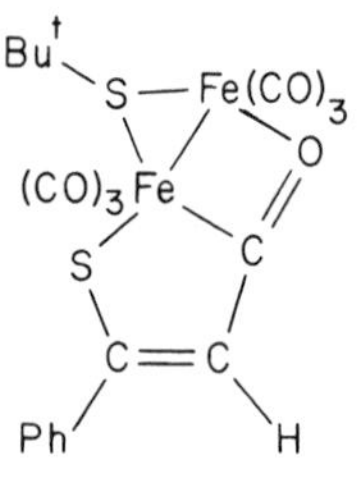

596

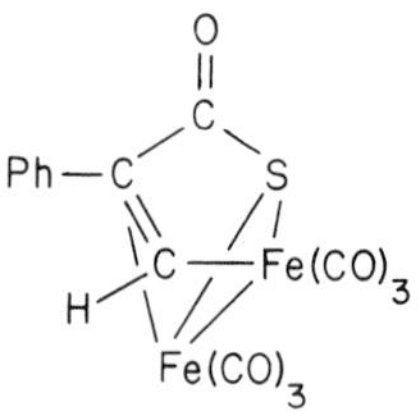

597

four-membered thia-metallacycle, 599, forms when [$Zr(C_5Me_5)_2(S)(L)$]

(598) (L = pyridine or 4-t-butylpyridine) is treated with RC≡CR (R = Et, Ph, or p-MeC$_6$H$_4$) (324).

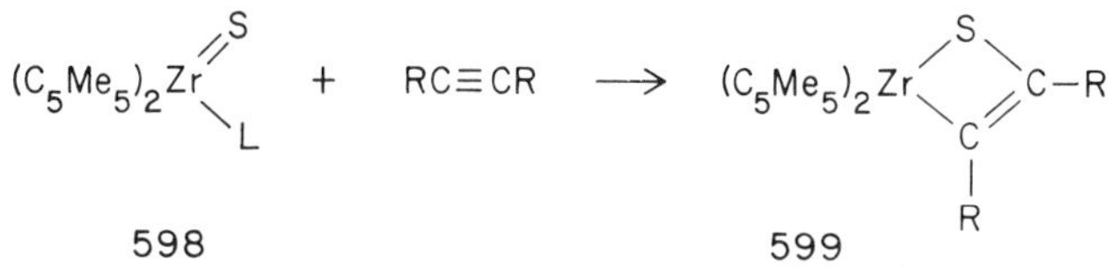

A new example of a type **H** complex, [NbCp$_2$(CR^1=CHR^2){CSCR^3= C(R^4)S}] **(601)**, was prepared from the reaction between [NbCp$_2$(CR^1= CHR^2)(η^2-CS$_2$)] **(600)** (R^1 = CF$_3$ or CN, R^2 = CF$_3$, CN, or H) and acetylenes, R^{3}C≡CR^4 (R^3 = CF$_3$, CN, CO$_2$Me, or CO$_2$Et, R^4 = CF$_3$, CN, CO$_2$Me, CO$_2$Et, or H) (325).

8. *Section VI, A, 3*

Alkylation converts the acetylide complex, [RuCp(C≡CSMe)(PMe$_3$)$_2$] **(602)**, to the vinylidene complex, {RuCp{C=C(Me)SMe}(PMe$_3$)$_2$]$^+$ **(603)**, (315) and the vinylidene complex, [RuCp{C=C(SMe)$_2$}(PMe$_3$)$_2$]$^+$ **(580)**, to the sulfonium complex, [RuCp{C=C(SMe)(SMe$_2$)}(PMe$_3$)$_2$]$^{2+}$ **(585)**, (318). The vinylidene complexes, [RuCp{C=C(SMe)$_2$}(PMe$_3$)$_2$]$^+$ **(580)** and [RuCp{C=C(Me)SMe}(PMe$_3$)$_2$]$^+$ **(603)**, as well as the sulfonium complex, [RuCp{C=C(SMe)(SMe$_2$)}(PMe$_3$)$_2$]$^{2+}$ **(585)**, can be reduced with Na/Hg to the acetylide complexes, [RuCp(C≡CSMe)(PMe$_3$)$_2$] **(602)**, [RuCp(C≡CMe)(PMe$_3$)$_2$] **(604)** and **602**, respectively (315,318). Protonation of **580** leads to the cis- and trans-isomers of the vinyl complex [RuCp{η^2-C(SMe)=CHSMe}(PMe$_3$)$_2$]$^{2+}$ **(605)** (315).

9. *Section VI, B, 1*

Cyclohexylisocyanide adds to the carbene carbon of [Cr{C(SPh)CH= CHPh}(CO)$_5$] **(581)**, causes a rearrangement of the ligands and the formation of an S-coordinated ketenimine complex, **606** (316).

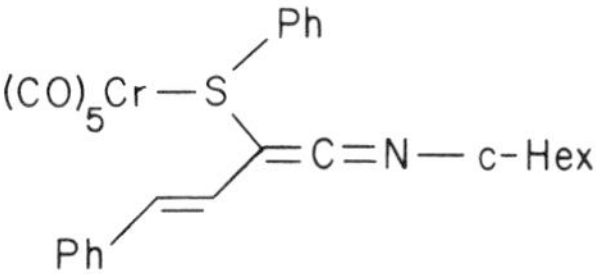

606

10. *Section VI, B, 4*

Reaction of [Fe(CO)₃{Fe(CO)₃—CH=CPh—S}] **(595)** with BuLi and [Et₃O][BF₄] leads to the formation of **607** and **608**. When **595** is treated with Na/Hg or BuLi and I₂, **608** is the only product (*326*).

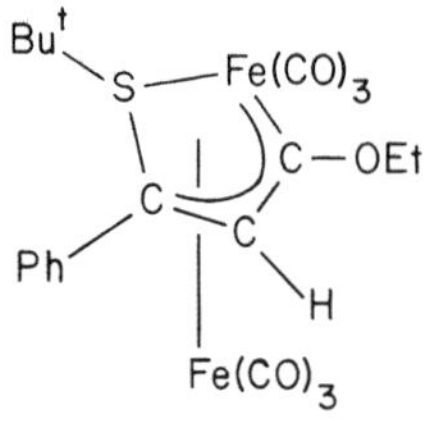

607

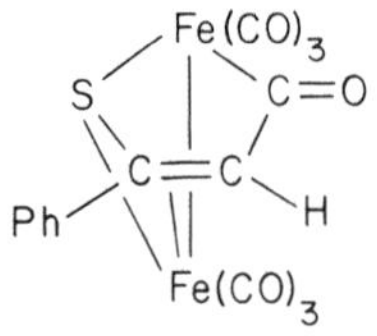

608

Reduction of the thiophene complex, **609** (R = H, Me) with Na[H₂Al(OCH₂CH₂OMe)₂] leads to **610**, which, upon treatment with base (NEt₃ or basic alumina), undergoes ring opening and oxidative addition of the thiophene ligand to give **611** (*327*). The ligand reverts to the organic

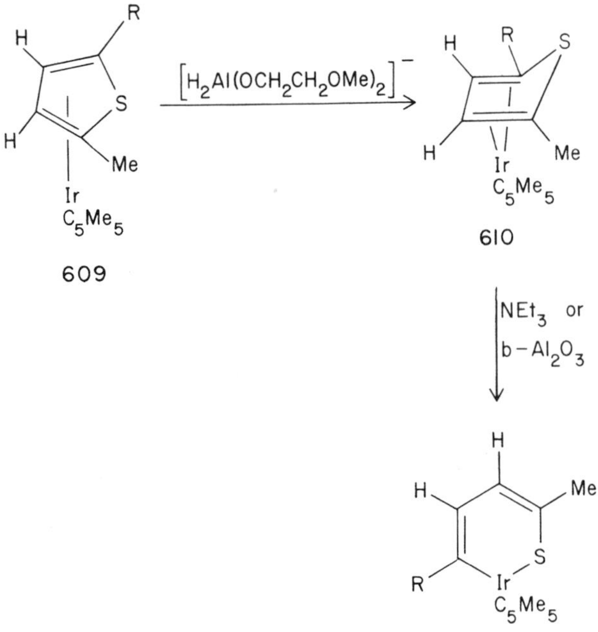

609

610

611

thia-cyclic ligand (as in **610**) upon treatment with [Cp(CO)$_2$Mo≡
Mo(CO)$_2$Cp] *(328)*. Oxidation of **611** with [Cp$_2$Fe]$^+$, however, leads to the
original η^5-coordinated thiophene complex, **609** *(327)*.

According to the present classification, the addition of PMe$_2$Ph to
[Os$_3$(CO)$_{10}$(μ-SCH$_2$CH=CH$_2$)(μ=H)] **(569)** which induces formal insertion of the organosulfur ligand into the metal hydride bond to form **612**,

612

(310) should be placed in a section with the reaction between **479** and
PEt$_3$, but the reaction is actually quite similar to the insertion of 2,3-hy-
drothiophene into the M—H bond of **32** to give **33** (Section II,C,1).

11. *Section VII*

Lorentz and co-workers report in two publications on the photochemically induced insertion of alkenes and dienes into the S—S bond of
[Fe$_2$(CO)$_6$S$_2$] which leads to [2 + 2] cycloaddition and the formation of
1,2-dithiolato ligands *(329,330)*. The addition of MeC≡CNMe$_2$ to
[M$_3$(CO)$_{10}$(μ_3-S)] (M = Fe or Ru) results in the ligand in **573** *(312)*, which
is reminiscent of that in **515**.

REFERENCES

1. B. Zwanenburg and A. J. H. Klunder, "Perspectives in the Organic Chemistry of
   Sulfur," p. v. Elsevier, Amsterdam, 1987.
2. S. G. Murray and F. R. Hartley, *Chem. Rev.* **81**, 365 (1981).
3. P. V. Broadhurst, *Polyhedron* **4**, 1801 (1985).
4. R. J. Angelici, F. B. McCormick, and R. A. Pickering, *in* "Fundamental Research in
   Organometallic Chemistry" (M. Tsutsui, Y. Ishii, and H. Yaozeng, eds.), p. 347. Van
   Nostrand, New York, 1982.
5. C. Bianchini, C. Mealli, A. Meli, and M. Sabat, *in* "Stereochemistry of Organometallic
   and Inorganic Compounds" (I. Bernal, ed.), Vol. 1, Chap. 3. Elsevier, Amsterdam,
   1986.
6. I. Omae, "Organometallic Intramolecular Coordination Compounds," p. 233. Elsevier,
   Amsterdam, 1986.
7. L. Markó and B. Markó-Monostory, *in* "The Organic Chemistry of Iron" (E. A.
   Koerner von Gustorf, F. W. Grevels, and I. Fischler, eds.), Vol. 2, p. 283. Academic
   Press, New York, 1981.

8. H. Alper and A. S. K. Chan, *Inorg. Chem.* **13**, 225 (1974).
9. H. Alper and C. K. Foo, *Inorg. Chem.* **14**, 2928 (1975).
10. H. Patin, G. Mignani, R. Dabard, and A. Benoit, *J. Organomet. Chem.* **168**, C21 (1979).
11. A. Benoit and J. Y. Le Marouille, *J. Organomet. Chem.* **218**, C67 (1981).
12. G. J. Kruger, L. Linford, H. G. Raubenheimer, and A. A. Chalmers, *J. Organomet. Chem.* **262**, 69 (1984).
13. E. K. Lhadi, H. Patin, and A. Darchen, *Organometallics* **3**, 1128 (1984).
14. G. J. Kruger, A. van A. Lombard, and H. G. Raubenheimer, *J. Organomet. Chem.* **331**, 247 (1987).
15. D. Seyferth, G. B. Womack, M. Cowie, and B. W. Hames, *Organometallics* **3**, 1891 (1984).
16. D. Seyferth, G. B. Womack, M. G. Gallaghar, M. Cowie, J. P. Fackler, and A. M. Mazany, *Organometallics* **6**, 283 (1987).
17. H. Patin, B. Misterkiewicz, J. Y. Le Marouille, and A. Mousser, *J. Organomet. Chem.* **314**, 173 (1986).
18. H. Patin, G. Mignani, A. Benoit, J. Y. Le Marouille, and D. Grandjean, *Inorg. Chem.* **20**, 4351 (1981).
19. H. Patin, G. Mignani, C. Mahé, J. Y. Le Marouille, T. G. Southern, A. Benoit, and D. Grandjean, *J. Organomet. Chem.* **197**, 315 (1980).
20. H. G. Raubenheimer, G. J. Kruger, and A. van A. Lombard, *J. Organomet. Chem.* **323**, 385 (1987).
21. H. Alper and A. S. K. Chan, *J. Am. Chem. Soc.* **95**, 4905 (1973).
22. P. H. Bird, U. Siriwardane, A. Shaver, O. Lopez, and D. N. Harpp, *J. Chem. Soc., Chem. Commun.*, 513 (1981).
23. G. Dettlaf, P. Hübener, J. Klimes, and E. Weiss, *J. Organomet. Chem.* **229**, 63 (1982).
24. J. Daub, G. Endress, U. Erhardt, K. H. Jogun, J. Kappler, A. Laumer, R. Pfiz, and J. Stezowski, *Chem. Ber.* **115**, 1787 (1982).
25. A. Benoit, J. Y. Le Marouille, C. Mahé, and H. Patin, *J. Organomet. Chem.* **233**, C51 (1982).
26. A. Lagadec, B. Misterkiewicz, A. Darchen, D. Grandjean, A. Mousser, and H. Patin, *Organometallics* **7**, 242 (1988).
27. D. Seyferth, G. B. Womack, L. C. Song, M. Cowie, and B. W. Hames, *Organometallics* **2**, 928 (1983).
28. A. Darchen, E. K. Lhadi, D. Grandjean, A. Mousser, and H. Patin, *J. Organomet. Chem.* **342**, C15 (1988).
29. U. Krüerke (ed.), "Gmelin's Handbuch der Anorganischen Chemie," Teil C1, p. 115. Springer-Verlag, Berlin, 1979.
30. H. Umland, F. Edelman, D. Wormsbächer, and U. Behrens, *Angew. Chem., Int. Ed. Engl.* **22**, 153 (1983).
31. H. Umland and U. Behrens, *J. Organomet. Chem.* **273**, C39 (1983).
32. H. Alper, *J. Organomet. Chem.* **84**, 347 (1975).
33. K. Hoffmann and E. Weiss, *J. Organomet. Chem.* **128**, 225 (1977).
34. G. N. Schrauzer and H. Kisch, *J. Am. Chem. Soc.* **95**, 2501 (1973).
35. H. Alper, *J. Organomet. Chem.* **73**, 359 (1974).
36. H. Alper, *Inorg. Chem.* **15**, 962 (1976).
37. H. Alper and A. S. K. Chan, *J. Organomet. Chem.* **61**, C59 (1973).
38. H. Alper, *J. Organomet. Chem.* **61**, C62 (1973).
39. A. R. Siedle, *J. Organomet. Chem.* **208**, 115 (1981).
40. H. Alper, *J. Organomet. Chem.* **80**, C29 (1974).
41. G. Mignani, H. Patin, and R. Dabard, *J. Organomet. Chem.* **169**, C19 (1979).

42. H. Patin, G. Mignani, and M. T. van Hulle, *Tetrahedron Lett.*, 2441 (1979).
43. E. D. Dobrzynski and R. J. Angelici, *J. Organomet. Chem.* **76**, C53 (1974).
44. H. Alper, N. D. Silavwe, G. I. Birnbaum, and F. R. Ahmed, *J. Am. Chem. Soc.* **101**, 6582 (1979).
45. H. Alper, F. W. P. Einstein, R. Nagai, J. F. Petrignani, and A. C. Willis, *Organometallics* **2**, 1291 (1983).
46. H. Alper, F. W. P. Einstein, F. W. Hartstock, and A. C. Willis, *J. Am. Chem. Soc.* **107**, 173 (1985).
47. H. Alper, F. W. P. Einstein, F. W. Hartstock, and A. C. Willis, *Organometallics* **5**, 9 (1986).
48. E. Lindner, I. P. Butz, S. Hoehne, W. Hiller, and R. Fawzi, *J. Organomet. Chem.* **259**, 99 (1983).
49. M. Pasquali, P. Leoni, C. Floriani, A. C. Villa, and C. Guastini, *Inorg. Chem.* **22**, 841 (1983).
50. R. L. Bennet, M. I. Bruce, I. Matsuda, R. J. Doedens, R. G. Little, and J. T. Veal, *J. Organomet. Chem.* **67**, C72 (1974).
51. R. L. Bennet, M. I. Bruce, and I. Matsuda, *Aust. J. Chem.* **28**, 2307 (1975).
52. J. Dupont, N. Beydoun, and M. Pfeffer, *J. Chem. Soc., Dalton Trans.*, 1715 (1989).
53. H. G. Raubenheimer, L. Linford, and A. van A. Lombard, *Organometallics* **8**, 2062 (1989).
54. S. L. Buchwald and R. B. Nielsen, *J. Am. Chem. Soc.* **110**, 3171 (1988).
55. H. D. Kaez, R. B. King, T. A. Manuel, R. D. Nichols, and F. G. A. Stone, *J. Am. Chem. Soc.* **82**, 4749 (1960).
56. R. B. King and F. G. A. Stone, *J. Am. Chem. Soc.* **82**, 4557 (1961).
57. A. E. Ogilvy, M. Draganjac, T. B. Rauchfuss, and S. R. Wilson, *Organometallics* **7**, 1171 (1988).
58. G. N. Glavee, L. M. Daniels, and R. J. Angelici, *Organometallics* **8**, 1856 (1989).
59. R. B. King, P. Treichel, and F. G. A. Stone, *J. Am. Chem. Soc.* **83**, 3600 (1961).
60. E. Boyar, A. J. Deeming, K. Henrick, M. McPartlin, and A. Scott, *J. Chem. Soc., Dalton Trans.*, 1431 (1986).
61. R. D. Adams and S. Wang, *J. Am. Chem. Soc.* **109**, 924 (1987).
62. S. S. Carleton, F. G. Kennedy, and S. A. R. Knox, *J. Chem. Soc., Dalton Trans.*, 2230 (1981).
63. K. Osakada, T. Chiba, Y. Nakamura, T. Yamamoto, and A. Yamamoto, *J. Chem. Soc., Chem. Commun.*, 1589 (1986).
64. D. C. Dittmer, K. Takahashi, M. Iwanami, A. I. Tsai, P. L. Chang, B. B. Blidner, and I. K. Stamos, *J. Am. Chem. Soc.* **98**, 2795 (1976).
65. E. J. Parker, J. R. Bodwell, T. C. Sedergran, and D. C. Dittmer, *Organometallics* **1**, 517 (1982).
66. S. L. Buchwald, R. B. Nielsen, and J. C. Dewan, *J. Am. Chem. Soc.* **109**, 1590 (1987).
67. R. D. Adams, *Acc. Chem. Res.* **16**, 67 (1983).
68. G. N. Schrauzer, H. N. Rabinowitz, J. K. Frank, and I. C. Paul, *J. Am. Chem. Soc.* **92**, 212 (1970).
69. P. G. Mente and C. W. Rees, *J. Chem. Soc., Chem. Commun.*, 419 (1972).
70. T. L. Gilchrist, P. G. Mente, and C. W. Rees, *J. Chem. Soc., Perkin Trans. 1*, 2165 (1972).
71. K. H. Pannell, A. J. Mayr, and D. VanDerveer, *Organometallics* **2**, 560 (1983).
72. K. H. Pannell, A. J. Mayr, R. Hoggard, and R. C. Petterson, *Angew. Chem., Int. Ed. Engl.* **19**, 632 (1980).
73. R. C. Petterson, K. H. Pannell, and A. Mayr, *Acta Crystallogr., Sect B* **36**, 2434 (1980).

74. K. H. Pannell, A. J. Mayr, and B. H. Carasco-Flores, *J. Organomet. Chem.* **354,** 97 (1988).

75. C. P. Morley, *Organometallics* **8,** 800 (1989).

76. C. P. Morley, R. R. Vaughn, and B. J. Wheatley, *J. Organomet. Chem.* **353,** C39 (1988).

77. F. Edelmann and U. Behrens, *J. Organomet. Chem.* **134,** 31 (1977).

78. O. Koch, F. Edelmann, B. Lubke, and U. Behrens, *Chem. Ber.* **115,** 3049 (1982).

79. Y. S. Yu and R. J. Angelici, *Organometallics* **2,** 1018 (1983).

80. J. R. Matachek, R. J. Angelici, K. A. Schugart, K. J. Haller, and R. F. Fenske, *Organometallics* **3,** 1038 (1984).

81. N. C. Schroeder, J. W. Richardson, Jr., S. L. Wang, R. A. Jacobson, and R. J. Angelici, *Organometallics* **4,** 1226 (1984).

82. W. R. Roper and A. H. Wright, *J. Organomet. Chem.* **233,** C59 (1982).

83. W. A. Schenk, D. Kuemmerle, and C. Burschka, *J. Organomet. Chem.* **349,** 183 (1988).

84. H. G. Raubenheimer, G. J. Kruger, L. Linford, C. F. Marais, R. Otte, J. T. Z. Hattingh, and A. Lombard, *J. Chem. Soc., Dalton Trans.,* 1565 (1989).

85. U. Schubert, J. Kron, and H. Hörnig, *J. Organomet. Chem.* **355,** 243 (1988).

86. C. E. L. Headford and W. R. Roper, *J. Organomet. Chem.* **244,** C53 (1983).

87. J. E. Guerchais, F. Le Floch-Perennou, F. Y. Pétillon, A. N. Keith, L. Manojlović-Muir, K. W. Muir, and D. W. A. Sharp, *J. Chem. Soc., Chem. Commun.,* 410 (1979).

88. J. L. Davidson, M. Shiralian, L. Manojlović-Muir, and K. W. Muir, *J. Chem. Soc., Chem. Commun.,* 30 (1979).

89. R. Rumin, L. Manojlović-Muir, K. W. Muir, and F. Y. Pétillon, *Organometallics* **7,** 375 (1988).

90. N. M. Agh-Ataby, J. L. Davidson, G. Douglas, and K. W. Muir, *J. Chem. Soc., Chem. Commun.,* 549 (1989).

91. A. S. Ward, E. A. Mintz, and M. P. Kramer, *Organometallics* **7,** 8 (1988).

92. E. A. Mintz and A. S. Ward, *J. Organomet. Chem.* **307,** C52 (1986).

93. K. R. Pörschke, *Chem. Ber.* **120,** 425 (1987).

94. Y. Stenstrøm, G. Klauck, A. Koziol, G. J. Palenik, and W. M. Jones, *Organometallics* **5,** 2155 (1986).

95. W. R. Roper and K. G. Town, *J. Chem. Soc., Chem. Commun.,* 781 (1977).

96. T. J. Collins and W. R. Roper, *J. Organomet. Chem.* **159,** 73 (1978).

97. G. N. Glavee and R. J. Angelici, *J. Am. Chem. Soc.* **111,** 3598 (1989).

98. D. Seebach, *Chem. Ber.* **105,** 487 (1972).

99. H. G. Raubenheimer and H. E. Swanepoel, *J. Organomet. Chem.* **141,** C21 (1977).

100. B. Dadamoussa, A. Darchen, P. L'Haridon, C. Larpent, H. Patin, and J. Y. Thepot, *Organometallics* **8,** 564 (1989).

101. G. J. Kruger, L. Linford, and H. G. Raubenheimer, *J. Chem. Soc., Dalton Trans.,* 2337 (1984).

102. H. G. Raubenheimer, S. Lotz, H. W. Viljoen, and A. A. Chalmers, *J. Organomet. Chem.* **152,** 73 (1978); H. G. Raubenheimer, S. Lotz, H. E. Swanepoel, H. W. Viljoen, and J. C. Rautenbach, *J. Chem. Soc., Dalton Trans.,* 1701 (1979); G. J. Kruger, G. Gafner, J. P. R. de Villiers, H. G. Raubenheimer, and H. Swanepoel, *J. Organomet. Chem.* **187,** 333 (1980).

103. S. Lotz, H. G. Raubenheimer, H. W. Viljoen, and J. C. Viljoen, *S. Afr. J. Chem.* **36,** 13 (1983).

104. J. C. Viljoen, S. Lotz, L. Linford, and H. G. Raubenheimer, *J. Chem. Soc., Dalton Trans.,* 1041 (1984).

105. H. G. Raubenheimer, G. J. Kruger, and H. W. Viljoen, *J. Organomet. Chem.* **319,** 361 (1987).

106. F. R. Kreissl and N. Ullrich, *Chem. Ber.* **122,** 1487 (1989).
107. H. Fischer, J. Schmid, and S. Zeuner, *Chem. Ber.* **120,** 583 (1987).
108. F. B. McCormick, W. B. Gleason, X. Zhao, P. C. Heah, and J. A. Gladysz, *Organometallics* **5,** 1778 (1986).
109. H. Werner, R. Feser, W. Paul, and L. Hofmann, *J. Organomet. Chem.* **219,** C29 (1981).
110. H. G. Raubenheimer, G. J. Kruger, and L. Linford, *Inorg. Chim. Acta* **150,** 173 (1988).
111. E. Wenkert, M. Shepard, and A. T. McPhail, *J. Chem. Soc., Chem. Commun.,* 1390 (1986).
112. U. Schubert, B. Heiser, L. Hee, and H. Werner, *Chem. Ber.* **118,** 3151 (1985).
113. H. D. Holden, B. F. G. Johnson, J. Lewis, P. R. Raithby, and G. Uben, *Acta Crystallogr. Sect. C: Cryst. Struct. Commun.* **39,** 1197 (1983).
114. H. D. Holden, B. F. G. Johnson, J. Lewis, P. R. Raithby, and G. Uben, *Acta Crystallogr., Sect. C: Cryst. Struct. Commun.* **39,** 1200 (1983).
115. D. E. Laycock and H. Alper, *J. Org. Chem.* **46,** 289, (1981).
116. H. Alper and H. N. Paik, *J. Organomet. Chem.* **121,** 225 (1976).
117. P. J. Pogorzelec and D. H. Reid, *J. Chem. Soc., Chem. Commun.,* 289 (1983).
118. J. Benecke and U. Behrens, *J. Organomet. Chem.* **363,** C15 (1989).
119. V. Crocq, J. C. Daran, and Y. Jeannin, *J. Organomet. Chem.* **373,** 85 (1989).
120. G. E. Herberich and H. Mayer, *J. Organomet. Chem.* **322,** C29 (1987).
121. K. Seitz and U. Behrens, *J. Organomet. Chem.* **294,** C9 (1985).
122. P. Ewing and L. J. Farrugia, *J. Organomet. Chem.* **373,** 259 (1989).
123. R. D. Adams, Z. Dawoodi, D. F. Foust, and B. E. Segmüller, *Organometallics* **2,** 315 (1983).
124. E. Lindner, C. P. Krieg, W. Hiller, and D. Hübner, *Chem. Ber.* **117,** 489 (1984).
125. F.J. Brown, *Prog. Inorg. Chem.* **27,** 1 (1980).
126. H. Fischer, *in* "The Chemistry of the Metal – Carbon Bond" (F. R. Hartley and S. Patai, eds.), p. 181. Wiley, Chichester, New York, 1982.
127. B. E. Boland-Lussier and R. P. Hughes, *Organometallics* **1,** 635 (1982).
128. G. Beck and W. P. Fehlhammer, *Angew. Chem., Int. Ed. Engl.* **27,** 1344 (1988).
129. W. P. Fehlhammer and G. Beck, *J. Organomet. Chem.* **369,** 105 (1989).
130. H. D. Holden, B. F. G. Johnson, J. Lewis, P. R. Raithby, and G. Uden, *Acta Crystallogr., Sect. C: Cryst. Struct. Commun.* **39,** 1203 (1983).
131. R. Bertani, M. Mozzon, and R. A. Michelin, *Inorg. Chem.* **27,** 2809 (1988).
132. J. Messelhäuser, I. P. Lorenz, K. Haug, and W. Hiller, *Z. Naturforsch., B: Anorg. Chem., Org. Chem.* **40,** 1064 (1985).
133. M. Rakowski DuBois, R. C. Haltiwanger, D. J. Miller, and G. Glatzmaier, *J. Am. Chem. Soc.* **101,** 5245 (1979).
134. O. A. Rajan, M. McKenna, R. C. Haltiwanger, and M. Rakowski DuBois, *Organometallics* **3,** 831 (1984).
135. C. W. Bird and E. M. Hollins, *J. Organomet. Chem.* **4,** 245 (1965).
136. A. Shaver, P. J. Fitzpatrick, K. Steliou, and I. S. Butler, *J. Am. Chem. Soc.* **101,** 1313 (1979).
137. S. Lotz, P. H. van Rooyen, and M. M. van Dyk, *Organometallics* **6,** 499 (1987).
138. A. R. Koray and M. L. Ziegler, *J. Organomet. Chem.* **169,** C34 (1979).
139. A. Lagadec, R. Dabard, B. Misterkiewicz, A. Le Rouzic, and H. Patin, *J. Organomet. Chem.* **326,** 381 (1987).
140. H. Patin, G. Mignani, C. Mahé, J. Y. Le Marouille, A. Benoit, and D. Grandjean, *J. Organomet. Chem.* **210,** C1 (1981).
141. H. G. Raubenheimer, S. Lotz, J. C. Viljoen, and E. du Plessis, *J. Chem. Res.* (Suppl.), 22 (1982).

142. H. Alper and J. L. Fabre, *Organometallics* **1**, 1037 (1982).
143. S. Lotz, J. L. M. Dillen, and M. M. van Dyk, *J. Organomet. Chem.* **371**, 371 (1989).
144. D. Seyferth, C. N. Rudie, J. S. Merola, and D. H. Berry, *J. Organomet. Chem.* **187**, 91 (1980).
145. H. G. Raubenheimer, G. J. Kruger, C. F. Marais, J. T. Z. Hattingh, R. Otte, and L. Linford, *in* "Advances in Metal Carbene Chemistry" (U. Schubert, ed.) p. 145. Kluwer Academic Publishers, Dordrecht, The Netherlands, 1989.
146. H. G. Raubenheimer, E. O. Fischer, U. Schubert, C. Krüger, and Y. H. Tsay, *Angew. Chem., Int. Ed. Engl.* **20**, 1055 (1981).
147. H. G. Raubenheimer, G. J. Kruger, A. Lombard, L. Linford, and J. C. Viljoen, *Organometallics* **4**, 275 (1985).
148. R. J. Angelici, *in* "Advances in Metal Carbene Chemistry" (U. Schubert, ed.) p. 123. Kluwer Academic Publishers, Dordrecht, The Netherlands, 1989.
149. T. Desmond, F. J. Lalor, G. Ferguson, and M. Parvez, *J. Chem. Soc., Chem. Commun.,* 75 (1984).
150a. N. C. Schroeder, R. Funchess, R. A. Jacobson, and R. J. Angelici, *Organometallics* **8**, 521 (1989).
150b. H. P. Kim and R. J. Angelici, *Adv. Organomet. Chem.* **27**, 51 (1987).
151. E. O. Fischer, D. Himmelreich, R. Cai, H. Fischer, U. Schubert, and B. Zimmer-Gasser, *Chem. Ber.* **114**, 3209 (1981).
152. E. O. Fischer and W. Röll, *Angew. Chem., Int. Ed. Engl.* **19**, 205 (1980).
153. D. L. Reger, E. Mintz, and L. Lebioda, *J. Am. Chem. Soc.* **108**, 1940 (1986).
154. H. G. Raubenheimer, S. Lotz, G. J. Kruger, A. Lombard, and J. C. Viljoen, *J. Organomet. Chem.* **336**, 349 (1987).
155. E. Lindner and W. Nagel, *Z. Naturforsch., B: Anorg. Chem., Org. Chem.* **32**, 1116 (1977).
156. F. R. Kreissl and H. Keller, *Angew. Chem., Int. Ed. Engl.* **25**, 904 (1986).
157. R. A. Doyle and R. J. Angelici, *J. Organomet. Chem.* **375**, 73 (1989).
158. J. P. Battioni, J. C. Chottard, and D. Mansuy, *Inorg. Chem.* **21**, 2056 (1982).
159. P. Legzdins and L. Sanchez, *J. Am. Chem. Soc.* **107**, 5525 (1985).
160. S. V. Evans, P. Legzdins, S. J. Rettig, L. Sanchez, and J. Trotter, *Organometallics* **6**, 7 (1987).
161. P. Legzdins, S. J. Rettig, and L. Sanchez, *Organometallics* **7**, 2394 (1988).
162. B. Gautheron, G. Tainturier, S. Pouly, F. Théobald, H. Vivier, and A. Laarif, *Organometallics* **3**, 1495 (1984).
163. P. Meunier, B. Gautheron, and A. Mazouz, *J. Organomet. Chem.* **320**, C39 (1987).
164. C. Giannotti and G. Merle, *J. Organomet. Chem.* **113**, 45 (1976).
165. L. Busetto, A. Palazzi, E. F. Serantoni, and L. R. Di Sanseverino, *J. Organomet. Chem.* **129**, C55 (1977).
166. A. F. Hill, W. R. Roper, J. M. Waters, and A. H. Wright, *J. Am. Chem. Soc.* **105**, 5939 (1983).
167. W. E. Buhro, M. C. Etter, S. Georgiou, J. A. Gladysz, and F. B. McCormick, *Organometallics* **6**, 1150 (1987).
168. F. B. McCormick, *Organometallics* **3**, 1924 (1984).
169. R. D. Adams, J. E. Babin, and M. Tasi, *Organometallics* **6**, 1717 (1987).
170. H. Fischer and R. Markl, *Chem. Ber.* **115**, 1349 (1982).
171. H. Fischer and S. Zeuner, *Z. Naturforsch., B: Anorg. Chem., Org. Chem.* **38**, 1365 (1983).
172. H. Fischer, *J. Organomet. Chem.* **222**, 241 (1981).
173. H. Fischer and S. Zeuner, *Z. Naturforsch., B: Anorg. Chem., Org. Chem.* **40**, 954 (1985).

174. H. Fischer, S. Zeuner, and J. Riede, *Angew. Chem., Int. Ed. Engl.* **23**, 726 (1984).
175. H. Fischer and S. Zeuner, *J. Organomet. Chem.* **252**, C63 (1983).
176. H. Werner, L. Hofmann, J. Wolf, and G. Müller, *J. Organomet. Chem.* **280**, C55 (1985).
177. J. Wolf, R. Zolk, U. Schubert, and H. Werner, *J. Organomet. Chem.* **340**, 161 (1988).
178. D. S. Gill, M. Green, K. Marsden, I. Moore, A. G. Orpen, F. G. A. Stone, I. D. Williams, and P. Woodward, *J. Chem. Soc., Dalton Trans.*, 1343 (1984).
179. F. R. Kreissl and N. Ulrich, *J. Organomet. Chem.* **361**, C30 (1989).
180. G. R. Clark, K. Marsden, W. R. Roper, and L. J. Wright, *J. Am. Chem. Soc.* **102**, 6570 (1980).
181. A. Mayr, G. A. McDermott, A. M. Dorries, and A. K. Holder, *J. Am. Chem. Soc.* **108**, 310 (1986).
182. E. Delgado, A. T. Emo, J. C. Jeffery, N. D. Simmons, and F. G. A. Stone, *J. Chem. Soc., Dalton Trans.*, 1323 (1985).
183. P. G. Byrne, M. E. Garcia, J. C. Jeffery, P. Sherwood, and F. G. A. Stone, *J. Chem. Soc., Dalton Trans.*, 1215 (1987).
184. N. S. Nametkin, V. D. Tyurin, A. I. Nekhaev, M. G. Kondrat'eva, and Yu. P. Sobolev, *Izv. Akad. Nauk SSSR, Ser. Khim.*, 217 (1982).
185. E. A. Chernyshev, O. V. Kuz'min, A. V. Lebedev, I. A. Mrachkovskaya, A. I. Gusev, N. I. Kirillova, and A. V. Kisin, *Zh. Obshch. Khim.* **56**, 367 (1986); *Chem. Abstr.* **106**, 50290x (1987).
186. E. A. Chernyshev, O. V. Kuz'min, A. V. Lebedev, A. I. Gusev, M. G. Los', N. V. Alekseev, N. S. Nametkin, V. D. Tyurin, A. M. Krapivin, N. A. Kubasova, and V. G. Zaikin, *J. Organomet. Chem.* **252**, 143 (1983).
187. A. I. Nekhaev, S. D. Alekseeva, N. S. Nametkin, V. D. Tyurin, G. G. Aleksandrov, N. A. Parpiev, M. T. Tashev, and H. B. Dustov, *J. Organomet. Chem.* **297**, C33 (1985).
188. B. I. Kolobkov, N. S. Nametkin, V. D. Tyurin, A. I. Nekhaev, G. G. Aleksandrov, M. T. Tashev, and H. B. Dustov, *J. Organomet. Chem.* **301**, 349 (1986).
189. E. Lindner, *J. Organomet. Chem.* **94**, 229 (1975).
190. H. G. Raubenheimer, G. J. Kruger, C. F. Marais, R. Otte, and J. T. Z. Hattingh, *Organometallics* **7**, 1853 (1988).
191. Y. Wakatsuki, H. Yamazaki, and H. Iwasaki, *J. Am. Chem. Soc.* **95**, 5781 (1973).
192. W. Paul and H. Werner, *Angew. Chem. Suppl.*, 396 (1983).
193. H. Werner, W. Paul, W. Knaup, J. Wolf, G. Müller, and J. Riede, *J. Organomet. Chem.* **358**, 95 (1988).
194. H. Werner and W. Paul, *Angew. Chem., Int. Ed. Engl.* **23**, 58 (1984).
195. L. Hofmann and H. Werner, *Chem. Ber.* **118**, 4229 (1985).
196. K. R. Grundy and W. R. Roper, *J. Organomet. Chem.* **113**, C45 (1976).
197. F. R. Kreissl and M. Wolfgruber, *J. Organomet. Chem.* **355**, 267 (1988).
198. I. F. Butler and A. E. Fenster, *J. Organomet. Chem.* **66**, 161 (1974).
199. R. P. Burns, F. P. McCullough, and C. A. McAuliffe, *Adv. Inorg. Chem. Radiochem.* **23**, 211 (1980).
200. F. Scott, G. J. Kruger, S. Cronje, A. Lombard, H. G. Raubenheimer, R. Benn, and A. Rufińska, *Organometallics* **9**, 1071 (1990).
201. H. Otto and H. Werner, *Chem. Ber.* **120**, 97 (1987).
202. E. Klei and J. H. Teuben, *J. Organomet. Chem.* **222**, 79 (1987).
203. M. R. Torres, A. Perales, and J. Ros, *Organometallics* **7**, 1223 (1988).
204. M. I. Bruce, M. J. Liddell, M. R. Snow, and E. R. T. Tiekink, *J. Organomet. Chem.* **352**, 199 (1988).
205. L. Busetto, A. Palazzi, and V. Foliadis, *Inorg. Chim. Acta* **40**, 147 (1980).

206. H. Werner and R. Werner, *J. Organomet. Chem.* **209**, C60 (1981).
207. M. I. Bruce, T. W. Hambley, M. R. Snow, and A. G. Swincer, *J. Organomet. Chem.* **273**, 361 (1984).
208. Y. J. Kim, K. Osakada, K. Sugita, T. Yamamoto, and A. Yamamoto, *Organometallics* **7**, 2182 (1988).
209. P. V. Yaneff, *Coord. Chem. Rev.* **23**, 183 (1977).
210. P. M. Treichel and D. C. Molzahn, *Inorg. Chem. Acta* **36**, 267 (1976).
211. R. D. Adams and N. M. Golembeski, *J. Am. Chem. Soc.* **101**, 1306 (1979).
212. R. D. Adams, N. M. Golembeski, and J. P. Selegue, *J. Am. Chem. Soc.* **101**, 5862 (1979).
213. R. D. Adams, N. M. Golembeski, and J. P. Selegue, *J. Am. Chem. Soc.* **103**, 546 (1981).
214. J. P. Selegue, *J. Am. Chem. Soc.* **104**, 119 (1982).
215. G. Grötsch and W. Malisch, *J. Organomet. Chem.* **262**, C38 (1984).
216. H. G. Raubenheimer, J. C. Viljoen, S. Lotz, and A. Lombard, *J. Chem. Soc., Chem. Commun.*, 749 (1981).
217. H. G. Raubenheimer, G. J. Kruger, and A. Lombard, *J. Organomet. Chem.* **240**, C11 (1982).
218. H. G. Raubenheimer, G. J. Kruger, and H. W. Viljoen, *J. Chem. Soc., Dalton Trans.*, 1963 (1985).
219. Y. Wakatsuki and H. Yamazaki, *J. Chem. Soc., Chem. Commun.*, 280 (1973).
220. O. Kolb and H. Werner, *J. Organomet. Chem.* **268**, 49 (1984).
221. M. Ebner and H. Werner, *Chem. Ber.* **119**, 482 (1986).
222. M. G. Mason, P. N. Swepston, and J. A. Ibers, *Inorg. Chem.* **22**, 411 (1983).
223. S. J. N. Burgmayer and J. L. Templeton, *Inorg. Chem.* **24**, 3939 (1985).
224. P. V. Broadhurst, B. F. G. Johnson, J. Lewis, and P. R. Raithby, *J. Chem. Soc., Chem. Commun.*, 140 (1982).
225. J. J. Maj, A. D. Rae, and L. F. Dahl, *J. Am. Chem. Soc.* **104**, 4278 (1982).
226. H. A. Harris, A. D. Rae, and L. F. Dahl, *J. Am. Chem. Soc.* **109**, 4739 (1987).
227. H. Fischer, U. Gerbing, and A. Tiriliomis, *J. Organomet. Chem.* **332**, 105 (1987).
228. H. Fischer, U. Gerbing, A. Tiriliomis, G. Müller, B. Huber, J. Riede, J. Hofmann, and P. Burger, *Chem. Ber.* **121**, 2095 (1988).
229. H. Fischer, J. Hofmann, U. Gerbing, and A. Tiriliomis, *J. Organomet. Chem.* **358**, 229 (1988).
230. H. Fischer and I. Pashalidis, *J. Organomet. Chem.* **348**, Cl (1988).
231. D. Touchard, P. H. Dixneuf, R. D. Adams, and B. E. Segmüller, *Organometallics* **3**, 640 (1984).
232. L. Carlton, J. L. Davidson, P. Ewing, L. Manojlović-Muir, and K. W. Muir, *J. Chem. Soc., Chem. Commun.*, 1474 (1985).
233. J. Dupont and M. Pfeffer, *J. Organomet. Chem.* **321**, C13 (1987).
234. R. D. Adams and S. Wang, *Organometallics* **6**, 739 (1987).
235. J. L. Davidson and D. W. A. Sharp, *J. Chem. Soc., Dalton Trans.*, 2283 (1975).
236. F. Y. Petillon, F. Le Floch-Perennou, J. E. Guerchais, and D. W. A. Sharp, *J. Organomet. Chem.* **173**, 89 (1979).
237. M. T. Ashby and J. H. Enemark, *Organometallics* **6**, 1318 (1987).
238. D. Seyferth, J. B. Hoke, and J. C. Dewan, *Organometallics* **6**, 895 (1987).
239. J. L. Le Quere, F. Y. Petillon, J. E. Guerchais, and J. Sala-Pala, *Inorg. Chim. Acta* **43**, 5 (1980).
240. J. Li, R. Hoffmann, C. Mealli, and J. Silvestre, *Organometallics* **8**, 1921 (1989).
241. F. Conan, J. Sala-Pala, J. E. Guerchais, J. Li, R. Hoffmann, C. Mealli, R. Mercier, and L. Toupet, *Organometallics* **8**, 1929 (1989).

242. W. A. Schenk and D. Kümmerle, *J. Organomet. Chem.* **303**, C25 (1986).
243. G. P. Elliott and W. R. Roper, *J. Organomet. Chem.* **250**, C5 (1983).
244. J. W. Ziller, D. K. Bower, D. M. Dalton, J. B. Keister, and M. R. Churchill, *Organometallics* **8**, 492 (1989).
245. M. McKenna, L. Wright, D. J. Miller, L. Tanner, R. C. Haltiwanger, and M. Rakowski DuBois, *J. Am. Chem. Soc.* **105**, 5329 (1983).
246. H. Fischer, U. Gerbing, J. Riede, and R. Benn, *Angew. Chem., Int. Ed. Engl.* **25**, 78 (1986).
247. H. Fischer and U. Gerbing, *J. Organomet. Chem.* **299**, C7 (1986).
248. H. Fischer, *J. Organomet. Chem.* **345**, 65 (1988).
249. E. O. Fischer, K. Weiss, and K. Burger, *Chem. Ber.* **106**, 1581 (1973).
250. J. G. Davidson, E. K. Barefield, and D. G. VanDerveer, *Organometallics* **4**, 1178 (1985).
251. S. Brandt and P. Helquist, *J. Am. Chem. Soc.* **101**, 6473 (1979).
252. E. J. O'Connor, S. Brandt, and P. Helquist, *J. Am. Chem. Soc.* **109**, 3739 (1987).
253. R. S. Iyer, G. H. Kuo, and P. Helquist, *J. Org. Chem.* **50**, 5898 (1985).
254. K. A. M. Kremer and P. Helquist, *J. Organomet. Chem.* **285**, 231 (1985).
255. G. Yoshida, H. Kurosawa, and R. Okawara, *J. Organomet. Chem.* **131**, 309 (1977).
256. W. A. Schenk, D. Rüb, and C. Burschka, *Angew. Chem., Int. Ed. Engl.* **24**, 971 (1985).
257. H. P. Kim, S. Kim, R. A. Jacobson, and R. J. Angelici, *Organometallics* **5**, 2481 (1986).
258. R. A. Doyle and R. J. Angelici, *Organometallics* **8**, 2207 (1989).
259. S. Myrvold, O. A. Nassif, G. Semelhago, A. Walker, and D. H. Farrar, *Inorg. Chim. Acta* **117**, 17 (1986).
260. L. Busetto, S. Bordoni, V. Zanotti, V. G. Albano, and D. Braga, in "Advances in Metal Carbene Chemistry" (U. Schubert, ed.), p. 141. Kluwer Academic Publishers, Dordrecht, The Netherlands, 1989.
261. I. B. Benson, J. Hunt, and S. A. R. Knox, *J. Chem. Soc., Dalton Trans.,* 1240 (1978).
262. H. G. Raubenheimer, G. J. Kruger, C. F. Marais, J. T. Z. Hattingh, L. Linford, and P. H. van Rooyen, *J. Organomet. Chem.* **355**, 337 (1988).
263. T. J. Collins and W. R. Roper, *J. Chem. Soc., Chem. Commun.,* 901 (1977).
264. L. Busetto, S. Bordoni, and V. Zanotti, *J. Organomet. Chem.* **339**, 125 (1988).
265. H. Le Bozec, J. L. Fillaut, and P. H. Dixneuf, *J. Chem. Soc., Chem. Commun.,* 1182 (1986).
266. H. P. Kim, S. Kim, R. A. Jacobson, and R. J. Angelici, *Organometallics* **3**, 1124 (1984).
267. W. Uedelhoven, K. Eberl, and F. R. Kreissl, *Chem. Ber.* **112**, 3376 (1979).
268. F. R. Kriessl, F. X. Müller, D. L. Wilkinson, and G. Müller, *Chem. Ber.* **122**, 289 (1989).
269. D. Wormsbächer, R. Drews, F. Edelmann, and U. Behrens, *Chem. Ber.* **115**, 1332 (1981).
270. R. Drews, F. Edelmann, and U. Behrens, *J. Organomet. Chem.* **315**, 369 (1986).
271. C. Knors, G. H. Kuo, J. W. Lauher, C. Eigenbrot, and P. Helquist, *Organometallics* **6**, 988 (1987).
272. L. Busetto, V. Zanotti, V. G. Albano, D. Braga, and M. Monari, *J. Chem. Soc., Dalton Trans.,* 1067 (1988).
273. H. P. Kim and R. J. Angelici, *Organometallics* **5**, 2489 (1986).
274. W. A. Schenk, D. Kuemmerle, and T. Schwietzke, *J. Organomet. Chem.* **349**, 163 (1988).
275. D. Touchard, J. L. Fillaut, P. H. Dixneuf, C. Mealli, M. Sabat, and L. Toupet, *Organometallics* **4**, 1684 (1985).
276. H. Umland and U. Behrens, *J. Organomet. Chem.* **287**, 109 (1985).

277. D. A. Lech, J. W. Richardson, Jr., R. A. Jacobson, and R. J. Angelici, *J. Am. Chem. Soc.* **106**, 2901 (1984).
278. J. W. Hachgenei and R. J. Angelici, *J. Organomet. Chem.* **355**, 359 (1988).
279. G. H. Spies and R. J. Angelici, *Organometallics* **6**, 1897 (1987).
280. S. Wang and J. P. Fackler, *Organometallics* **8**, 1578 (1989).
281. M. M. Singh and R. J. Angelici, *Inorg. Chem.* **23**, 2691 (1984).
282. W. A. Herrmann, J. Weichmann, R. Serrano, K. Blechschmitt, H. Pfisterer, and M. L. Ziegler, *Angew. Chem., Int. Ed. Engl.* **22**, 314 (1983).
283. M. Herberhold, W. Ehrenreich, and W. Bühlmeyer, *Angew. Chem., Int. Ed. Engl.* **22**, 315 (1983).
284. W. A. Herrmann, J. Rohrmann, and A. Schäfer, *J. Organomet. Chem.* **265**, C1 (1984).
285. W. A. Herrmann, C. Hecht, M. L. Ziegler, and B. Balbach, *J. Chem. Soc., Chem. Commun.*, 686 (1984).
286. M. Herberhold, W. Jellen, and H. H. Murray, *J. Organomet. Chem.* **270**, 65 (1984).
287. T. Fassler, D. Buchholz, G. Huttner, and L. Zsolnai, *J. Organomet. Chem.* **369**, 297 (1989).
288. R. D. Adams and S. Wang, *Organometallics* **4**, 1902 (1985).
289. D. Seyferth and G. B. Womack, *J. Am. Chem. Soc.* **104**, 6839 (1982).
290. D. Seyferth, G. B. Womack, R. S. Henderson, M. Cowie, and B. W. Hames, *Organometallics* **5**, 1568 (1986).
291. C. J. Ruffing and T. B. Rauchfuss, *Organometallics* **4**, 524 (1985).
292. D. Seyferth, G. B. Womack, and L. C. Song, *Organometallics* **2**, 776 (1983).
293. D. Seyferth and G. B. Womack, *Organometallics* **5**, 2360 (1986).
294. D. Seyferth and R. S. Henderson, *J. Organomet. Chem.* **182**, C39 (1979).
295. J. Messelhäuser, K. U. Gutensohn, I. P. Lorenz, and W. Hiller, *J. Organomet. Chem.* **321**, 377 (1987).
296. R. Weberg, R. C. Haltiwanger, and M. Rakowski DuBois, *Organometallics* **4**, 1315 (1985).
297. T. B. Rauchfuss, D. P. S. Rodgers, and S. R. Wilson, *J. Am. Chem. Soc.* **108**, 3114 (1986).
298. J. R. Halbert, W. H. Pan, and E. I. Steifel, *J. Am. Chem. Soc.* **105**, 5476 (1983).
299. C. M. Bolinger, T. B. Rauchfuss, and S. R. Wilson, *J. Am. Chem. Soc.* **103**, 5620 (1981).
300. C. M. Bolinger, J. E. Hoots, and T. B. Rauchfuss, *Organometallics* **1**, 223 (1982).
301. C. M. Bolinger and T. B. Rauchfuss, *Inorg. Chem.* **21**, 3947 (1982).
302. D. M. Giolando, T. B. Rauchfuss, A. L. Rheingold, and S. R. Wilkinson, *Organometallics* **6**, 667 (1987).
303. M. Rakowski DuBois, D. L. DuBois, M. C. VanDerveer, and R. C. Haltiwanger, *Inorg. Chem.* **20**, 3064 (1981).
304. C. M. Bolinger, T. B. Rauchfuss, and A. L. Rheingold, *J. Am. Chem. Soc.* **105**, 6321 (1983).
305. C. J. Casewit, R. C. Haltiwanger, J. Noordik, and M. Rakowski DuBois, *Organometallics* **4**, 119 (1985).
306. G. Godziela, T. Tonker, and M. Rakowski DuBois, *Organometallics* **8**, 2220 (1989).
307. R. G. W. Gingerich and R. J. Angelici, *J. Am. Chem. Soc.* **101**, 5604 (1979).
308. R. J. Angelici and R. G. W. Gingerich, *Organometallics* **2**, 89 (1983).
309. R. A. Doyle and R. J. Angelici, *J. Am. Chem. Soc.* **112**, 194 (1990).
310. R. D. Adams and M. P. Pompeo, *Organometallics* **9**, 1718 (1990).
311. R. D. Adams and M. P. Pompeo, *Organometallics* **9**, 2651 (1990).
312. R. D. Adams, G. Chen, J. T. Tanner, and J. Yin, *Organometallics* **9**, 595 (1990).
313. W. D. Jones and L. Dong, *J. Am. Chem. Soc.* **113**, 559 (1991).

314. J. W. Park, L. M. Henling, W. P. Schaefer, and H. Grubbs, *Organometallics* **9**, 1650 (1990).
315. D. C. Miller and R. J. Angelici, *Organometallics* **10**, 79 (1991).
316. R. Aumann and J. Schröder, *Chem. Ber.* **123**, 2053 (1990).
317. D. C. Miller and R. J. Angelici, *J. Organomet. Chem.* **394**, 235 (1990).
318. D. C. Miller and R. J. Angelici, *Organometallics* **10**, 89 (1991).
319. M. G. Choi, L. M. Daniels, and R. J. Angelici, *J. Organomet. Chem.* **383**, 321 (1990).
320. R. Aumann, J. Schröder, and H. Heinen, *Chem. Ber.* **123**, 1369 (1990).
321. D. Touchard, J. L. Fillaut, P. H. Dixneuf, R. D. Adams, and B. E. Segmüller, *J. Organomet. Chem.* **386**, 95 (1990).
322. R. Kergoat, M. M. Kubicki, L. C. Gomes de Lima, H. Scordia, J. E. Guerchais, and P. L'Haridon, *J. Organomet. Chem.* **367**, 143 (1989).
323. Th. Fässler and G. Huttner, *J. Organomet. Chem.* **376**, 367 (1989).
324. M. J. Carney, P. J. Walsh, and R. G. Bergman, *J. Am. Chem. Soc.* **112**, 6426 (1990).
325. J. Amaudrut, J. Sala-Pala, J. E. Guerchais, and R. Mercier, *J. Organomet. Chem.* **391**, 61 (1990).
326. Th. Fässler, G. Huttner, D. Günauer, S. Fiedler, and B. Eber, *J. Organomet. Chem.* **381**, 409 (1990).
327. J. Chen, L. M. Daniels, and R. J. Angelici, *J. Am. Chem. Soc.* **112**, 199 (1990).
328. J. Chen and R. J. Angelici, *Organometallics* **9**, 879 (1990).
329. A. Kramer and I. P. Lorenz, *J. Organomet. Chem.* **388**, 187 (1990).
330. A. Kramer, R. Lingnau, I. P. Lorenz, and H. A. Mayer, *Chem. Ber.* **123**, 1821 (1990).

ADVANCES IN ORGANOMETALLIC CHEMISTRY, VOL. 32

Hydroformylation Catalyzed by Ruthenium Complexes

PHILIPPE KALCK, YOLANDE PERES, and JEAN JENCK

Laboratoire de Chimie des Procédés
Ecole Nationale Supérieure de Chimie
31077 Toulouse Cédex, France

I

INTRODUCTION

The hydroformylation reaction or oxo synthesis, discovered in 1938 by O. Roelen in the laboratories of Ruhr Chemie in Germany, is of great importance as a synthetic tool. Starting from an alkene and using syn-gas ($H_2/CO = 1/1$), aldehydes with one more carbon are obtained, according to Eq. (1). More than 5 million tons of aldehydes are produced in the

$$R \diagdown\diagup + H_2 + CO \xrightarrow{[Cat.]} R \diagdown\diagup\diagdown CHO + R \diagup\diagdown_{CHO} \qquad (1)$$

world, so this oxo synthesis is of great industrial importance. The method is chiefly devoted to the hydroformylation of propene, the products being required for various uses including plastifiers, solvents, and starting materials for detergents.

The first catalysts used by organic chemists or developed in industrial plants were based on cobalt precursors. The pressures required are very high, 120–180 bar (1 bar ≅ 10^5 Pa), and the temperature generally used is 200°C. The selectivities in aldehydes are approximately 90%, with hydrogenation of the alkene being the main side reaction, and the linearity is low (<70%). This last feature is the key to the process since the main industrial uses require linear products. In the Shell process the phosphine ligand PBu$_3$ is introduced, leading to alcohols as products. The catalytic activity of rhodium was recognized at an early stage, but the temperature and pressure conditions were almost the same as those for cobalt. Usually rhodium is 10^3 to 10^4 times more active than cobalt, but it is 10^3 times more expensive! Around 1965 a rhodium complex where the coordination sphere contains triphenylphosphine ligands were discovered in Wilkinson's laboratory. Most present processes in industry, or for organic synthesis, employ this [HRh(CO)(PPh$_3$)$_3$] complex. In the plants it works almost exclusively for propene hydroformylation, for which good conversion rates to butanals are reached ($\sim$98%), with the linearity being about 92%.

Although ruthenium is significantly less expensive than rhodium and although its use has been recommended since 1960 (*1*) for the oxo synthesis, complexes of this metal have not been developed as catalysts. However, many papers and patents have referred to the results obtained employing various ruthenium complexes. The purpose of this article is to analyze the work done involving ruthenium compounds, restricting the scope only to the hydroformylation reaction and not to the carbonylation reaction, which would demand to too lengthy an article. In this review we examine successively mononuclear ruthenium complexes, ruthenium clusters as precursors, photochemical activation, and supported catalysis.

II

HYDROFORMYLATION WITH MONONUCLEAR RUTHENIUM COMPLEXES

Apart from the paper by Pichler *et al.* (*2*) where the use of ruthenium salts is mentioned for the oxo synthesis, the first paper where a well-defined complex was employed was reported by Wilkinson and group in a short communication in 1965 (*3*). The complex is [Ru(CO)$_3$(PPh$_3$)$_2$], and this species converts pent-1-ene at 100 bar of total pressure of CO and H$_2$ (1/1) at 100°C for 15 hours into 80% hexanals. A more detailed study appeared in 1968 (*4*). Under essentially the same experimental conditions (except 110°C), [Ru(CO)$_3$(PPh$_3$)$_2$] induced a linear to branched aldehyde ratio of 65/35. Two other complexes were tested: [RuCl$_3$(PPh$_3$)$_3$]·MeOH, which gave only 44% conversion, and [RuCl$_2$(PPh$_3$)$_3$], which is quite ineffective since it gave the insoluble complex [RuCl$_2$(CO)$_2$(PPh$_3$)$_2$]. The important observation was made that [Ru(CO)$_3$(PPh$_3$)$_2$] was recovered unchanged after reaction.

This complex was also identified starting from other mononuclear complexes as precursors, namely, $[RuH_2(PPh_3)_4]$, $[RuH_4(PPh_3)_3]$, $[RuH_2(CO)_2(PPh_3)_2]$, $[RuH_2(CO)(PPh_3)_2]$, $[RuH(NO)(PPh_3)_3]$, $[Ru(CO_2-CH_3)_2(PPh_3)_2]$, $[Ru(CO_2CF_3)_2(PPh_3)_2]$, and $[Ru(CO_2CMe_3)_2(PPh_3)_2]$ (5). These various compounds, under the oxo conditions, evidently generate the same active species. Moreover, the same linear to branched ratio was found after catalysis. The isomeric ratio can be slightly improved to 82/18 by using $[Ru(CO)_3(Ph_2PCH_2CH_2PPh_2)]$, but the conversion rate is lowered to 11% instead of 80–100%. The two complexes $[RuH_2(PPh_3)_4]$ and $[RuH_4(PPh_3)_3]$ yield significant amounts of hydrogenated products.

Investigations of the mechanism have shown, *inter alia,* that the reaction is first order in ruthenium concentration, the linear to branched aldehyde ratio is almost unchanged when the catalyst concentration is varied, and the rate of the hydroformylation is significantly increased by high partial pressures of hydrogen but dramatically inhibited by high partial pressures of carbon monoxide. Moreover, the optimum temperature was found to be around 120°C; below 80°C the conversion rate is very slow, whereas above 150°C the ruthenium species are transformed into less active or inactive ruthenium complexes. Starting from the $[RuH_2(CO)_2(PPh_3)_2]$ complex, no induction period was observed, the reaction giving rise to the same distribution of products. The rate-determining step is certainly the activation of dihydrogen to form the dihydridoruthenium species [Eq. (2)].

$$[Ru(CO)_3(PPh_3)_2] + H_2 \rightleftharpoons [RuH_2(CO)_2(PPh_3)_2] + CO \qquad (2)$$

Among the steps in Scheme 1, those related to the hydrogen transfer (step 2) or the CO insertion (step 3) must be very fast since no isomerization nor hydrogenation processes are concomitant with hydroformylation. A mechanism analogous to that proposed for rhodium, that is, occurring on a single metal center, can account for all these observations (Scheme 1).

Addition of an excess of triphenylphosphine to the complex induces a marked improvement of the selectivity in linear aldehyde. Presumably, instead of having $[RuH_2(CO)_2(PPh_3)$ (alkene)], in the presence of an excess of PPh$_3$ the complex $[RuH_2(CO)(PPh_3)_2(alkene)]$ is formed so that the electron density on the central atom is increased. Under such conditions the hydridic character of the hydride ligand is more pronounced, giving rise to higher ratios in the linear alkyl species. However, the first step is very sensitive to the amount of triphenylphosphine present, since an excess of this ligand results in a dramatic decrease in the rates. In our opinion a second catalytic cycle can be proposed (Scheme 2).

In the presence of a large excess of triphenylphosphine there is strong competition between all the ligands so that coordination of the alkene in Step 1 is retarded, and, possibly, CO insertion (step 3) as well as the reductive elimination are less favored than in the previous mechanism

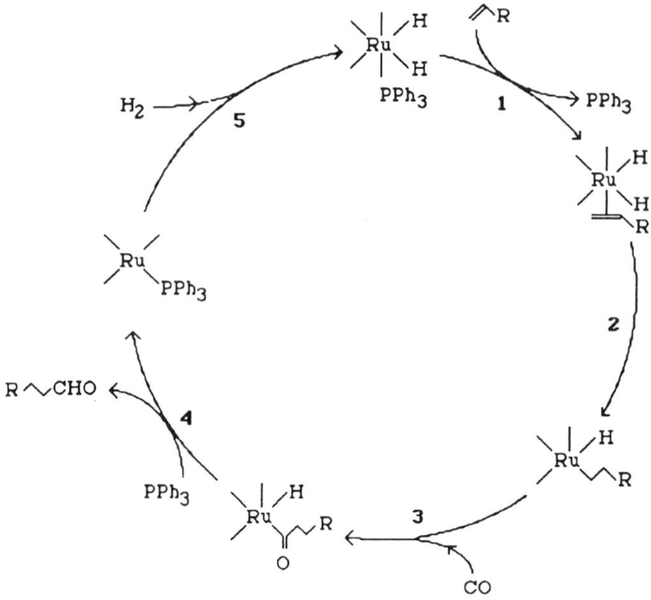

PHILIPPE KALCK *et al.*

SCHEME 1

(Scheme 1). The situation here is quite different from that which prevails for rhodium for which the excess of PPh_3 leads to higher selectivities in linear aldehyde without a decrease in the reaction rate. As is usually observed (*1*), the reactivity of the alkenes is strongly related to steric demands. Moreover, cyclohexene and butadiene do not react at all.

The reactivity of the cyclopentadienyl-containing ruthenium complex $[(\eta^5\text{-}C_5H_5)Ru(CO)_2]_2$ has been explored by Ugo and co-workers (*6*). The analogous iron complex has a very low level of catalytic activity. In contrast, the ruthenium complex is active between 50 and 200 bar (total pressure, $P_{H_2} = P_{CO}$), in the range 120–150°C. The studies have been carried out with oct-1-ene, the conversions reached under the best conditions being about 80% in 40 hours. Hydroformylation, isomerization, and hydrogenation of the alkene were observed as well as hydrogenation of aldehydes to the corresponding alcohols. Thus, this complex is a poor hydroformylation catalyst: indeed, at 50 bar and 135°C, 15% of aldehyde was obtained with an n/iso ratio of about 77/23, but extensive isomerization of oct-1-ene occurred (45%) accompanied by hydrogenation (5%). In order to prevent isomerization, it was found necessary to increase the total pressure, but this takes place to the detriment of aldehyde formation and the selectivity in linear aldehyde. For instance, at 200 bar and 135°C, the conversion was 60%, yielding aldehydes 9%, isomerization 12%, octane

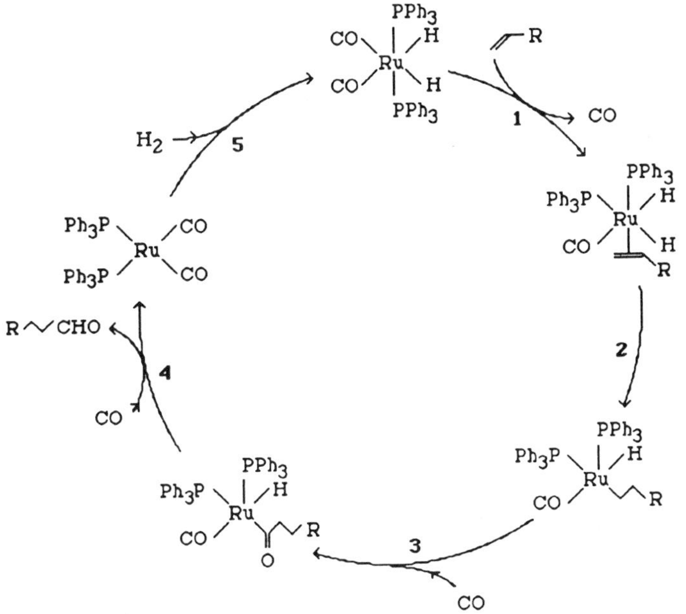

SCHEME 2

12%, and n/iso ratios of approximately 2.8. Carbon monoxide has an inhibiting effect on the catalytic activity although the n/iso ratio was improved to 4 when the partial CO pressure was increased. Below 150°C the cyclopentadienyl ligand remained attached to ruthenium. The authors proposed that the first step was the formation of a hydrido mononuclear ruthenium species according to Eq. (3).

$$[(\eta^5\text{-}C_5H_5)Ru(CO)_2]_2 + H_2 \rightleftharpoons 2[(\eta^5\text{-}C_5H_5)Ru(CO)_2H] \tag{3}$$

Creation of a vacant site by CO dissociation is favored at low CO pressures, whereas the CO insertion step seems not to be very sensitive to pressure. Above 150°C the cyclopentadienyl ruthenium complex was destroyed, and many complexes were formed, with $[Ru_3(CO)_{12}]$ as the major component. This mixture of compounds is of poor activity, and $[Ru_3(CO)_{12}]$ tested under the same conditions presents the same low level of catalytic activity.

In a subsequent study, Ugo and co-workers (7) started from the monomeric species $[(\eta^5\text{-}C_5H_5)Ru(CO)_2X]$ (X = Cl, Br, I) in order to generate more easily the hydride intermediate $[(\eta^5\text{-}C_5H_5)Ru(CO)_2H]$, employing a tertiary amine to remove HX [Eq. (4)]. The activity of the ruthenium

$$[(\eta^5\text{-}C_5H_5)Ru(CO)_2X] + H_2 \rightleftharpoons [(\eta^5\text{-}C_5H_5)Ru(CO)_2H] + HX \tag{4}$$

species was compared with that of the iron analogs. The latter complexes are less active, and we focus our attention on the catalytic performances of the ruthenium compounds which are the subject of this review. The hydride species, formed according to Eq. (4), was detected by ^{1}H NMR. Higher conversion rates were observed than when starting from the dinuclear complex [see Eq. (3)]. Hydrogenation of the alkene to the alkane and isomerization still occurred, but the aldehydes formed were hydrogenated to the corresponding alcohols. In the absence of tertiary amine, high levels of condensation products were observed as well.

Some attempts to modulate the electron density on the metal center by introduction of pentamethyl- or pentaphenyl-cyclopentadienyl ligands resulted in decreased activities, presumably due essentially to the appreciable steric hindrance introduced in the coordination sphere of the Ru center. As the dimer precursor $[(\eta^5\text{-}C_5H_5)Ru(CO)_2]_2$ leads to more isomerization of the alkene than the monomeric complexes, the equilibrium of Eq. (3) is certainly not fully shifted toward the formation of the $[(\eta^5\text{-}C_5H_5)Ru(CO)_2H]$ species. The authors have checked that the dimer itself catalyzes the isomerization of oct-1-ene to oct-2-ene.

To solve the problem of separating the organic products from the catalyst, an interesting approach was developed by Borowski, Cole-Hamilton, and Wilkinson (8). The monosulfonated triphenylphosphine ligand dpm $(Ph_2PC_6H_4SO_3Na)$ was coordinated to various metal precursors, among them ruthenium, to impart to the resulting complex good water solubility. A simple decantation allows separation of the products. The complex $[RuHCl(dpm)_3]$ was shown to be active for hydroformylation of hex-1-ene at 60 bar total pressure and 90°C. The authors observed about 30% conversion to aldehydes in 24 hours as well as formation of large amounts of internal alkene. The n/iso ratio was found to be 75/25. The monosulfonated triphenylphosphine ligand as well as its carboxylated analog were used in the presence of an amphiphilic molecule to facilitate transfer between the organic phase and the ruthenium-containing aqueous phase (9). For instance, the combination of $[RuCl_3 \cdot 3H_2O]$ plus 2 mol of (4-$CO_2H)C_6H_4PPh_2$, in the presence of $[NMe_3(C_{16}H_{33})]Br$, transformed 14% of dodec-1-ene into the C_{13} aldehydes with a n/iso ratio equal to 7.

More recently, Taqui Khan and co-workers (10) introduced the potentially tetradentate ethylenediaminetetraacetate ligand in the ruthenium coordination sphere in order to obtain an efficient water-soluble catalyst precursor. Indeed, starting from the ruthenium(III) aquo EDTA species $[Ru(EDTA)(H_2O)]^-$, carbonylation gives the paramagnetic carbonyl complex $[Ru(EDTA)(CO)]^-$ which is able to induce the heterolytic activation of dihydrogen (Scheme 3). The hydroformylation of hex-1-ene performed at 50 bar $(CO/H_2 = 1/1)$ and 130°C in a 80/20 ethanol–water solvent

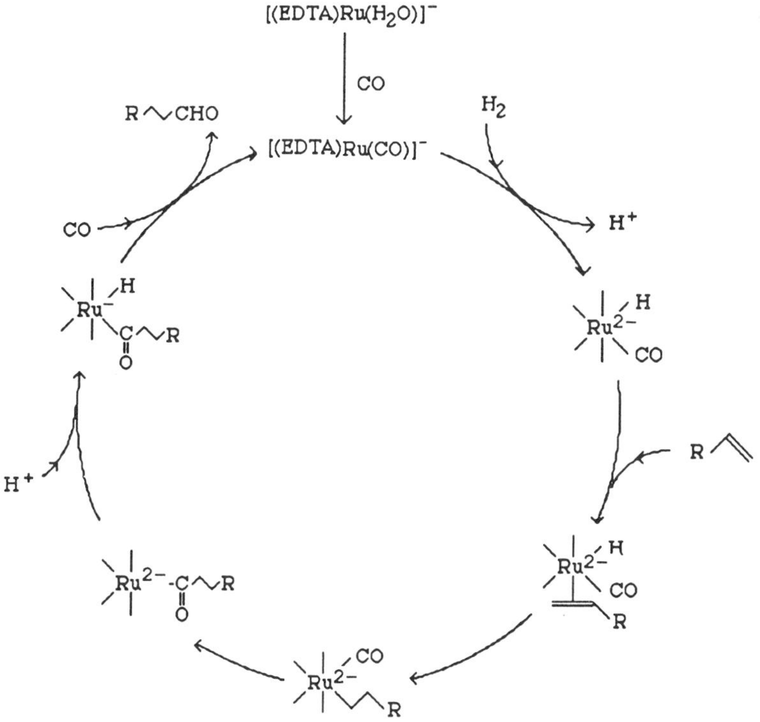

SCHEME 3.   Catalytic cycle proposed for $[(EDTA)Ru(H_2O)]^-$. The coordination positions occupied by EDTA have been omitted for clarity.

mixture gave 100% conversion to heptanal, in a contact time of 12 hours. In other words, a complete selectivity in the linear aldehyde was achieved, with turnover frequencies being approximately 0.2 minute^{-1}. According to the kinetic equations written by the authors, the catalytic cycle shown in Scheme 3 can be proposed. The high regioselectivity of this reaction may be due to the steric hindrance imposed on the coordinated alkene by the presence of EDTA coordinated to the metal center.

The same catalyst precursor, generated from [(EDTA)RuCl] which is also water soluble, was used for the hydroformylation of allylic alcohol under the same reaction conditions (11). At 50 bar and 130°C, in water as solvent, 4-hydroxybutanal was produced [Eq. (5)], together with about 2% of formaldehyde. However, the reaction proceeded further to give butane-1,4-diol by hydrogenation and γ-butyrolactone as well as dihydrofuran by cyclization [Eq. (6)]. The same catalytic cycle as that proposed in Scheme 3 can be considered. A kinetic investigation revealed a first-order dependence on the ruthenium complex concentration and on the allyl alcohol

$$HO\diagdown\diagup\diagdown + H_2 + CO \rightarrow HO\diagdown\diagup\diagdown\diagup CHO \qquad (5)$$

$$HO\diagdown\diagup\diagdown\diagup CHO \quad \xrightarrow{\;H_2\;} \quad HO\diagdown\diagup\diagdown\diagup OH$$
$$\xrightarrow{\text{cyclization}} \quad \text{(lactone)} + \text{(dihydrofuran)} \qquad (6)$$

concentration as well. The reaction rate is also directly proportional to the total $CO + H_2$ pressure. The activation energy was calculated to be 23 kcal/mol. The limiting step was the hydride transfer, as found for hex-1-ene (see Scheme 3), and the turnover frequency was approximately 0.6 $minute^{-1}$.

The complex $K[Ru(EDTA-H)Cl] \cdot 2H_2O$ was shown by Taqui Khan et al. (12) to catalyze the water gas shift reaction [Eq. (7)] and thus to

$$CO + H_2O \rightleftharpoons CO_2 + H_2 \qquad (7)$$

produce hydrogen from carbon monoxide (10–40 bar) and water. The same precursor utilizes the reagents CO and H_2 in the reaction medium to transform allyl alcohol, as previously described. The same isomeric distribution was obtained (e.g., 4-hydroxybutanal 35%, γ-butyrolactone 25%, dihydrofuran 25%, butane-1,4-diol 1%, with traces of formaldehyde). From these preliminary results it can be concluded that the water gas shift process proceeds faster than the hydroformylation reaction.

An interesting catalytic system was described by Pinke in a patent of 1975 (13). This system combines a complex of a group 8 transition metal with an aluminum hydride. Such catalysts can perform the hydroformylation of mixtures of terminal and internal alkenes to the corresponding linear aldehyde and alcohol. One example is related to the "$[RuCl_2(PBu_3)_3] + AIHEt_2$" system which transforms, at 120 bar ($H_2/CO = 1/2$) and 150°C over 24 hours, heptene into octanal and octanol. Linear products were detected exclusively.

Instead of using triphenylphosphine, Oswald et al. in a patent ascribed to Exxon (14) have used slightly more basic ligands. Among the hydroformylation catalysts, two examples are given related to ruthenium. The complexes were generated in situ by adding six phosphorus ligands to $[RuH(OAc)(PPh_3)_3]$. The two ligands of interest are $Ph_2P(CH_2CH_2SiPr_3)$ and $(Ph_2PCH_2CH_2)_2SiMe_2$. Presumably, under the oxo conditions the $[H_2Ru(CO)_2L_2]$ complexes are formed. Indeed, working on various alkenes at 100°C and 25 bar with a 1/1 hydrogen/carbon monoxide mixture, selectivities of approximately 65% in aldehydes were obtained with a n/iso ratio of about 2.3. In the last example, using the diphosphine ligand, a low conversion of but-1-ene was observed.

III

CATALYSIS INITIATED BY CLUSTERS

This section surveys the use of various di-, tri-, and polynuclear ruthenium complexes as precursors for the homogeneous hydroformylation of alkenes. Several arbitrary assumptions have been made so as to include dinuclear starting complexes which are strictly not cluster compounds. Moreover, no distinction is made between neutral and anionic precursors. Also, in several cases, particularly in the patents, information is lacking concerning the intermediate species involved in the catalytic cycles. Interestingly, half of the described systems come from patents, and there are few fundamental studies which clearly establish the implication of cluster species during the catalysis.

A. Neutral Species

Schultz and Bellstedt (15) have investigated the catalytic activity of $[Ru_3(CO)_{12}]$ directly generated under CO pressure from hydrated ruthenium oxide. No ruthenium pentacarbonyl was detected in the off gases after catalysis. Analysis of the mixtures revealed the presence of $[H_4Ru_3(CO)_{12}]$ and $[H_4Ru_3(CO)_{10}]$. Hydroformylation of propene was examined at 300 bar with a CO/H_2 mixture of 1/2. At 90°C, 14.7% propane, 75% butanals, and 4.7% butanols were obtained. By-products were butylbutyrates, formates, dipropylketones, as well as traces of 2-ethylhexanals and 2-ethylhexanols. Increasing the temperature to 150°C markedly increases the hydrogenated products, affording 23.3% propane, 35.3% butanols, and formates (8.7%). However, the n/iso ratio does not vary significantly and is generally about 2/1, whereas it should be at least 9/1 to compete with the rhodium phosphine system. Similarly to the effect of increasing the temperature, when the total pressure is increased in the range 100 to 1000 bar, propane formation and hydrogenation of aldehydes to alcohols are favored. In addition, the selectivity in butanal and butanol formation depends strongly on the amounts of ruthenium since for concentrations higher than about 3×10^{-2} % (mole of Ru/mole of C_3H_6) almost no C_4 aldehydes are present and mainly butanols are formed. Presumably the two reactions of hydroformylation and hydrogenation into alcohol should be successive since for longer reaction times the formation of alcohol is increased. In the same paper, the authors show (15) that $[RuCl_3 \cdot 3H_2O]$ generates under 300 bar of CO carbonyl complexes similar to those derived from $[RuO_2(H_2O)_x]$. However, a low selectivity into C_4

products was obtained, and numerous by-products including heavy compounds were formed.

The catalytic activity of $[Ru_3(CO)_{12}]$ was also recognized by Wilkinson and co-workers (5) during their first study of ruthenium phosphine-containing complexes, described in the first part of this review (vide supra). When compared with $[Ru(CO)_3(PPh_3)_2]$ and related mononuclear complexes, the carbonyl $[Ru_3(CO)_{12}]$ by itself gives at 120°C and 100 bar a modest catalytic activity for the hydroformylation of hex-1-ene (24% conversion in 20 hours), but a rather interesting n/iso ratio of 82/18 (5). Addition of 3 mol of phosphorus ligand to $[Ru_3(CO)_{12}]$ induces a dramatic effect on the catalytic activity and selectivity. Indeed, alkyl phosphines, particularly PBu_3, and phosphites reduced the conversions to 10% whereas less basic phosphorus ligands such as PPh_3, $P(OPh)_3$, or $P(O-2-naphthyl)_3$ led to 90–97.6% yields. Triphenylphosphine or phosphite ligands appeared to combine high conversion rates and selectivities into C_7 aldehydes, although approximately 50% heptanal and 36–42% 2-methylhexanal formed. The 2-naphthylphosphite ligand induced the formation of a few aldehydes (10%) together with large amounts of alcohols (n 28% and iso~27%, respectively) and heavy compounds (30%). From the reaction mixture various polynuclear ruthenium compounds were recovered, the predominant species being $[H_4Ru_4(CO)_8L_4]$ (5).

The catalytic activity of $[Ru_3(CO)_{12}]$ as well as $[H_4Ru_4(CO)_8(DIOP)_2]$, where DIOP is 2,3-isopropylidene-2,3-dihydroxy-1,4-bis(diphenylphosphino)butane, was also investigated by Piacenti and co-workers (16,17). At 150°C and partial pressures P_{CO} 50 bar and P_{H_2} 45 bar, $[Ru_3(CO)_{12}]$ gave 83% aldehydes, a small quantity of alcohols, and approximately 13% alkanes, with a selectivity in linear aldehyde of 81%. Moreover, at very low carbon monoxide partial pressures ($P_{CO} = 7$ bar, $P_{H_2} = 85$ bar) not only were hydrogenated products obtained (5% aldehydes, 36% alcohols, and 59% alkanes) but the linearity also decreased significantly (72%). In addition, the nonconverted alk-1-ene was extensively isomerized, showing a composition similar to that found after hydroformylation of alk-2-ene (here cis-pent-2-ene). The authors did not propose any catalytic mechanism and supposed that a second catalytically active species appears at low P_{CO} values in minor quantities. This species is responsible (17) for the isomerization of the alkene, the exchange between the hydrido ligands and hydrogen from the gaseous phase, and the hydrogenation of both alkene and aldehydes. The complex $[H_4Ru_4(CO)_8(DIOP)_2]$ showed a peculiar behavior. For the hydroformylation of pent-1-ene-5-d_3 under the same conditions (150°C, $P_{CO} = 50$ bar, $P_{H_2} = 45$ bar), lower conversions were obtained, the hydrogenation to alcohols and alkanes was increased, but the linearity was improved to 85%.

Another interesting example was reported in a patent by Cooper (*18*). He prepared ketones using $[Ru_3(CO)_{12}]$ as the catalyst precursor from essentially ethylene and propylene, for example, at 100°C and 120 bar ($H_2/CO = 2/1$). In addition to the expected butyraldehydes, 23% heptanone was obtained with no production of butanols. With an H_2/CO mixture of 3/1 and in acetic acid as solvent, the selectivity to ketone products can reach 50%. Similarly, at 150°C and 120 bar, H_2/CO of 3/1 in glacial acetic acid, ethylene gave approximately 52% penta-3-one and 48% propanal.

Hydroformylation of fluoroalkenes catalyzed by various transition metal complexes, among them $[Ru_3(CO)_{12}]$, has been examined by Ojima et al. (*19–21*). The presence of very strong electron-withdrawing groups on the alkene dramatically modifies the regioselectivity of the hydride transfer onto the carbon–carbon double bond so that the branched aldehyde is preferentially produced. At 100°C, 130 bar ($CO/H_2 = 1/1$), with $[Ru_3(CO)_{12}]$, 3,3,3-trifluoropropene was converted in 16 hours to 62% aldehydes (n/iso 15/85) and 38% alkane. Addition of triphenylphosphine (no indication of the P/Ru ratio was given) lowered the conversion but improved significantly the selectivity into the branched aldehyde (n/iso 8/92). Simultaneously, hydrogenation was almost completely suppressed. Similarly, pentafluorostyrene (*16*) can be hydroformylated as well as fluoroethylene. The perfluoroaldehydes can be used for the synthesis of various biologically active fluoroamino acids and related compounds (see Ref. *20*, and references quoted therein).

The carbonyl $[Ru_3(CO)_{12}]$ is a good cocatalyst for the low pressure hydroformylation of internal alkenes using the classic rhodium phosphine $[HRh(CO)(PPh_3)_3]$ system in the presence of an excess of triphenylphosphine (P/Rh = 200) (*22*). Starting from a mixture of hex-2- and hex-3-ene, the addition of $[Ru_3(CO)_{12}]$ (Rh/Ru = 1/1) increased both the reaction rate and the n/iso ratio of heptanals. More recently, Poilblanc and co-workers (*23*) have prepared a mixed ruthenium–rhodium complex formulated as $[ClRh(\mu\text{-}CO)(\mu\text{-}dppm)_2Ru(CO)_2]$ (dppm is $Ph_2PCH_2PPh_2$). This species shows catalytic activity in the hydroformylation of pent-1-ene at 40 bar ($H_2/CO = 1/1$) and 75°C. Conversion to hexanals was 90% in 24 hours and the linearity reached 70%. No further report has appeared to determine the role of the two metals in this catalysis.

A very interesting synergistic effect has been found by Hidaï et al. (*24–26*) by using $[Co_2(CO)_8]-[Ru_3(CO)_{12}]$ mixtures. Particularly relevant is the hydroformylation of cyclohexene (110°C, 80 bar, $H_2/CO = 1/1$, 4 hours). Under these conditions, $[Co_2(CO)_8]$ or $[Ru_3(CO)_{12}]$ alone gave very poor yields (14 and 3%, respectively) of cyclohexylmethanal, whereas for an Ru/Co ratio of 9.9 an initial rate 27 times as fast as that with $[Co_2(CO)_8]$ alone was observed. Even a 1/1 ratio did not result in the simple additive effect of the two individual compounds, but the reaction rate with $[Co_2(CO)_8]$ was increased by a factor of 6. With $[Co_2(CO)_8]$ the solvent does not play any significant role, but for the bimetallic catalyst it was shown that the initial rate was strongly dependent on the nature of the solvent. Interestingly, alcohols such as ethanol or methanol were the best among the solvents investigated. Presumably the most important synergistic effect could result from the dinuclear reductive elimination of aldehydes by interaction of cobalt–acyl and ruthenium–hydride species.

Finally, we mention the catalytic activity of a dinuclear species formulated as $[Ru_2(OOCR)_4X]$ (R = alkyl from C_2 to C_{15}, X = Cl, Br, I) generated by treatment under nitrogen of $[RuCl_3 \cdot 3H_2O]$ with sodium carboxylates in water at 90°C (27). Presumably such complexes, for which the oxidation state is formally 2.5 (28), are in fact polynuclear species which involve ruthenium(II) and ruthenium(III) sites. Hydroformylation of propene with the isobutyrate ruthenium complex afforded, at 170°C and 200 bar ($CO/H_2 = 1/1$) for 30 minutes, 98% selectivity in butyraldehydes, with a linear to branched ratio of 1.14. A few hydrogenated products can be formed at higher temperatures, essentially butanols, but markedly increasing the pressure, for instance, to 280 bar (170°C), allows a full selectivity in aldehyde to be reached. The bromo and iodo analogs have been also studied.

All the work mentioned in this section is silent on the mechanistic aspects of the reactions, except for two species observed by Schultz and Bellstedt (15) which are hydrido tri- or tetranuclear complexes, namely, $[H_4Ru_4(CO)_{12}]$ and $[H_4Ru_3(CO)_{10}]$.

B. Anionic Species

Anionic ruthenium complexes appear to play a very significant role as starting materials. Indeed, as early as 1981, Chang patented a new synthetic method to obtain linear aldehydes or alcohols using dianionic complexes (29). Experiments were carried out showing that neutral and monoanionic compounds induced selectivities comparable to those observed with $[Ru_3(CO)_{12}]$ or mononuclear species, whereas dianionic complexes allowed n/iso ratios as high as 99 to be obtained! For instance (30), at 145 bar ($CO/H_2 = 1/1$), 160°C, in THF, for 1 hour $[H_2Ru_4(CO)_{13}]$ transformed 5% pent-1-ene into 82.2% hexanal and 17.8% 2-methylpentanal (n/iso 4.62). Under the same experimental conditions, the polynuclear monoanionic complex (PPN)[HRu_4(CO)_{13}], where PPN is $N(PPh_3)_2$, gave a mixture of 75.7% hexanal and 24.3% 2-methylpentanal (n/iso 3.12, conversion 14.0%). The corresponding dianionic compound

(PPN)$_2$[Ru$_4$(CO)$_{13}$] converted 27.3% of pent-1-ene to 97.5% hexanal and 2.5% of the branched isomer (n/iso 39.0). Similarly, (PPN)$_2$[Ru$_6$(CO)$_{18}$] presented a high level of selectivity with an n/iso ratio of 99 in 30 minutes. However, reaction times dramatically influence the selectivity, as this ratio is approximately 31 in 1.5 hours 15 in 5 or 9 hours, and 9 in 29 hours. No clear explanation can account for this decrease in selectivities except, perhaps, that the active species is rapidly transformed into other reactive but mediocre selective species.

Salts of [Ru(CO)$_4$]$^{2-}$ or [H$_2$Ru$_4$(CO)$_{12}$]$^{2-}$ can also conveniently be used. These dianionic complexes can be directly generated by mixing [Ru$_3$(CO)$_{12}$] or [H$_4$Ru$_4$(CO)$_{12}$] with another dianionic compound such as [Na$_2$Fe(CO)$_4$] or [Na$_2$W$_2$(CO)$_{10}$] (*30,31*). The addition of a second metal complex also seems to provide a "stabilization" of the ruthenium active species since, for instance, in 3 hours a selectivity of 91% has been maintained for a conversion of 68% of pent-1-ene. Moreover, addition of [Co$_2$(CO)$_8$] to the dianionic complex (PPN)$_2$(H$_2$Ru$_4$(CO)$_{12}$] (molar ratio 1/1) increases the reactivity and the stability of the system. In fact 31% of oct-1-ene is converted in 1 hour to *n*-nonanal with 97% selectivity.

In a subsequent patent (*32*), Chang has shown that the anionic ruthenium precursors can be regenerated by treating the deactivated solutions with a very strong reducing agent such as potassium bis(*sec*-butyl)borohydride, potassium benzophenone, or potassium hydride (~2 equiv). For instance, a first run carried out with [PPN]$_2$[Ru$_6$(CO)$_{18}$] gave 84% selectivity in nonanal. After distillation of the solvent and the oxo products, treatment of the residue under argon with potassium benzophenone gave a catalyst precursor which was introduced in a second hydroformylation test to give 31% conversion in 2 hours and 94.1% selectivity in linear aldehyde. Such a reactivation can transform systems having a poor selectivity into active species. This is the case for [PPN][H$_3$Ru$_4$(CO)$_{12}$] (first run 84% selectivity toward linear product, second run 95.7% selectivity) and for the mixed cluster [PPN] [CoRu$_3$(CO)$_{13}$] (58.2 and 96.4%, respectively). However, all these catalytic systems present a modest level of activity since turnover frequencies are between 3 and 12 hour^{-1}.

From these data it is difficult to identify the true nature of the anionic species generated by the reduction, although, from the work of Shore *et al.* (*33,34*), it should be a dianionic complex. Indeed, treatment of [Ru$_3$(CO)$_{12}$] with sodium benzophenone in Na/Ru molar ratios of 1/1, 1.5/1, 2/1, and 3/1 was shown to afford, respectively, [Ru$_6$(CO)$_{18}$]$^{2-}$, [Ru$_4$(CO)$_{13}$]$^{2-}$, [Ru$_3$(CO)$_{11}$]$^{2-}$, and [Ru$_4$(CO)$_{12}$]$^{4-}$. The preparation of more highly charged complexes was also patented, and it was claimed that these species are active for the hydroformylation reaction. Starting from $[Ru_3(CO)_{12}]$ or an anionic compound, by treatment with the correct stoichiometry of sodium benzophenone, the three complexes $[Ru_4(CO)_{11}]^{6-}$, $[Ru_6(CO)_{17}]^{4-}$, and $[Ru_6(CO)_{16}]^{6-}$ have been prepared (*34*).

Patents based on the monoanionic complexes $[HRu_3(CO)_{11}]^-$, $[HRu_2Co(CO)_{11}]^-$, and $[HRu_2Fe(CO)_{11}]^-$, among others, have been disclosed by National Distillers Corp. (*35*) to give a good activity in polar solvents. For instance, in DMSO at 135°C and 64 bar of CO/H_2 (1/1) for 4 hours, 8-methoxy-1,6-octadiene was transformed to 9-methoxynon-7-enal with a selectivity of 93%. An analogous approach has been used by Knifton from Texaco, who introduced a very interesting and new method to stabilize the active anionic ruthenium species. Quaternary phosphonium or ammonium bases or salts are added to the system. These serve not only as stabilizing cations but also as a polar medium. In addition, low melting salts allow separation of the organic products from the catalytic system, which can be recycled without loss of activity (*36–40*).

The ruthenium/phosphonium salt system is very efficient in converting internal alkenes to the corresponding aldehydes and alcohols. Generally the alkanols are obtained, so that this catalytic system promotes three different functions: carbon–carbon double bond migration, hydroformylation of the alkene, then reduction of the aldehyde function to the alcohol. Thus, starting from $[RuO_2(H_2O)_x]$ and PBu_4Br at 180°C, 83 bar CO/H_2 (1/2) for 4 hours, oct-2-ene was 98% converted. Among the products 72% were oxo compounds and 66% were C_9 alcohols, with a linearity of 49% (*36,39*). The complexes $[Ru(acac)_3]$ or $[Ru_3(CO)_{12}]$ can be conveniently used as ruthenium sources, but $RuCl_3$ gave poor results. The reactant solutions show the $[HRu_3(CO)_{11}]^-$ anion to be the major species with small quantities of $[Ru(CO)_3Br_3]^-$ and $[Ru(CO)_5]$ present, as well as other unidentified species (*39*). The hydrido triruthenium anion is likely to be responsible for the hydroformylation reaction and the reduction of aldehyde, but the contribution of all the species present in solution was not elucidated. The ruthenium carbonyl compounds were shown to be stable in the quaternary phosphonium salt matrix, even under low pressure and high temperature conditions, so that the aldehydes and/or alcohols can be separated from the crude mixture by fractional distillation and the residual "melt" catalyst recycled without loss of activity (*36,39*). The highest linearity (69%) was obtained starting from $[RuO_2(H_2O)_x]$ and $[PBu_4][CH_3COO]$, although the conversion was only 56% in 4 hours; the implication of ion pair interactions with the $[HRu_3(CO)_{11}]^-$ cluster anion was invoked to explain this better selectivity (*39*).

Addition of two tributylphosphine ligand equivalents per ruthenium did not significantly alter the reactivity pattern, but an excess ($P/Ru = 5$)

diminished considerably the yield, which was reduced from 99 to 4%. Whereas $[Co_2(CO)_8]$ itself under the experimental conditions (particularly at low total pressure) is almost inactive, the system $[RuO_2]/[Co_2(CO)_8]/$ PBu_3 (1/1/1.5) gave 53% conversion of internal octene to 9% C_9 alcohols and 25% C_9 aldehydes (46% linearity). Higher concentrations in catalyst, about 15 times more (39), induced higher yields of alcohols. Various diphosphine or even triphosphine ligands (37), tertiary amines, and bidentate nitrogen-containing ligands (38) can be added to ruthenium to improve the linearity of the oxo products. For instance, tris(2-diphenylphosphinoethyl)phosphine gave 63% linearity, and 2,2'-dipyridyl afforded 69% linearity. Infrared analysis of the solutions after catalysis revealed that $[HRu_3(CO)_{11}]^-$ remains the major species. In addition, several CO bands typical of ruthenium carbonyl bipyridyl clusters were detected.

Knifton has also shown (36–38,40) that nitrogen- or phosphorus-ligand modified ruthenium complexes, in a phosphonium salt matrix, can conveniently catalyze the hydroformylation of terminal alkenes with high selectivities in linear oxo products. Usually selectivities better than 80% were achieved. In the best case (160°C, 95 bar, $CO/H_2 = 1/2$) a linearity in nonanol of 94% was obtained starting from $[Ru_3(CO)_{12}]$, 2,2'-bipyridine, and $[PBu_4]Br$. The main products were alcohols and not aldehydes. However, it is often difficult to reduce the isomerization of oct-1-ene as well as its hydrogenation. The $[Ru_3(CO)_{12}]/2,2'$-bipyridine (bipy) system has been extensively explored. Two equilibria have been proposed to account for the infrared data and the effects of the bipy ligand [eqs. (8) and (9)].

$$[Ru_3(CO)_{12}] + bipy \rightleftharpoons [Ru_3(CO)_{10}(bipy)] + 2CO \qquad (8)$$

$$[HRu_3(CO)_{11}]^- + bipy \rightleftharpoons [HRu_3(CO)_9(bipy)]^- + 2CO \qquad (9)$$

This study, performed on the hydroformylation of oct-1-ene, has shown that below 140°C nonanals are the predominant products with linearities of approximately 97% (99% in one run at 100°C), whereas above 180°C nonanol was obtained almost exclusively with high octene conversions (>98% at 200°C) but poor linearities (65%). At high temperatures a 10-fold excess of bipy increases the nonanol linearity (to 76%). This parameter is not very sensitive to the CO or H_2 partial pressures as the total pressure is above about 95 bar. The author (40) seems to prefer coordination of the alkene to a ruthenium center or hydride transfer to form an alkyl ruthenium cluster as the two possible rate-determining steps. Thus, by optimization of the experimental conditions it is possible to reach high linearities (n/iso > 100) for the hydroformylation of terminal alkenes by the $[Ru_3(CO)_{12}]$/bidentate nitrogen- or phosphorus-containing ligand/phosphonium salt system. However, little information concerning the mechanism is available.

Two research groups have explored the use of the CO/H_2O couple instead of CO/H_2 in the hydroformylation reaction. In fact, this reaction was discovered by Reppe using $[Fe(CO)_5]$ (*1*) and requires the presence of a base to produce hydrido species. Investigation of this reaction by Pettit *et al.* (*41*) has shown that the two hydrido iron species are formed [eq. (10)]:

$$[Fe(CO)_5] \xrightarrow[CO_2]{OH^-} [HFe(CO)_4]^- \xrightarrow[OH^-]{H_2O} [H_2Fe(CO)_4] \qquad (10)$$
$$\qquad\qquad\qquad\qquad\quad \mathbf{1} \qquad\qquad\qquad\qquad \mathbf{2}$$

The anionic complex $[HFe(CO)_4]^-$ is produced at pH 12, whereas $[H_2Fe(CO)_4]$ is obtained at lower pH values (10.7) and is the species which reacts with ethylene.

Further exploration of this reaction catalyzed by other metal complexes (*42*) has revealed that several clusters are more active both for the water gas shift reaction [Eq. (11)] and for the hydroformylation reaction [Eq. (12)].

$$CO + H_2O \rightleftharpoons CO_2 + H_2 \qquad (11)$$

$$RCH{=}CH_2 + 2CO + H_2O \rightarrow RCH_2CH_2CHO + CO_2 \qquad (12)$$

The aldehyde is generally reduced to the alcohol under the experimental conditions [Eq. (13)]. The complexes $[Ru_3(CO)_{12}]$ as well as

$$RCH_2CH_2CHO + CO + H_2O \rightarrow RCH_2CH_2CH_2OH + CO_2 \qquad (13)$$

$[H_4Ru_4(CO)_{12}]$ were shown to transform propene into butyraldehyde under a partial CO pressure of 28 bar ($P_{C_3H_6} = 12$ bar) at 100°C for 10 hours in the presence of a large excess of aqueous trimethylamine. These catalytic systems are more active for the water gas shift reaction (turnover frequency 340 hour^{-1}) than for the production of oxo compounds($\sim$ 12 hour^{-1}). However, no indications about the nature of the ruthenium species produced in basic media were given.

Laine (*43–47*) used potassium hydroxide to promote the catalytic activity of $[Ru_3(CO)_{12}]$ and $[H_4Ru_4(CO)_{12}]$ for the hydroformylation of pent-1-ene. Under 64 bar of CO pressure, at 135 or 150°C, high selectivities for straight-chain aldehydes were obtained, for example, 97%. As the subsequent reduction of aldehydes to alcohols is lower, important aldol condensations occurred owing to the presence of a base in solution. Analysis of the reaction mixtures has shown that the anionic $[H_3Ru_4(CO)_{12}]^-$ cluster is likely to be the active species. Since this complex was recognized as the major component of the low pressure ruthenium-catalyzed water gas shift reaction, we believe that the two catalytic cycles should have as a common point a hydride species which interacts with the alkene to give an alkyl ruthenium cluster intermediate. In our opinion, it is at least a dihydride species which does not release dihydrogen, which is directly implicated in the hydroformylation cycle.

When the total concentration in ruthenium was decreased, the catalytic activity fell off indicating that cluster catalysis was occurring (46). Moreover, Laine has found a synergistic effect between ruthenium and iron. Indeed, whereas the turnover frequencies displayed by $[Fe_3(CO)_{12}]$ and $[Ru_3(CO)_{12}]$ for the hydroformylation of pent-1-ene were, respectively, 38 and 40 hour^{-1}, a 1:1 mixture of $[Fe_3(CO)_{12}]$ and $[Ru_3(CO)_{12}]$ gave a value of 230 hour^{-1}.

More recently, in our laboratory (48) we have shown that the dinuclear bridged ruthenium complex $[Ru_2(\mu\text{-}CH_3COO)_2(CO)_4(PPh_3)_2]$ is an active catalyst for the low pressure (10–30 bar) hydroformylation of oct-1-ene. The reaction requires a basic promoter (aqueous KOH or NEt_3) and the CO/H_2 (1/1) syn-gas mixture. No isomerization nor hydrogenation was observed, but the selectivity into linear aldehyde remains at a modest level (n/iso ~ 3). It is suggested that the main role of the basic promoter is to generate an anionic active intermediate, formulated as $[Ru_2(\mu\text{-}CH_3COO)_2(CO)_3H(PPh_3)_2]^-$.

C. Mechanistic Studies

The literature related to mechanistic studies on the ruthenium clusters involved in the hydroformylation reaction is scarce. The main evidence concerning the retention of the trinuclear structure comes from Süss-Fink and co-workers (49–53). In 1980, Süss-Fink (49) published his first observation that under low pressure (57 bar, $C_2H_4/H_2/CO = 1/1/2$) at 100°C, the anionic cluster $[NEt_4][HRu_3(CO)_{11}]$ in DMF catalyzed the hydroformylation of ethylene to propanal (yield 74% in 5 hours). Similarly, propene was transformed to the C_4 aldehydes with a high selectivity in n-butanal (95.5%). The anionic trinuclear cluster was found to be unchanged after catalysis (49). Under H_2 pressure (40 bar), this complex gives the tetranuclear anion $[H_3Ru_4(CO)_{12}]^-$ which restores $[HRu_3(CO)_{11}]^-$ under CO pressure (40 bar) [see Eq. (14)] (50). The tetranuclear anionic precursor

$$[HRu_3(CO)_{11}]^- \underset{CO}{\overset{H_2}{\rightleftharpoons}} [H_3Ru_4(CO)_{12}]^- \qquad (14)$$

gave strictly the same result as the trinuclear anion for the hydroformylation of ethylene and propene. In our opinion, the use of a syn-gas with twice as much CO as H_2 should be a favorable factor to the retention of the $[HRu_3(CO)_{11}]^-$ structure. Such a CO/H_2 mixture also precludes the hydrogenation of propene or butanals (51).

Deuteration experiments for the carbonylation of ethylene have given greater insight into the mechanism (52). Starting from a $C_2H_4/D_2/CO$ mixture (13:13:26 bar) the deuterioformylation of $CH_2{=}CH_2$ gave $CH_2D{-}CH_2{-}CDO$ with 97% incorporation of deuterium on the C-1 (formyl) carbon atom, 12% on C-2, and 80% on C-3 (methyl). The anionic cluster has been characterized by an X-ray crystal determination (53) as follows:

which is more simply represented as

After the first catalytic cycle, the species $[DRu_3(CO)_{11}]^-$ is produced; incorporation of $CH_2{=}CH_2$, CO, and D_2 regenerated this species and gave CH_2DCH_2CDO. However, owing to the reversibility of the cycle some $CH_2{=}CHD$ is formed, together with $[HRu_3(CO)_{11}]^-$, so that some incorporation of deuterium in the methylene position (12%) is observed. Moreover, deuteriation of the methyl position (CH_2D) is not complete. However, the main catalytic cycle is probably as shown in Scheme 4. Moreover, the μ^2-$\eta^2(C,O)$-propionyl anionic intermediate formed during the hydroformylation of $CH_2{=}CH_2$ by $[HRu_3(CO)_{11}]^-$ was trapped by deuteration (CF_3COOD) so that the following neutral species was obtained:

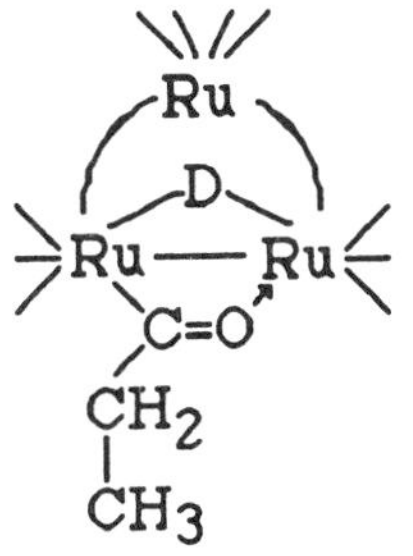

$$\text{(structure: } Ru_3 \text{ cluster with } \mu\text{-D, } C=O, CH_2, CH_3 \text{ ligands)}$$

Thus, all reactions in the catalytic cycle involve trinuclear ruthenium intermediates.

Furthermore, Kaesz et al. (54) have shown that $[HRu_3(CO)_{11}]^-$ treated with ethylene under 3 bar then protonated by HBF_4 gave the analogous neutral species $[Ru_3(\mu\text{-}H)(\mu\text{-}O{=}CEt)(CO)_{10}]$. Moreover, further exposure with ethylene (3 bar) and CO (1 bar) resulted in the formation of $[Ru_3(CO)_{12}]$ as well as a dinuclear species formulated as $Ru_2(\mu\text{-}O{=}CEt)_2(CO)_6$. The latter complex was not detected by Süss-Fink (55).

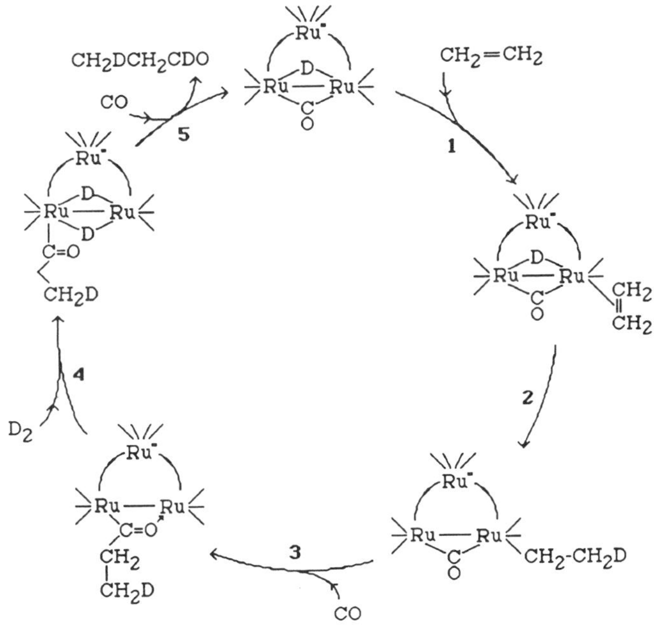

SCHEME 4. Adapted from Ref. 52.

IV

PHOTOCHEMICAL ACTIVATION AND SUPPORTED CATALYSIS

A. *Photochemical Activation*

To the best of our knowledge, very few studies have been made on photochemically promoted hydroformylation reactions. The ruthenium precursor $[Ru(CO)_3(PPh_3)_2]$ was studied by Gordon and Eisenberg (56). Irradiation of ethylene or propene under CO/H_2 pressure (nearly 1 bar) gave alkanes, aldehydes, and alcohols. High proportions of alkanes were obtained so that, for instance, for styrene only ethylbenzene was produced. No reaction occurred when the sample was not photolyzed. Thus, the system is photoassisted but in no case photoinitiated. Although small amounts of $[RuH_2(CO)_2(PPh_3)_2]$ were detected at early reaction times, under the conditions used for hydroformylation the $[Ru(CO)_3(PPh_3)_2]$ precursor was converted completely to $[Ru(CO)_4(PPh_3)]$. This complex requires photolysis to react with dihydrogen, giving rise to $[RuH_2(CO)_3(PPh_3)]$. The catalytic cycle proposed is given by Scheme 5.

B. *Supported Catalysis*

Many efforts have been undertaken to graft transition metal complexes onto various supports in order to retain the performance of the soluble catalyst precursors and to allow easy separation of the catalysts from the reaction products. Most studies have been concerned with polymers, particularly with functionalized styrene–divinylbenzene resins. This approach to immobilize homogeneous catalysts has been reviewed, with all the strategies to anchor metal complexes on organic or inorganic supports examined (57–59).

Pittman *et al.* (60) has studied the complex $[HRh(CO)(PPh_3)_3]$ supported through the exchange of one or more triphenylphosphine ligand(s) by diphenylphosphinated styrene–divinylbenzene copolymers (Pol-PPh$_2$) and observed that this system induced high linear to branched selectivities. A similar study was carried out (61) for $[(Pol-PPh_2)_2Ru(CO)_3]$ complexes in order to examine if such a catalyst would exhibit modified selectivities with regard to the homogeneous analog $[(PPh_3)_2Ru(CO)_3]$ for the hydroformylation of pent-1-ene. Under 80 bar of CO/H_2 (1/1) at 140°C and for 20 hours, the anchored ruthenium complex gave n/iso ratios of 3.5 to 3.8 for phosphine loadings of approximately 29% and P/Ru ratios of 3.1 and 6.7. These results are to be compared with those obtained starting from $[Ru(CO)_3(PPh_3)_2]$ itself (n/iso 2.3) or in the presence of an excess of

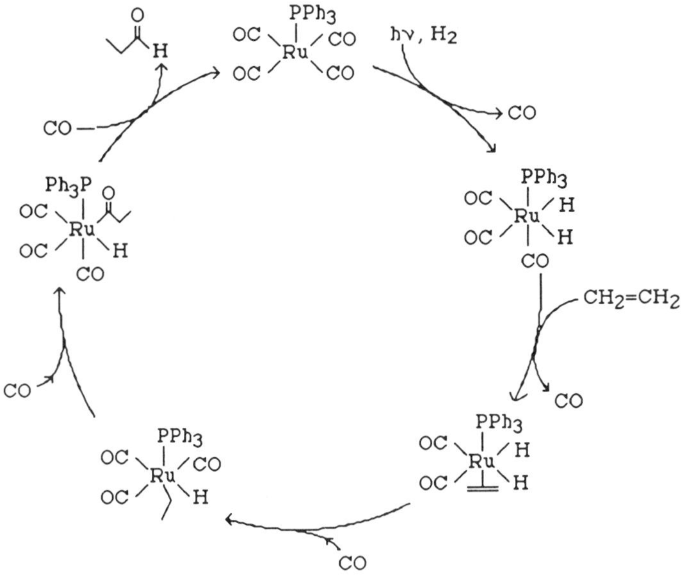

SCHEME 5. Adapted from Ref. 56.

triphenylphosphine ligand (P/Ru = 3.3, n/iso 3.3). However, in molten PPh$_3$, the homogeneous catalyst allowed a ratio of 5.1 of n/iso to be obtained. The reactions proceeded more slowly for the polymer-anchored catalysts than for the homogeneous complexes. Moreover, the substitution of the two phosphine ligands to give the [(Pol-PPh$_2$)$_2$Ru(CO)$_3$] complex introduced cross-linkages into the polymer so that the internal mobility of the resin is reduced and thus its swelling ability. Thus, the diffusion of reactants into the polymer to reach the metal sites is made more difficult, and slower reaction rates are obtained. Higher n/iso ratios were observed for higher phosphine loadings owing to the presence of the following equilibrium [Eq. (15)]:

$$[(Pol\text{-}PPh_2)RuH_2(alkene)(CO)_2] + Pol\text{-}PPh_2 \rightleftharpoons$$
$$[(Pol\text{-}PPh_2)_2RuH_2(alkene)(CO)] + CO \quad (15)$$

When the equilibrium is shifted to right the hydrido ligands present a more hydridic character, so that more linear alkyl species are formed.

A similar approach has been carried out by Kim et al. (62) who disclosed the use of sulfonated styrene–divinylbenzene resins plus a tris(dialkylaminophosphine) to obtain a phosphonium attached salt which can exchange triphenylphosphine ligands in [Ru(CO)$_3$(PPh$_3$)$_2$] or [RuCl$_2$-

(PPh$_3$)$_3$]. The preparation of these supported complexes are patented, and the hydroformylation of alkenes was claimed. In a complementary study of the use of low melting phosphonium salts, Alexander and Knifton (*63*) described the use of insoluble polymeric phosphonium salts to coordinate ruthenium and to catalyze the hydroformylation of alkenes. Particularly interesting is the polystyryl(methyltributyl) phosphonium polymer. Optionally [Co$_2$(CO)$_8$] can be added to the ruthenium precursor, generally [RuO$_2$(H$_2$O)$_x$], which was shown (*39*) to generate the cluster anion [HRu$_3$(CO)$_{11}$]$^-$. Starting from oct-1-ene, linearities in C$_9$ aldehydes and alcohols as high as 84% were obtained.

More recently, Lieto and co-workers (*64*) have shown that anionic ruthenium entities, [HRu$_3$(CO)$_{11}$]$^-$ and [H$_3$Ru$_4$(CO)$_{12}$]$^-$, as well as the mixed metal cluster [H$_3$RuOs$_3$(CO)$_{12}$]$^-$, can be supported on ammonium styrene–divinylbenzene membranes or beads with retention of their structure during the hydroformylation of hex-1-ene (60 bar, 100°C, H$_2$/CO = 1/5, 48 hours). Beside some metal leaching in solution, giving rise to nonactive metal species and to appreciable isomerization, the supported clusters were characterized by higher performances than their soluble partners. In particular, [H$_3$Ru$_4$(CO)$_{12}$]$^-$ when immobilized onto a support generated greater selectivities in linear aldehyde (75% instead of 55%). The authors inferred that the chemically inert polymer backbone induces an ordering for the catalytic sites which is not present in the homogeneous phase. For our part, we believe that, even if the polymers are macroporous, the sites around the grafted clusters generate a sufficient steric hindrance to promote formation of the linear alkyl-metal species.

In a patent for Exxon (*65*), Mitchell claimed the use of various unsaturated phosphorus- or nitrogen-containing ligands, generally quaternarized, to prepare, among other metals, ruthenium complexes derived from Ru$_3$(CO)$_{12}$. The phosphorus atoms can be bonded to supports such as silica, titania, or even semiconductor titania to give, for example, [TiO$_2$/(Ti—O—PPh$_2$)$_2$Ru$_3$(CO)$_{10}$] or similar substituted complexes. Although experimental examples were very scarce in this paper, the hydroformylation of alkenes was claimed.

The use of metal phthalocyanine compounds has also been described (*66,67*). The catalysts can either be supported on an inert carrier or used in a liquid–liquid two-phase system. Various functionalized phthalocyanine ruthenium complexes have been mentioned for the reaction of interest. For instance, ruthenium phthalocyanine disulfonate transformed hept-3-ene into 3-*n*-propylpentanol (80°C, 18 hours, 120 bar) in a two-phase system. Further details (*67*) on the preparation of metal complexes supported on silica- and alumina-type supports have appeared. Generally, a mixture of metal phthalocyanine sulfonate and hydrated alumina produced a paste which was shaped in spheres, then calcinated at 500°C under nitrogen. Solid supports such as zeolites have also been claimed (67).

Transition metal carbonyl clusters intercalated with lamellar materials such as graphite or smectites have been prepared and have been shown to be useful in catalysis (68). For instance, sodium montmorillonite was impregnated with $[Ru(NH_3)_6]Cl_3$ in water under reflux. After filtration and drying, the resulting complex was dissolved in methanol and reduced with carbon monoxide at 80°C, 100 bar, for 16 hours. Infrared analysis has shown the formation of a sodium montmorillonite-intercalated $[Ru_3(CO)_{12}]$ complex. This system was claimed to be active for the hydroformylation of propene either with CO/H_2 or CO/H_2O in the presence of a basic promoter.

The cationic complex $[Ru(NH_3)_6]^{3+}$ was similarly used to introduce ruthenium into a zeolite (69). A Faujasite-type zeolite (Na—X) was charged with 3.6% of ruthenium and subjected to a mixture of $H_2/CO/C_2H_4$ at 100 bar, 200°C, for 46 hours. Turnover frequencies of 20 hour^{-1} were obtained, and propanal, propanol, and 2-methylbutanal (resulting from the dimerization of ethylene) were produced in the ratio 1/3/1. Some metal leached to give presumably a mononuclear species which afforded in a further run, in homogeneous solution, exclusively propanal and propanol (1/1). The remaining ruthenium-encaged complex gave the same product distribution as in the first run, and only negligible amounts of metal were lost on reuse. However, if more ruthenium was introduced into the zeolite, the CO reduction produced also metal aggregates which catalyze the Fischer–Tropsch reaction.

Finally, adsorption of $[Ru_3(CO)_{12}]$ on γ-Al_2O_3 has been carried out to give, after calcination at 500°C, a Ru/Al_2O_3 catalyst. Similarly $Ru/K/Al_2O_3$ has been prepared (70). The reactivity of these two systems was tested at 275°C, 90 bar, for the hydroformylation of propene. The Ru/Al_2O_3 system gave turnover frequencies of 18.4 hour^{-1} in propane and 11.2 hour^{-1} in C_4 oxygenated products. However, the $Ru/K/Al_2O_3$ catalyst was more selective since the two values obtained were, respectively, 0.9 and 21.4 hour^{-1}. Infrared studies in a high-pressure cell revealed the presence of a ruthenium-formyl surface species.

V

CONCLUSION

When compared with the rhodium complexes, ruthenium systems appear to be characterized by modest levels of catalytic activity. However, by a careful control of the experimental conditions, particularly by the use of polar solvents or phosphonium matrices, anionic ruthenium mononuclear or clusters species allow high selectivities in linear aldehydes ($\geq 97\%$) to be obtained. Most of the studies are reported in patents. More detailed mechanistic studies are required in order to achieve a better understanding of the catalysts, and such research may well lead in the future to more efficient systems.

REFERENCES

1. A. Mullen, *in* "New Syntheses with Carbon Monoxide," (J. Falbe, ed.) Springer-Verlag, Berlin, 1980.
2. H. Pichler, B. Firnhaber, and D. Kioussis, *Brennstoff. Chem.* **44**, 337 (1963).
3. D. Evans, J. A. Osborn, F. H. Jardine, and G. Wilkinson, *Nature (London),* **208,** 1203 (1965).
4. D. Evans, J. A. Osborn, and G. Wilkinson, *J. Chem. Soc. A,* 3133 (1968).
5. R. A. Sanchez-Delgado, J. S. Bradley, and G. Wilkinson, *J. Chem. Soc., Dalton Trans.,* 399 (1976).
6. E. Cesarotti, A. Fusi, R. Ugo, and M. Zanderighi, *J. Mol. Catal.* **4**, 205 (1978).
7. A. Fusi, E. Cesarotti, and R. Ugo, *J. Mol. Catal.* **10**, 213 (1981).
8. A. F. Borowski, D. J. Cole-Hamilton, and G. Wilkinson, *Nouv. J. Chim.,* 137 (1978).
9. M. J. H. Russell and B. A. Murrer, Fr. Patent 2,489,308 to Johnson Matthey Ltd. (03.09.1981).
10. M. M. Taqui Khan, S. B. Halligudi, and S. H. R. Abdi, *J. Mol. Catal.* **48**, 313 (1988).
11. M. M. Taqui Khan, S. B. Halligudi, and S. H. R. Abdi, *J. Mol. Catal.* **45**, 215 (1988).
12. M. M. Taqui Kahn, S. B. Halligudi, and S. H. R. Abdi, *J. Mol. Catal.* **48**, 7 (1988).
13. P. A. Pinke, U.S. Patent 4,210,608 to U.O.P. Inc. (08.12.1975).
14. A. A. Oswald, T. G. Jermansen, A. A. Westner, and I. D. Huang, U.S. Patent 2,511,613 to Exxon Research and Engineering Company (21.08.1981).
15. H. F. Schulz and F. Bellstedt, *Ind. Eng. Chem. Prod. Res. Dev.* **12**, 176 (1973).
16. G. Braca, G. Sbrana, F. Piacenti, and P. Pino, *Chim. Ind. (Milan)* **52**, 1091 (1970).
17. M. Bianchi, G. Menchi, P. Frediani, U. Matteoli, and F. Piacenti *J. Organomet. Chem.* **247**, 89 (1983).
18. J. L. Cooper, U.S. Patent 4,602,116 to Eastman Kodak Company (09.04.1985).
19. T. Fuchikami and I. Ojima, *J. Am. Chem. Soc.* **104**, 3527 (1982).
20. I. Ojima, K. Kato, M. Okabe, and T. Fuchikami, *J. Am. Chem. Soc.* **109**, 7714 (1987).
21. I. Ojima and T. Fuchikami, U.S. Patent 4,370,504 to Sagami Chemical Research Center (1983); see also Japan Patent 127,592 (22.07.1984).
22. R. R. Hignett and P. J. Davidson, Fr. Patent 2,395,246 to Johnson Matthey & Co. Ltd. (21.06.1978).
23. B. Chaudret, B. Delavaux, and R. Poilblanc, *Nouv. J. Chim.* **7**, 679 (1983).
24. M. Hidaï, A. Fukuoka, Y. Koyasu, and Y. Uchida, *J. Chem. Soc., Chem. Commun.,* 516 (1984).
25. M. Hidaï, A. Fukuoka, Y. Koyasu, and Y. Uchida, *J. Mol. Catal.* **35**, 29 (1986).
26. M. Hidaï and H. Matsuzaka, *Polyhedron* **7**, 2369 (1988).
27. J. L. Cooper, U.S. Patent 4,474,995 to Eastman Kodak Company (26.08.1982).
28. F. A. Cotton and G. Wilkinson, "Advanced Inorganic Chemistry," 4th Ed., Wiley, New York, 1980.
29. B. H. Chang, U.S. Patent 4,453,019 to Ashland Oil Inc. (02.09.1982).

30. B. H. Chang, U.S. Patent 4,506,101 to Ashland Oil Inc. (02.09.1982).
31. B. H. Chang, U.S. Patent 4,539,306 to Ashland Oil Inc. (02.09.1982).
32. B. H. Chang, U.S. Patent 4,547,595 to Ashland Oil Inc. (02.10.1982).
33. S. G. Shore and C. C. Nagel, *J. Chem. Soc., Chem. Commun.*, 530 (1980); U.S. Patent 4,349,521 to Ohio State University Research Foundation (09.1982).
34. S. G. Shore and A. Bhattacharyya, U.S. Patent 4,496,532 to Ohio State University Research Foundation (27.07.1983).
35. National Distillers Corp., U.S. Patent 4,633,021 (29.11.1985).
36. J. F. Knifton, J.-J. Lin, R. A. Grigsby, Jr., and W. H. Brader, Jr., U.S. Patent 4,451,679 to Texaco Inc. (21.10.1982).
37. J. F. Knifton, U.S. Patent 4,451,680 to Texaco Inc. (21.10.1982).
38. J. F. Knifton and R. A. Grigsby, Jr., U.S. Patent 4,469,895 to Texaco Inc. (21.10.1982).
39. J. F. Knifton, *J. Mol. Catal.* **43**, 65 (1987).
40. J. F. Knifton, *J. Mol. Catal.* **47**, 99 (1988).
41. H. C. Kang, C. H. Mauldin, T. Cole, W. Slegeir, K. Cann, and R. Pettit, *J. Am. Chem. Soc.* **99**, 8323 (1977).
42. R. Pettit, U.S. Patent 4,306,084 to the University of Texas (06.12.1978).
43. R. M. Laine, R. G. Rinker, and P. C. Ford, *J. Am. Chem. Soc.* **99**, 252 (1977).
44. R. M. Laine, *J. Am. Chem. Soc.* **100**, 6451 (1978).
45. R. M. Laine, Fr. Patent 2,443,282 to S.R.I. International (06.12.1978).
46. R. M. Laine, Symposium on Metal Clusters in Catalysis, American Chemical Society, San Francisco Meeting, August 24–29 (1980).
47. R. M. Laine, *Ann. N.Y. Acad. Sci.* **333**, 124 (1980); R. M. Laine, *J. Mol. Catal.* **14**, 137 (1982).
48. J. Jenck, P. Kalck, E. Pinelli, M. Siani, and A. Thorez, *J. Chem. Soc., Chem. Commun.*, 735 (1988).
49. G. Süss-Fink, *J. Organomet. Chem.* **193**, C20 (1980).
50. G. Süss-Fink and J. Reiner, *J. Mol. Catal.* **16**, 231 (1982).
51. G. Süss-Fink and G. F. Schmidt, *J. Mol. Catal.* **42**, 361 (1987).
52. G. Süss-Fink and G. Herrmann, *J. Chem. Soc., Chem. Commun.*, 735 (1985).
53. B. F. G. Johnson, J. Lewis, P. R. Raithby, and G. Süss-Fink, *J. Chem. Soc., Dalton Trans.*, 1356 (1979).
54. C. E. Kampe, N. M. Boag, and H. D. Kaesz, *J. Am. Chem. Soc.* **105**, 2896 (1983).
55. G. Süss-Fink, personnal communication.
56. E. M. Gordon and R. Eisenberg, *J. Organomet, Chem.* **306**, C53 (1986).
57. D. C. Bailey and S. H. Langer, *Chem. Rev.* **81**, 109 (1981).
58. F. Ciardelli, G. Braca, C. Carlini, G. Sbrana, and G. Valentini, *J. Mol. Catal.* **14**, 1 (1982).
59. C. U. Pittman, Jr., "Comprehensive Organometallic Chemistry" (G. Wilkinson, F. G. A. Stone, and E. W. Abel, eds.), Vol. 8, p. 553–611. Pergamon, 1982.
60. C. U. Pittman, Jr., and R. M. Hanes, *J. Am. Chem. Soc.* **98**, 5402 (1976).
61. C. U. Pittman, Jr., and G. M. Wilemon, *J. Org. Chem.* **46**, 1901 (1981).
62. L. Kim, T. E. Paxson, and S. C. Tang, Eur. Patent 0.002,557 to Shell International Research (19.12.1977).
63. D. C. Alexander and J. F. Knifton, U.S. Patent 4,511,740 to Texaco Inc. (01.09.1983).
64. H. Marrakchi, M. Haimeur, P. Escalant, J. Lieto, and J. P. Aune, *Nouv. J. Chim.* **10**, 159 (1986).
65. H. L. Mitchell III, U.S. Patent 4,473,505 to Exxon Research and Engineering Co. (27.09.1982).
66. E. H. Homeier, U.S. Patent 3,984,478 to Universal Oil Products Co. (27.05.1975).

67. E. H. Homeier and R. W. Johnson, U.S. Patent 4,234,455 to Universal Oil Products Co. (09.04.1979).
68. J. A. Hinnenkamp, U.S. Patent 4,413,146 to National Distillers and Chemical Corporation (25.11.1981).
69. P. F. Jackson, B. F. G. Johnson, J. Lewis, R. Ganzerla, M. Lenarda, and M. Graziani, *J. Organomet. Chem.* **190**, C1 (1980).
70. I. L. C. Freriks, P. C. de Jong-Versloot, A. G. T. G. Kortbeek, and J. P. van den Berg, *J. Chem. Soc., Chem. Commun.,* 253 (1986).

ADVANCES IN ORGANOMETALLIC CHEMISTRY, VOL. 32

X-Ray Structural Analyses of Organomagnesium Compounds

PETER R. MARKIES, OTTO S. AKKERMAN, and
FRIEDRICH BICKELHAUPT

Scheikundig Laboratorium
Vrije Universiteit
1081 HV Amsterdam, The Netherlands

WILBERTH J. J. SMEETS, and ANTHONY L. SPEK

Vakgroep Kristal-en Structuurchemie
Universiteit van Utrecht
3584 CH Utrecht, The Netherlands

I

INTRODUCTION

A. *Scope of the Review*

Organomagnesium compounds are well-established tools in organic chemistry. Grignard reagents are known from the beginning of the 1900s,

147

and many synthetic applications have been discovered since. After 1930, other reagents, especially organolithium reagents, were developed as alternatives. Being complementary to the "old" organomagnesium chemistry, organolithium chemistry showed deviating selectivities in some reactions and even opened up new synthetic possibilities by the use of halogen–metal exchange and direct metallation reactions. A general disadvantage of organolithium compounds, however, is their lower stability and higher cost. Therefore, organomagnesium chemistry remains important and is still standard in many synthetic procedures.

Compared to the long history of synthetic applications of organomagnesium compounds, the investigation of their structure and bonding covers a much shorter period. In many textbooks they are still treated as simple "carbanionic" reagents, without a more detailed discussion of their nature. For the delayed development of this aspect of organomagnesium chemistry, three major factors are responsible.

First, working with these air-sensitive compounds is frustrated by many experimental difficulties, mainly owing to a high reactivity toward oxygen and water. Special precautions must be taken to obtain reliably pure reagents for the structural investigations and to prevent premature change or even decomposition.

A second important factor is the growing armada of physical and spectroscopical techniques which have become available only recently. When the investigation of the nature of organomagnesium compounds drew new interest in the 1960s, the number of available experimental techniques was still rather limited and consisted mainly of the measurement of conductivities and degrees of association of organomagnesiums in solution. More powerful techniques like NMR spectroscopy or X-ray crystallography were in an early stage of development. After the modern techniques had become more accessible, they gave a large momentum to the fundamental research on organomagnesium compounds. NMR spectroscopy proved to be very suitable for the study of solutions. With CIDNP techniques, the occurrence of intermediates with a radical nature was detected during the Grignard formation reaction. A disadvantage in comparison with organolithium chemistry was the lack of magnesium isotopes suitable for NMR spectroscopy; only recently is progress being made with ^{25}Mg. Reliable thermochemical measurements of heats of hydrolysis or other reactions were facilitated by the development of adequate instrumentation and provided information about the relative stabilities of organomagnesium reagents.

Thirdly, as crystallographic methods developed to become routine techniques, their importance increased tremendously. They provide a very direct and detailed picture of the composition of organometallic reagents, as compared with the more indirect information obtained from the other methods. Recent reviews have appeared in this series on the crystal structures of organolithium compounds and the heavier alkali metal compounds. The study of organolithium structures especially has become a large field of research, with many structures reported during the 1980s. This was stimulated both by the important synthetic applications of organolithium compounds and by the relative accessibility of lithium for theoretical calculations. Reports on the structures of heavier alkali metal compounds are scarcer; this is mainly due to the experimental difficulties which are encountered in their synthesis and crystallization.

To our knowledge, a review specifically devoted to organomagnesium structures has not been published earlier. Related papers deal mainly with inorganic magnesium compounds, for example, salts, including those with organic or inorganic ligands. This article mainly concentrates on organomagnesium compounds having a Mg—C bond, which are also the most interesting reagents from a synthetic viewpoint. This field is still growing in importance, as can be concluded from the relatively large number of recent publications. A systematic survey of all data is of importance from various points of view:

1. The crystal structures visualize the bonding and the range of coordination modes possible for organomagnesium compounds. They can serve as models for the complexes present in solution that were studied by association measurements or NMR techniques.
2. Understanding of chemical bonding is needed to understand which bond types are possible and to draw relations to the physical behavior of organomagnesium compounds. Comparison with other main group metals, especially lithium, is of interest.
3. Structural investigations can be of benefit for the synthetic use of organomagnesium compounds, since reaction mechanisms will be better understood. Crystal structures will aid in visualizing the reaction pathway and in understanding regio- or stereoselectivities. Models for the transition states will be based on a more realistic foundation.
4. Structural data will serve as reference material for calculations in theoretical chemistry and in the parameterization of magnesium in semiempirical methods. Traditionally, theory has been concentrating on organolithium calculations. As more sophisticated programs and computer hardware become available, more attention will also be directed toward heavier metals and larger molecules. In the future, calculations will increasingly stimulate experimental structural investigations by indicating "key compounds" which show interesting features and are amenable to computation.

5. A new development is low temperature, high precision crystal structure determination to determine electron distributions in crystalline solids. If reliable charge distributions in organomagnesium complexes were measured in this way, valuable information would be obtained on the polarity of Mg—C or Mg—O bonds. So far, application of this method to an organomagnesium compound has not been reported; this may in part be due to the difficulty of growing perfect crystals.

The Cambridge Structural Database (1989 release), a Chemical Abstracts computer search, and recent issues of relevant journals were surveyed on organomagnesium structures, and 64 structures containing a Mg—C bond were found. They are discussed in classes which are defined by coordination number, degree of association, and the type of Mg—C bonds. Many structures are relatively recent (1980 or later) and are therefore not included in the well-known review on group 2 chemistry in the series *Comprehensive Organometallic Chemistry*. No complete review on organomagnesium chemistry has appeared since. In the recent past, more specialized topics such as magnesium-anthracene derivatives and polyether complexes have emerged; they are discussed in separate sections in this article. Some structures without a Mg—C bond are also included in supplementary tables because they are of interest for comparison with organomagnesium structures. The survey of this latter class of compounds is not intended to be complete.

B. *Recent Developments in Structural Organomagnesium Chemistry*

An important reason for us to collect the material in this article was the presentation, against an integral background, of 20 recent structures from our own investigations of the late 1980s. Our primary area of interest is the coordination of arylmagnesium compounds, both inter- and intramolecular. A number of compounds were synthesized covering the range from simple organomagnesium species to large intramolecularly coordinated complexes. For most of them, crystals suitable for X-ray structure determination were obtained and an X-ray study carried out. The results, from which a short summary is given in this section, form a considerable extension of organomagnesium structures. Full papers describing these new structures appeared only recently or are in preparation.

Relatively little is known about the crystal structures of unsolvated organomagnesium compounds. From this viewpoint, our polymeric structures of [diphenylmagnesium]$_n$ and of a [dineopentylmagnesium·neopentylmagnesium bromide]$_n$ adduct contribute to fill a gap in the existing literature. The structures of diphenylmagnesium and di(*p*-tolyl)magnesium as their THF adducts were also determined. Diphenylmagnesium gives the expected monomeric bis(tetrahydrofuranate) complex. In the case of di(*p*-tolyl)magnesium, a remarkable structure was found containing both the monomeric and the dimeric complex in a 2 : 1 stoichiometry. The complexation of bis(*p-tert*-butylphenyl)magnesium with diglyme and tetraglyme lead to the isolation of complexes containing pentacoordinated magnesium. In the latter case, two oxygen atoms of the ether ligands remain uncomplexed. This may suggest that a point of saturation for intermolecular coordination of a diarylmagnesium complex has been reached.

Intramolecular coordination strongly influences the structures of organomagnesium compounds. Coordinating solvent molecules are successively substituted by intramolecular ligands, and eventually the complex becomes completely saturated by "intramolecular solvation." In addition, coordination numbers can be induced which are unusually high in organomagnesium chemistry (up to six). This was demonstrated in a series of derivatives of diphenylmagnesium and phenylmagnesium bromide. Bis(2,6-dimethoxymethylphenyl)magnesium proved to be hexacoordinated, with the central magnesium atom completely surrounded by the aryl substituents. Bis(*o*-anisyl)magnesium had a remarkable dimeric structure, with one μ-bridging anisyl group forming a four-membered coordinative ring and the second connecting both magnesium atoms as a 1,3-bridging ligand. In 1-oxa-6-magnesadibenzo(4,5,7,8)cyclodecadieen·(THF)$_2$, another unexpected effect was found. The conformational constraints imposed by the formation of two six-membered coordinative rings induces a pseudo-trigonal bipyramidal geometry, which has two vacant positions filled by THF molecules.

The induction of a hexacoordinated state by substituents leading to five-membered coordinative rings is demonstrated in a series of four *o*-CH$_2$O(CH$_2$CH$_2$O)$_n$Me-substituted phenylmagnesium bromide structures ($n = 0-3$). Again, in the lower members of this series, extra THF molecules serve to fill the vacant coordination sites.

Interesting chemistry was encountered with organomagnesium derivatives of 1,3-xylyl crown ethers. The crystal structures of 2-bromomagnesio-1,3-xylyl-15-crown-4 and 2-bromomagnesio-1,3-xylyl-18-crown-5 were determined. During their formation from the corresponding aryl bromides and magnesium metal, a remarkable ether cleavage reaction occurred as a result of intramolecular coordination. In the case of the crown-4 Grignard reagent, the reaction has a remarkable selectivity which could be understood once its crystal structure had been determined. 1,3-Xylyl-18-crown-5 formed complexes with a remarkable rotaxane structure on complexation with diphenylmagnesium or bis(*p-tert*-butylphenyl)magnesium. The complexation of the large (double) macrocycle 1,3,16,18-dixylyl-30-crown-8 with bis(*p-tert*-butylphenyl)magnesium gave a 1:2 complex. In this case, the diarylmagnesium units are bonded to the outside of the crown either ring, and rotaxane structures apparently do not occur. This illustrates the rather sharp boundary conditions for the occurrence of rotaxane or "threaded" structures: the cavity has to be just right for the size of the metal center.

Surprisingly, a smaller crown ether, 1,3-xylyl-15-crown-4, was metallated both by diphenylmagnesium and phenylmagnesium bromide. Such reactivity, though common in organolithium chemistry, is unique for organomagnesium compounds. It is obviously induced by the crown ether ring, which facilitates a reaction path with a special low-energy transition state. The metallation reaction yields an asymmetric crown ether-substituted diphenylmagnesium derivative, 2-phenylmagnesio-1,3-xylyl-15-crown-4, which was characterized by its crystal structure. The 2-bromo-substituted analogs of the crown ethers mentioned above give halogen–metal exchange reactions with organomagnesium reagents. Via this route, 2-(*p-tert*-butylphenylmagnesio)-1,3-xylyl-18-crown-5 was synthesized and subsequently characterized by its crystal structure. This reaction is analogous to the lithium–halogen exchange which normally has no counterpart in organomagnesium chemistry.

C. *Some Aspects of Bonding in Organomagnesium Compounds*

Having a (Ne) s^2 electronic configuration and a strongly electropositive character, magnesium has a tendency to become Mg^{2+}. This tendency is not fully realized in organomagnesium compounds, but does result in strongly polarized bonds with other elements, including those to carbon. In comparison to lithium, the electronegativity of magnesium is higher (Li 0.98, Mg 1.31); accordingly, the Mg—C bond is more covalent. This is in line with the lower reactivity and the higher thermal stability of organomagnesium compounds as compared with lithium analogs. However, the divalency of magnesium imparts to the metal atom a relatively large positive charge as compared with monovalent lithium, since magnesium is attached to two negatively charged groups. The net effect may even impart a larger positive charge on magnesium as compared to the monovalent lithium; this aspect calls for a computational investigation. As a result, many (electrostatic) arguments valid for the explanation of organolithium structures can be equally applied to organomagnesium chemistry. An important difference between organolithium and organomagnesium coordination chemistry naturally remains: the stoichiometry of negative to positive centra for magnesium is 2 : 1, whereas for lithium it is 1 : 1. This leads to a higher tendency of organolithium compounds to form three-dimensional clusters.

An isolated diorganylmagnesium molecule in the gas phase usually adopts a linear structure with a central sp-hybridized magnesium atom and two negatively charged ligands at opposite positions. In practice, this situation is very difficult to achieve: only a few organomagnesium compounds can be transferred into the gas phase as thermal decomposition normally precedes vaporization. In a gas-phase electron diffraction study of dineopentylmagnesium (Section II), a linear molecule was indeed found.

The tendency of aggregation of organomagnesium compounds is high, owing to the highly unsaturated character of magnesium. For dineopentylmagnesium, the monomer and dimer are in equilibrium even in the gas phase, as was shown by a mass spectroscopic investigation. In the liquid and solid state, aggregates are formed which probably in most cases are polymeric. These polymers are constituted of chains of magnesium atoms connected by bridging (organic) ligands. The central structural element in these oligomers is the four-membered Mg_2C_2 ring, which represents a favorable electrostatic array of two positively (magnesium) and two negatively (carbon) charged atoms. The magnesium atoms connect two such rings in a spiro fashion, forming a linear chain; they are (pseudo) tetrahedrally surrounded by four organic ligands, and the R groups bridge between two magnesium atoms with a three-center two-electron bond. Angles necessarily deviate from the ideal tetrahedral geometry: inside the four-membered rings, they tend to be close to 90°. The formation of aggregates can be prevented by sterically demanding groups, as was recently shown by a solid-state structure of $(Me_3Si)_3C—Mg—C(SiMe_3)_3$ (Section III,B,1), which is linear and analogous to that of dineopentylmagnesium in the gas phase.

In the presence of polarizable electron-rich organic compounds, especially Lewis bases such as ethers or tertiary amines, organomagnesium compounds form complexes: the polymeric structures which are present in the solvent-free species are broken down to form lower aggregates or monomers. In apolar solvents like petroleum ether, benzene, or toluene, most organomagnesium reagents retain their polymeric nature and have therefore a (very) low solubility. For this reason, organomagnesium chemistry is almost always performed in ethereal solvents. By establishing Lewis acid (organomagnesium)–Lewis base (ether) interactions, the magnesium atom tries to attain coordinative saturation. As in the aggregated state, the central magnesium atom normally achieves (pseudo) tetrahedral coordination which, however, is not so much indicative of electronic saturation but rather dictated by steric hindrance; there is insufficient space around magnesium (bonding radius $1.4 \pm 0.05\text{Å}$) in $R_2Mg \cdot 2L$ to accommodate more than two of the rather bulky Lewis bases L in addition to the two anionic ligands R required by Mg(II). In general, the coordination geometry deviates from the ideal tetrahedral geometry by larger C—Mg—C (or C—Mg—X) angles ($120-149°$) and smaller L—Mg—L angles ($95°$). A qualitative analysis suggests that several factors may contribute to this effect. In the first place, the inherent linear *sp* hybridization of the strongly bound R_2Mg unit is only partially perturbed by the (weaker!) coordination of the Lewis base L; from this point of view, the large R—Mg—R angle may be seen as a sort of "memory effect" (which, incidentally, is even more pronounced in R_2Zn analogs). Another way of looking at the problem takes Bent's rule as the starting point: of the four bonds emanating from magnesium, those to carbon (electronegativity 2.5) are much less polarized toward the ligand than those to the heteroatom (e.g., oxygen, electronegativity 3.5); consequently, magnesium invests more *s* character (wider angle!) in the bonds to carbon than in those to ether oxygen. Finally, in some cases steric factors will increase the tendency to have larger R—Mg—R angles because many R groups are bulkier in the direct vicinity of the magnesium than an average ether, especially when R is an aryl group.

In order to investigate the role which orbital interactions play in the coordination of ether molecules to organomagnesium compounds, we analyzed the angular orientation of the ether C—O—C triangle toward the magnesium atom, using the data of Mg—O interactions from our own crystal structures. In this series, both inter- and intramolecular organomagnesium complexes are included, and the Mg—O distance (R) varied from 2.0 Å (strong Mg—O bond) to 2.7 Å (very weak Mg—O interaction).

There exist two descriptions of the lone pairs in ethers (or, for that matter, in water). The oxygen is considered to be either (approximately) sp^3 hybridized, which makes the two lone pairs degenerate and places them in a plane perpendicular to the C—O—C plane; or it is considered sp^2-p hybridized involving a sp^2 lone pair along the bisector of C—O—C and one in a p_z orbital perpendicular to the C—O—C plane. Chakrabarti and Dunitz have investigated the spatial orientation of alkali and alkaline earth cations toward ethers (*1*). In a similar fashion, we have analyzed the geometric relation between carbon-bonded magnesium and the ether oxygen in our organomagnesium structures (Table I). This relation can be defined by two angles (Fig. 1). The angle between the Mg—O bond and the bisector of C—O—C is α; Mg—A $= R \sin \alpha$ defines the distance of Mg from this bisector. The angle β is the dihedral angle between the plane O—Mg—A and the plane perpendicular to the C—O—C plane; as the

TABLE I

ORIENTATION OF R—O—R ETHER GROUPS TOWARD THE MAGNESIUM ATOM, AS
DEFINED BY THE PARAMETERS Mg—O BOND DISTANCE, $R \sin \alpha$, AND β

Compound	Structure	Mg—O(Å)	α (°)	$R \sin \alpha$ (°)	β (°)
EtMgBr·(Et$_2$O)$_2$	**5c**	2.026	6	0.20	32
		2.053	9	0.30	1
Ph$_2$Mg·(THF)$_2$	**5l**	2.030(4)	11	0.39	8
(p-Tolyl)$_2$Mg·(THF)$_2$	**5m**	2.050	6	0.22	13
		2.031(6)	9	0.32	4
[(p-Tolyl)$_2$Mg·THF]$_2$	**6d**	2.020(5)	14	0.48	20
Ph$_2$Mg·(1,3-xylyl-18-crown-5)	**26**	2.520(4)	28	1.18	19
		2.222(4)	47	1.63	0
		2.204(3)	48	1.66	3
		2.516(4)	35	1.43	14
(p-tert-BuPh)$_2$Mg·tetraglyme	**30**	2.234(6)	41	1.47	9
		2.117(5)	35	1.21	4
		2.198(6)	35	1.27	4
[(o-Anisyl)$_2$Mg·THF]$_2$	**37**	2.052(5)	3	0.12	84
		2.166(4)	35	1.25	40
		2.063(5)	27	0.94	5
		2.100(5)	16	0.57	5
[2,6-Bis(CH$_3$OCH$_2$)phenyl]$_2$Mg	**38**	2.282(3)	13	0.53	28
		2.338(3)	12	0.47	32
		2.327(3)	12	0.47	27
		2.311(3)	10	0.40	35
1-Oxa-6-magnesiadibenzo(4,5,7,8)-	**39**	2.242(4)	20	0.77	1
cyclodecadieen·(THF)$_2$		2.221(4)	7	0.29	33
		2.095(3)	9	0.31	24
2-MgBr-1,3-xylyl-15-crown-4	**40**	2.33(1)	37	1.42	7
		2.13(1)	45	1.55	1
		2.12(1)	28	1.02	4
		2.49(1)	52	1.97	7
2-MgBr-1,3-xylyl-18-crown-5	**41**	2.170(4)	0	0.36	21
		2.235(4)	46	1.61	6
		2.331(4)	50	1.79	5
		2.126(4)	3	0.12	56
2-MgPh-1,3-xylyl-15-crown-4	**42**	2.619(3)	57	2.20	3
		2.183(3)	35	1.25	10
		2.209(3)	27	1.01	4
		2.335(3)	43	1.59	0
2-[(p-tert-BuPh)Mg]-1,3-xylyl-	**43**	2.713(4)	55	2.23	2
18-crown-5		2.337(4)	51	1.83	4
		2.146(4)	8	0.28	26
		2.317(3)	5	0.22	84
[1-MgBr-2-(CH$_3$OCH$_2$)ben-	**44**	2.037(6)	27	0.92	10
zene·THF]$_2$		2.105(7)	8	0.30	26
1-MgBr-2-2,5,8-trioxanonyl	**46**	2.129(6)	10	0.36	23
benzene·(THF)		2.148(5)	14	0.51	3
		2.222(5)	47	1.64	3
		2.178(5)	41	1.42	7
1-MgBr-2-2,5,8,11-tetraoxadodecyl	**47**	2.151(7)	33	1.17	1
benzene		2.186(7)	31	1.11	3
		2.197(6)	46	1.59	4
		2.331(6)	60	2.01	2

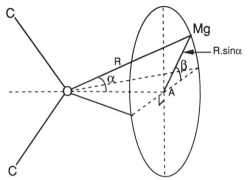

FIG. 1. Geometric relation of magnesium with respect to the ether oxygen in polar coordinates ($R \sin \alpha$ versus β).

latter contains the two sp^3-hybridized oxygen lone pairs, β is a measure indicating how far magnesium is removed from the lone pair plane.

The pictorial presentation of Fig. 2 ($R \sin \alpha$ against β in polar coordinates) leads to interesting results which in some regards differ from those obtained for the free ions (*1*). Contrary to the free ions, carbon-bonded magnesium shows a larger range of α, or, in other words, there seems to be no particular tendency for magnesium to be located on or near the bisector, which is the direction of the sp^2 lone pair and/or of the ether dipole axis; the maximum for α is about 60°. Second, β is large (up to 84°) *only* when the distance $R \sin \alpha$ to the bisector is small, which means that magnesium is located in a rather flat ellipsoid along the plane of the lone pairs (only *one* quadrant is shown in Fig. 2). When interpreting these findings, one must be aware, as Dunitz has pointed out, that the geometrical constraints of the ligands will not always permit magnesium and ether oxygen to attain their energetically most favorable relative orientations (*1*). Nevertheless, the remarkable fixation of magnesium to the lone pair plane, especially at large angles α, and the absence of any preference for location *on* the ether bisector seem to indicate that magnesium likes to be bonded to the sp^3-hybridized oxygen lone pairs and that electrostatic factors, though clearly present, are not dominant; there seems to be a subtle balance between orbital overlap and electrostatics. This is probably also the reason for the strong prevalence of tetracoordination in most organomagnesium compounds. Inorganic compounds of Mg^{2+}, such as the dihalides (Section IV,A), have a stronger tendency to be hexacoordinated, in spite of the smaller radius of Mg^{2+} (0.65 Å) compared with that of Mg in $RMgX \cdot L_n$ (1.4 Å). Thus, not surprisingly, the coordination chemistry of Mg^{2+} appears to be governed by charge effects much more than that of organometallic magnesium.

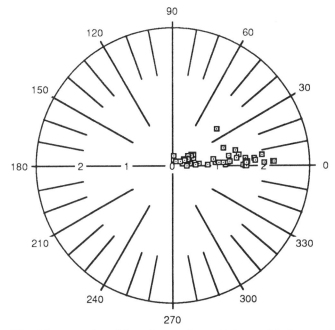

FIG. 2.   Pictorial presentation of the polar coordinates $R \sin \alpha$ and β of magnesium in the compounds in Table I (view of Fig. 1 in the direction A → O).

II

STRUCTURES OF ORGANOMAGNESIUM COMPOUNDS IN THE GAS PHASE

Though not belonging to the topic of this article in a strict sense, a brief discussion of gas-phase structure is imperative for reasons of complementarity and comparison, and it is facilitated by the scarcity of structures. By the use of gas-phase electron diffraction (GED), it is possible to determine the structure of molecules in the gas phase. Results obtained with this technique on organomagnesium compounds are of special interest for theoretical chemists, since calculations refer to molecules in the gas phase, too. In many cases, the structure of an organomagnesium compound in the gas phase will be different from that in the crystalline state. These differences often stretch beyond the effects of crystal packing.

A major drawback of GED is that only small molecules can be resolved completely; for larger molecules, the degrees of freedom must be limited during the calculations by assuming and optimizing values for many of the structural parameters. A second barrier for the use of GED may be the low volatility of many substrates. This presents a large problem for organo-magnesium compounds: many complexes lose their coordinated solvents on heating, leaving behind a nonvolatile, polymeric residue. Thermal instability of organomagnesium compounds imposes another limitation to the use of this technique. Until now, only four organomagnesium compounds have been measured: dicyclopentadienylmagnesium (**1**), bis(pentamethylcyclopentadienyl)magnesium (**2**), dineopentylmagnesium (**3**), and cyclopentadienylneopentylmagnesium (**4**), all determined by the same research group. They could be measured at nozzle temperatures of 125, 160, 160, and 90°C, respectively.

For **1**, a ferrocene-like structure with parallel cyclopentadienyl rings was found with Mg—C and C—C distances of 2.339 and 1.423 Å, respectively (*2*). The distance Mg—Cp centroid is 2.008 Å. The best agreement between calculated and experimental results was obtained from a model with eclipsed Cp hydrogen atoms, but a staggered conformation cannot be ruled out. No significant bending of the hydrogen atoms out of the Cp plane was detected. Recently, the GED structure of the pentamethyl derivative **2** was determined (together with those of the corresponding calcium and ytterbium compounds) (*3*). A geometry was found analogous to that of the parent compound. Important bond distances are as follows: Mg—C 2.341 Å, Mg—Cp centroid 2.011 Å, C_{Cp}—C_{Cp} 1.428 Å, and C_{Cp}—C_{Me} 1.520 Å. The methyl groups are bent 5° outward from the Cp plane.

The GED structure determination of **3** showed a centrosymmetric molecule with a linear C—Mg—C center (*4*). On evaporation of **3**, which is trimeric in benzene solution and probably polymeric in the crystalline state, dissociation into monomeric molecules occurs. The structure obtained is in accordance with the considerations in Section I,C. Important geometric parameters of the structure are the distances Mg—C (2.126 Å) and C—C (1.541 Å) and the angle Mg—C—C (118°). By fusion of **3** and **2**, the mixed crystalline compound **4** (m.p. 70–72°C) was obtained. The structure of this compound, which is stable toward comproportionation to the two symmetrical parent compounds, again showed a linear geometry at magnesium (Cp—Mg—Np 180°) (5). Relevant bond distances (Mg—C_{Cp} 2.328 Å, Mg—C_{Np} 2.12 Å) and angles (Mg—C_{Np}—C_{Np} 126°) were comparable to those in the symmetrical compounds.

With these four structures, the possibilities of GED contributing to organomagnesium structural chemistry may not be completely exhausted. Other compounds suitable for this technique (because of their volatility in high vacuum) have become known. In principle, diffraction measurements can also be performed on solutions. The use of this technique is limited to very simple systems. To our knowledge, only one recent example has been reported: MgI_2 solutions in Et_2O and THF were studied by large angle X-ray scattering (6). In diethyl ether (44°C, 2.8 M solution) a dimeric structure is proposed with the central Mg_2I_2 unit forming a square, four-membered ring and a terminal iodine bonded to each magnesium. A uniform Mg—I bond length of 2.65 Å and two I—I distances [3.75 (diagonal) and 5.30 Å] were found. In a THF solution (1.7 M), magnesium iodide is reported to be monomeric with an appreciable degree of dissociation into Mg—I^+ and I^-. In both species, the Mg—I bond distance is 2.56 Å. Both in diethyl ether and in THF solution, solvent molecules will complete the coordination sphere of the magnesium atoms.

In this context, we mention also the gas-phase structures of $MgBr_2$ (7) and MgI_2 (8); both are linear, and the bond lengths are Mg—Br 2.34 Å and Mg—I 2.52 Å.

III

STRUCTURES OF ORGANOMAGNESIUM COMPOUNDS IN THE SOLID STATE

A. Tetrahedral Coordination of Magnesium

1. Simple Monomeric and Dimeric Organomagnesium Compounds

In organomagnesium chemistry, tetrahedral coordination can be regarded as the normal situation. In almost all structure determinations of simple Grignard or diorganylmagnesium compounds, the magnesium atoms are surrounded by four ligands. Because of different states of association, three basic types of complexes can be distinguished: monomeric (5), dimeric (6), and polymeric (7). In the general structures 5–7, R presents an organic group, X an anionic substituent (Grignard, X = halogen; diorganylmagnesium, X = R), and L the coordinating ligands (ethers or amines). The monomeric structure 5, in which the central magnesium atom is coordinated to two heteroatom ligands, presents the most common type of

5   6

7

complex found in organomagnesium chemistry: a large number of crystal structures (**5a–5m**) fall into this category (Table II). Because of their similarity, only a few structures are actually shown in this review. Dimeric structures of type **6** result if organomagnesium compounds are crystallized from a weakly coordinating solvent like triethylamine or diisopropyl ether. For steric reasons the coordinative power of these solvents is too low to break organomagnesium complexes down into monomers. In addition, the ability of X to act as a bridging group is of importance (alkoxide, amide > halogen > alkyl, aryl group). In the absence of coordinating solvents, organomagnesium compounds will generally form polymeric structures of type **7**. In these structures, a chain of tetracoordinated magnesium atoms is connected by μ-bridging R groups via three-center two-electron bonds. In solvent-free Grignard compounds, either alternation of Mg_2R_2 (R = alkyl, aryl) and Mg_2X_2 (X = halogen) four-membered rings on the one hand or mixed bridges Mg_2RX on the other is conceivable.

The crystallographic study of organomagnesium compounds started around 1964 with the investigation of the simple Grignards phenylmagnesium bromide and ethylmagnesium bromide, which were found to be of type **5**. From the first compound, the bis(tetrahydrofuranate) (**5a**) (*9*) and bis(diethyl etherate) (**5b**) (*10*) were characterized. The precision of the structures obtained was rather low, owing to a low crystal quality and the limited experimental techniques of those days. Ethylmagnesium bromide bis(diethyl etherate) (**5c**) gave better crystals (*11*), and representative values for the relevant bond angles and distances in the simple monomeric Grignard complex were obtained (Mg—Br 2.48 Å, Mg—O 2.03 and 2.05 Å, Mg—C 2.15 Å, Br—Mg—C 125.0°).

TABLE II

RELEVANT BOND ANGLES AND BOND LENGTHS FOR TYPE 5 COMPOUNDS

	Compound (RMgX·L₂)			Bond angles (°)		Bond lengths (averaged, Å)			
Structure	R	X	L₂	C—Mg—X	L—Mg—L	Mg—C	Mg—X	Mg—L	Ref.
5a	Ph	Br	(THF)₂	n.a.[a]	n.a.	n.a.	n.a.	n.a.	9
5b	Ph	Br	(Et₂O)₂	n.a.	n.a.	n.a.	2.44	2.03	10
5c	Et	Br	(Et₂O)₂	125	101	2.15	2.48	2.04	11
5d	Me	Me	(Quinuclidine)₂	129	108	2.19		2.24	12
5e	Me	Me	TMEDA	130	82	2.17		2.24	13
5f	Ph	Ph	TMEDA	119	83	2.17		2.20	14
5g	R	R	(p-Dioxane)₂	124	100	2.16		2.04	15
R = methyl-o-carboranyl									
5h	Et	Br	(−)-Sparteine	115	84	2.24	2.48	2.15	16
5i	t-Bu	Cl	(−)-Sparteine	115	84	2.19	2.33	2.17	17
5j	Et	Br	(+)-α-Isosparteine	112	84	2.22	2.51	2.18	18
5k	Et	Br	(+)-6-Bz-sparteine	111	83	2.34	2.51	2.16	19
5l	Ph	Ph	(THF)₂	122	94	2.13		2.03	b
5m	p-Tol	p-Tol	(THF)₂	124	97	2.13		2.04	b

[a] n.a., Not available.
[b] This work.

5c   5d

The structures of the diorganylmagnesium complexes dimethylmagne-sium·(quinuclidine)$_2$ (**5d**) (*12*), dimethylmagnesium·TMEDA (**5e**) (*13*), and diphenylmagnesium·TMEDA (**5f**) (*14*) are very similar, so only **5d** is depicted. The lengths of the Mg—C and Mg—N bonds lie in the intervals 2.16–2.22 and 2.20–2.26 Å, respectively. It is noteworthy that the average Mg—N bond distances are larger than those of Mg—C or Mg—O, be-cause normally the bonding radius of N (0.75 Å) is intermediate between that of C (0.77 Å) and O (0.73 Å). Relative to the ideal tetrahedral geometry, a tendency toward deformation to larger C—Mg—C angles (**5d**, 129.0°; **5e**, 130.0°; **5f**, 119.2°) and smaller N—Mg—N angles (**5d**, 108.2°; **5e**, 81.5°; **5f**, 82.5°) is observed. For **5e** and **5f**, this angle contraction is imposed by the formation of a five-membered ring on complexation of TMEDA. In the case of **5d**, other effects must play a role, because, for steric reasons, a large angle between the bulky quinuclidine ligands would be expected and a small angle between the methyl groups. Here, as in many other organomagnesium structures to follow, the factors discussed in Sec-tion I,C seem to be operative. A more complex molecule, but with basically the same geometry as **5d–5f**, is di(methyl-*ortho*-carboranyl)magne-sium·bis(*para*-dioxane) (**5g**) (*15*). In this compound, the two organyl groups are large C_2B_{10} carboranyl clusters, bonded to magnesium with one

5g

of their carbon atoms. The magnesium atom is further coordinated by two *p*-dioxane molecules, which are acting as monodentate ligands. The bond distances and angles in **5g** have normal values (Mg—C 2.156 Å, Mg—O 2.038 Å, C—Mg—C 123.5°, O—Mg—O 99.5°).

A series of Grignard complexes with optically active sparteine ligands was characterized in order to study the induction of asymmetry in the polymerization of acrylates. In these compounds, the sparteine ligands act as a bidentate ligand, giving type **5** complexes. The structures of EtMgBr·[(−)-sparteine] **(5h)** *(16)*, *t*-BuMgCl·[(−)-sparteine] **(5i)** *(17)*, EtMgBr·[(−)-α-isosparteine] **(5j)** *(18)*, and EtMgBr·[(+)-6-benzylsparteine] **(5k)** *(19)* proved to be very similar, so only **5h** is depicted here. Some

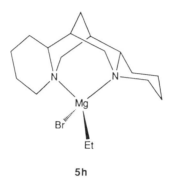

5h

variation occurs in the bond lengths toward magnesium (Mg—C 2.19–2.34 Å, Mg—N 2.13–2.19 Å). The angles around the central magnesium atom were quite constant (N—Mg—N 83.1–84.0°, Br/Cl—Mg—C 111.4–115.1°).

To facilitate a comparison of all type **5** structures, their relevant geometric parameters are collected in Table II, which also contains average values for each parameter. Remarkable is the consistent tendency of deviation from the ideal tetrahedral arrangement. In line with the discussion in Section I,C, the C—Mg—X angles are larger and the L—Mg—L angles smaller than the ideal tetrahedral values. Geometric constraints of the bidentate ligand are presumably responsible for the small angles of entries **5e**, **5f**, and **5h–5k**. Although the material is too limited to draw general conclusions, it seems that the nature of the heteroatom (O or N) is of less importance for the L—Mg—L angle than steric factors. On the other hand, the C—Mg—X angle shows a tendency to be smaller when X is a halogen than when X is carbon, which is in line with electronegativity of the substituent being important (Bent's rule).

Some years after the presentation of the monomeric structure of ethylmagnesium bromide, a centrosymmetric dimer of the same compound was reported **(6a)** *(20)*. In this complex, which has the composition (ethylmag-

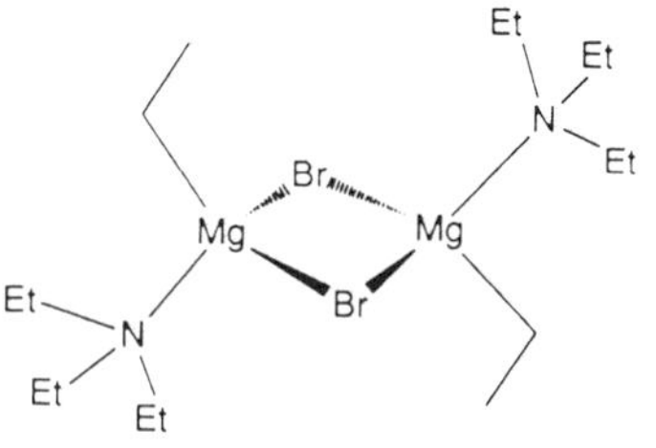

6a

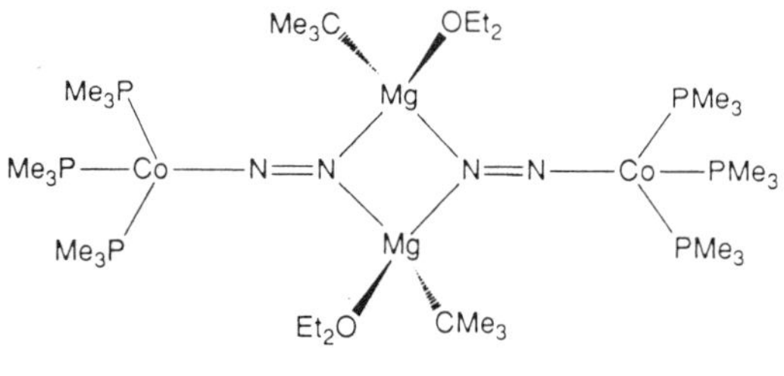

6c

nesium bromide · triethylamine)$_2$, the bromine atoms occupy the bridging positions. Each magnesium atom has terminal bonds to one organic group and one coordinating solvent molecule. The two Mg—Br bonds are almost equal (2.566 and 2.567 Å) and about 0.1 Å longer than in **5c**. The terminal bonds (Mg—N 2.15 and 2.18 Å, Mg—C 2.18Å) are practically identical to those in type **5** complexes. For the analogous diisopropyl ether complex of ethylmagnesium bromide (**6b**) (*21*), a similar structure was found. The terminal bonds of **6b** are short (Mg—O 2.019 Å, Mg—C 2.09 Å). The Mg—Br—Mg and Br—Mg—Br angles (86.8 and 93.2°) show that the central four-membered Mg$_2$Br$_2$ ring is almost square. The structure of a bimetallic cobalt diazenide organomagnesium complex (**6c**), as published by Klein *et al.*, can also be regarded as a type **6** complex (*22*). In this compound, the magnesium atoms are connected by two (Me$_3$P)$_3$Co—N=N— bridging ligands (Mg—N 2.097 Å). The central Mg$_2$N$_2$ ring of the centrosymmetric molecule is almost square (Mg—N 2.097 Å, Mg—N—Mg 95.8°, N—Mg—N 84.2°). The terminal positions R and L are occupied by *tert*-butyl and diethyl ether ligands, respectively.

During our own investigations, we obtained crystal structures of the tetrahydrofuranate complexes of diphenylmagnesium and bis(*p*-tolyl)magnesium. The publication of these structures is in preparation (*22a*). The first compound (**5l**) crystallized as the expected monomeric bis(tetrahydrofuranate) in analogy to **5f**. Some characteristic structural parameters of **5l**

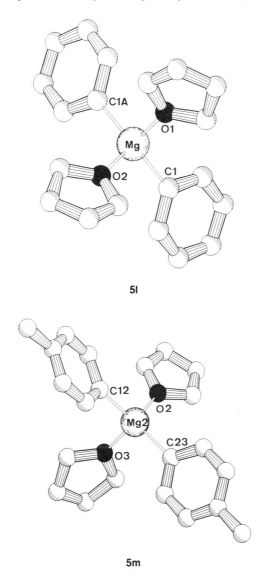

5l

5m

are as follows: Mg—C 2.127(4) Å, Mg—O 2.030(4) Å, C—Mg—C 122.4(1)°, and O—Mg—O 94.2(1)°. A more surprising crystal structure was found for bis(p-tolyl)magnesium, since it contained both the monomer **5m** and the dimer **6d** in a 2:1 stoichiometry. The formation of crystals with this composition must be facilitated by the simultaneous presence of both species during the crystallization process (concentrated solution!) and,

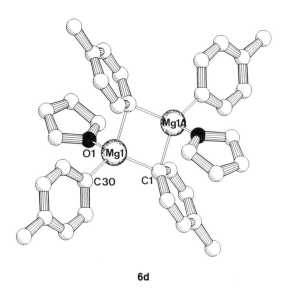

6d

of course, an energetically favorable crystal packing. The structure of **5m** has much in common with that of its diphenylmagnesium analog **5l**. The centrosymmetric complex **6d** has a novel feature, since it is the first simple diarylmagnesium species with a μ-bridging aryl group. This three-center two-electron bond is not symmetrical: a long [2.313(7) Å] and a short [2.245(7) Å] Mg—C contact is present. The dimer can thus be regarded as being composed of two strongly associated monomers. Both bridge Mg—C bonds are elongated, as compared with the terminal Mg—C bond [2.130(7) Å]. Inside the four-membered Mg_2C_2 ring, a large [C—Mg—C 102.5(2)°] and a small [Mg—C—Mg 77.5(2)°] angle are found. The terminal ligands have normal bond lengths [Mg—O 2.020(5) Å, Mg—C 2.130(7) Å]. The bridging of the aryl groups is unsymmetrical: both bridging aryl groups are perpendicular to the Mg_2C_2 plane, but their planes are inclined toward the Mg $\cdots$ Mg axis. The stronger Mg—C bond lies relatively close to the plane of the aryl group [28.3(3)°] and therefore has high σ character. For the weaker Mg—C bond [46.4(3)°], some p interaction between magnesium and the aromatic ring must be involved.

2. *Polymeric Solvent-Free Organomagnesium Compounds*

In the absence of coordinating solvents like ethers, most organomagnesium compounds tend to form polymeric structures. Microcrystalline solids or viscous liquids are generally obtained on desolvation of organomagnesium ether complexes, which are almost insoluble in apolar (non-ethereal) solvents; thus, the growth of single crystals is difficult. The first studies of solvent-free organomagnesium compounds were therefore performed using the powder diffraction technique, resulting in two similar (type **7**) structures for the relatively simple compounds dimethylmagnesium (**7a**) and diethylmagnesium (**7b**) (*23*). In both compounds, a linear chain of tetracoordinated magnesium atoms is connected by μ-bridging alkyl groups via three-center two-electron bonds. As compared with normal two-center bonds, the Mg—C distances are slightly elongated (**7a**, 2.24 Å; **7b**, 2.26 Å). Contrary to the structure of **6d**, no asymmetry in the M—C bond lengths is observed. The Mg_2C_2 rings in the polymeric chain are clearly rhomboid, with small Mg—C—Mg angles (**7a**, 75°; **7b**, 72°) and large C—Mg—C angles (**7a**, 105°; **7b**, 108°).

The use of powder diffraction for structure determination is limited to simple molecules, since for the resolution of larger molecules too many structural assumptions have to be made. In those cases, it will be necessary to obtain single crystals. Until recently, however, no examples were known in the literature, undoubtedly owing to experimental difficulties. In principle, three approaches for obtaining good quality crystals are available:

1. Solid and volatile organomagnesium compounds can be slowly sublimed in a static high vacuum. For instance, we used this technique recently to obtain crystals of ether-free diphenylzinc. Since only a few organomagnesium compounds can be sublimed, the perspectives of this approach are limited. We attempted the sublimation of dineopentylmagnesium, which will surely have a crystal structure different from that in the gas phase. Unfortunately, only low quality crystals have been obtained so far.

2. Organomagnesium compounds with sufficient solubility can be crystallized from apolar solvents (toluene, alkanes). This method is limited to compounds having large apolar organic groups which promote solubility and sterically oppose oligomerization. An adduct $[(Np_2Mg)_2 \cdot (NpMgBr)_2]_n$ crystallized from a dineopentylmagnesium solution which inadvertently contained a small amount of magnesium bromide (**7c**) (*24*). The structure of **7c** proved to be the first crystal structure of a simple, solvent-free alkylmagnesium halide compound. In the type **7** polymeric chain, an alternating pattern of three Mg_2Np_2 rings and one Mg_2Br_2 ring was found, connected in an almost perpendicular spiro fashion. Remarkable features are the long Mg—Br bonds [2.808(4) and 2.818(6) Å] and the unsymmetrical Mg—Np—Mg bridges which have one short (2.20–2.29 Å) and one long (2.33–2.42 Å) Mg—C bond. The long Mg—Br distance can be interpreted to be a consequence of initial dissociation of the polymeric structure into $Np_6Mg_4^{2+}$ cations and Br^- anions. The asymmetric bridging

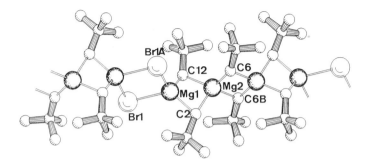

7c (from Ref. *24* with permission of Elsevier Science Publishers. B.V., Amsterdam)

of the neopentyl groups, which was not found in **7a** and **7b**, may in part be related to their larger steric size, as compared with the methyl or ethyl groups. It is also noteworthy that, contrary to **7a**, **7b**, and **7d**, the magnesium atoms are not arranged along a single straight line, but pairwise on two parallel straight lines separated by a distance of 1.10 Å.

3. By a high dilution technique, the ether complex of an organomagnesium compound can in certain cases be dissolved in a large quantity of an apolar solvent. Using this technique, we crystallized diphenylmagnesium (20 mmol), starting from its diethyl etherate, from benzene (2 liters). From the initially clear solution, crystals with composition $[Ph_2Mg]_n$ (**7d**) formed over a period of several weeks (*22a*). In the polymeric structure, the magnesium atoms are bridged by μ-phenyl groups, forming a linear chain

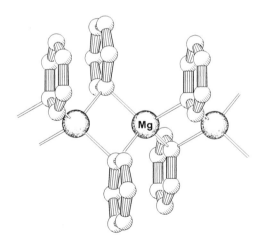

7d

of perpendicularly connected four-membered Mg_2C_2 rings [Mg—C 2.261(2) Å]. In contrast with complex **6d**, the three-center two-electron bonding of the bridging phenyl groups is symmetrical.

3. *Cyclic Organodimagnesium Compounds*

On symmetrization of di-Grignard reagents to the analogous halogen-free compounds, both polymeric and cyclic species may be expected. If the first possibility occurs, crystallization will be difficult, and, indeed, no such species have been characterized structurally. The growth of single crystals from the cyclic complexes proved to be easier: three crystal structures have been published so far. A common remarkable phenomenon in these structures is that they are not composed of the smallest possible, monomeric units; rather, large-membered rings are formed out of two or three monomeric units. Although, here again, crystal packing effects might play a role, clearly the thermodynamic preference for these large ring structures is decisive; this follows from the predominant occurrence of dimeric species in solution (as determined by association measurements) which presumably have the same monocyclic structure. In all structures, the magnesium atoms themselves have a normal coordination geometry (type **5** complex). The first report concerns $[(CH_2)_4Mg \cdot (p\text{-dioxane})]_2 \cdot [p\text{-dioxane}]$ (**8**) with a

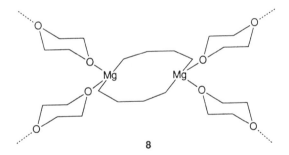

8

dimeric (dimagnesacyclodecane) structure (*25*). Each magnesium atom is coordinated by two *p*-dioxane molecules, which are bridging to other residues in a polymeric network. As usual, the coordination at magnesium deviates from the ideal tetrahedral angles with an enlargement of the C—Mg—C angle (C—Mg—C 128°, O—Mg—O 90°). This explains why the five-membered ring structure (monomeric complex) is unfavorable and thus not encountered, either in the solid state or in solution (*27*). The structure of $[(CH_2)_5Mg \cdot (THF)_2]_2$ (**9**) (*26*) has much in common with that of **8**: a dimagnesacyclododecane with each magnesium solvated by two THF molecules. Relative to **8**, the C—Mg—C angle is even further

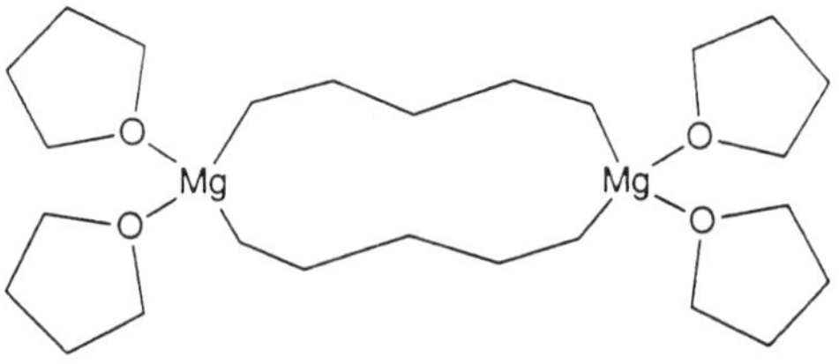

9

enlarged (C—Mg—C 141.5°, O—Mg—O 90.8°). A thermodynamic analysis in the case of **9** has shown that enthalpy (ring strain in the six-membered monomer) and entropy (loss of free particles on oligomerization) play an opposing (!) role in making the twelve-membered dimer the preferred species (*26*).

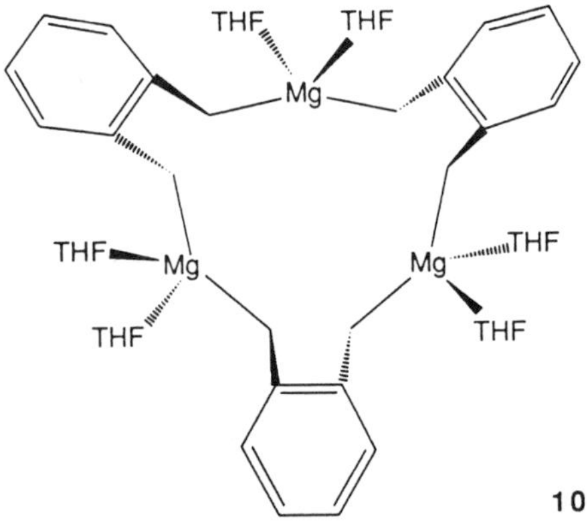

10

Trimeric [*o*-xylidenemagnesium · (THF)$_2$]$_3$ (**10**) (*27*), has a trimagnesacyclopentadecane structure. The coordination geometry of the magnesium atoms shows the same trends as in the previous type **5** compounds (C—Mg—C 126.8 and 130.5°, O—Mg—O 94.1 and 94.0°). The formation of a trimer is possibly preferred because of the rigidity of an *o*-xylidene unit which may prevent the complete relief of strain in the dimer.

Recently, the crystal structure of [1,8-naphthalenediyl-Mg · THF]$_4$ (**11**) was determined (*28*). In this case, not a simple ring structure but a tetrameric cluster is formed. This is facilitated by the special bonding mode of the 1,8-naphthalenediyl groups, which have one C—Mg single bond and one carbon atom μ-bridging between two magnesiums. The central cluster is formed by a slightly elongated tetrahedral array of four magnesium

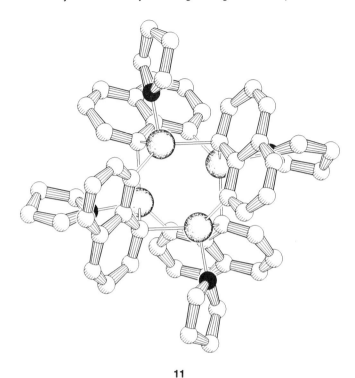

11

atoms, which are all tetracoordinated. Each magnesium has one Mg—C σ bond [2.15(2) Å] and is involved in two three-center two-electron bonds [2.28(2), 2.29(2) Å] to two other naphthalenes. One THF molecule serves to complete the coordination sphere [Mg—O 2.07(2) Å]. By association measurements at room temperature, the compound was shown to be tetrameric in THF solution as well. This behavior is unique, since almost all organomagnesium compounds are monomeric in this strongly basic solvent. It should be added that the R value (0.15) is rather high owing to free disordered THF molecules in the crystal.

B. Nontetrahedral Coordination of Magnesium

1. Di- and Tricoordination of Organomagnesium Compounds

Although the tetrahedral coordination geometry is preferred by organomagnesium compounds, many examples have been structurally characterized which show deviating coordination numbers. In most cases, this deviating behavior can be rationalized. Low coordination numbers can be realized by using sterically demanding groups. In these species, the central magnesium atoms are shielded; this prevents the coordination of a normal number of solvent molecules. In a recent crystal structure, even dicoordination of magnesium was realized in the solid state: in crystalline bis[tris(trimethylsilyl)methyl]magnesium (**12**) (*29*), a linear centrosymmetric structure was encountered; its magnesium center cannot form coordinative bonds with THF. A linear C—Mg—C angle (180°) is found, as expected for a purely *sp*-hybridized magnesium. Surprisingly, the Mg—C bonds are only slightly shortened (2.116 Å) as compared with the mean value for a type **5** complex (2.15 Å). This structure is the first solid-state example in which the magnesium atom has only Mg—C single bonds, and no further ligands. A tricoordinated magnesium is present in bis(trimethylsilyl)methylmagnesium chloride monodiethyl etherate (**13**)

12

13

(*30*). The size of the trimethylsilyl groups prevents the coordination of a second ether molecule to form a normal type **5** complex.

Another example of tricoordinate magnesium was found in the crystal structure of *sec*-butylmagnesium hexamethyldisilazane (**14**) (*31*). A selec-

14

tive, kinetically controlled deprotonation of hexamethyldisilazane by the mixed dialkylmagnesium reagent *n*-butyl-*sec*-butylmagnesium yielded **14** and *n*-butane. Since the reaction was performed in an apolar solvent (*n*-hexane), the product was necessarily solvent free. A dimeric structure was found with a highly symmetric four-membered Mg_2N_2 ring in its center. All Mg—N bond lengths are identical, and the anionic nitrogen bridges symmetrically between the two magnesium atoms (Mg—N 2.118 Å). The central ring is only slightly distorted from square (N—Mg—N 92.9°, Mg—N—Mg 87.1°). One terminal *sec*-butyl group is bonded to each magnesium atom with a rather short bond (Mg—C 2.08 Å). In **14**, the steric size of the trimethylsilyl- and *sec*-butyl groups prevents the formation of associates higher than the dimer as such aggregation would involve $Mg_2(s\text{-}Bu)_2$ rings.

2. *Penta- and Hexacoordination of Magnesium*

A remarkable exception from the tendency to form type **5** structures in normal Grignard reagents is the crystal structure of methylmagnesium bromide tris(tetrahydrofuranate) (**15**) (*32*). A possible explanation is the small size of the methyl group. In **15**, the central magnesium has an

15

approximately trigonal bipyramidal coordination. The bromine and methyl groups are disordered and occupy mutually exchangeable positions (*R* value 0.13). They occupy equatorial positions (C—Mg—Br 126°) together with a THF molecule (O—Mg—Br/Me 112° and 122°, Mg—O 2.06 Å). Both axial positions are occupied by THF molecules (Mg—O 2.04 and 2.24 Å).

Increased coordination numbers for magnesium are also found for an ethylmagnesium chloride/magnesium chloride adduct. A crystal structure determination revealed the formation of a tetrameric aggregate with the composition $[EtMgCl \cdot MgCl_2 \cdot (THF)_3]_2$ (**16**) (*33*). In this cluster, two magnesium atoms carrying one terminal ethyl group (Mg—C 2.19 Å) and one THF molecule (Mg—O 2.14 Å) have a distorted trigonal bipyramidal geometry. The other two magnesium atoms, which are only bonded to chloro and terminal THF ligands, are hexacoordinate (Mg—O 2.11 and

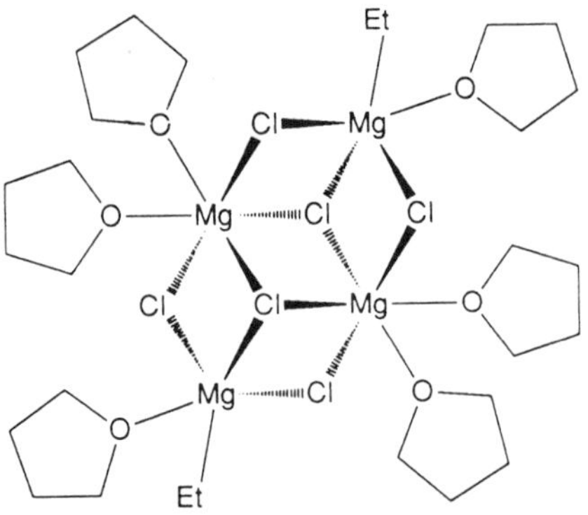

16

2.04 Å). The cluster is held together by doubly (μ_2, Mg—Cl 2.44 Å) or triply (μ_3, Mg—Cl 2.60 Å) bridging chloro ligands.

The complexes of bis(phenylethynyl)magnesium with THF and TMEDA were shown to be hexacoordinate (approximately octahedral). This coordination number is quite unusual for diorganylmagnesium compounds. It is probably facilitated here by the relatively high electronegativity and the low steric demand of the phenylethynyl group; as a consequence, the positive charge on magnesium will be higher and its steric hindrance less than normal, making magnesium an exceptionally strong Lewis acidic center. The crystal structure determination of (Ph—C≡C)$_2$Mg·(THF)$_4$ (**17**) has to be regarded as incomplete (*34*). Complex **17** was shown to possess two structural isomers, with the organyl groups occupying trans and cis positions relative to each other. In the crystal structure of (Ph—C≡C)$_2$Mg·(TMEDA)$_2$ (**18**) (*35*), the pheny-

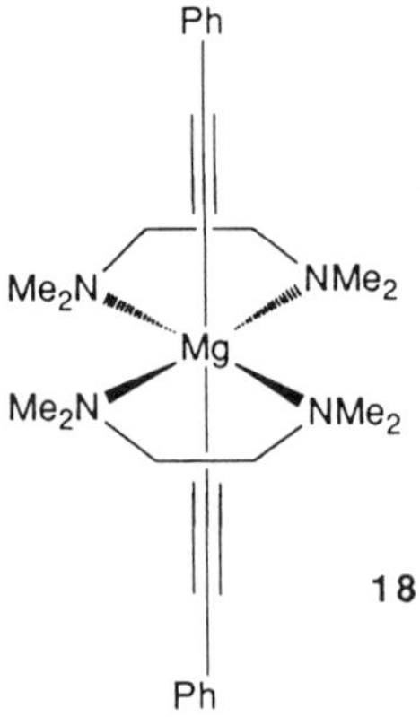

18

lethynyl groups have a trans configuration with a C—Mg—C angle of 180°. The complex has a pseudooctahedral coordination geometry, with small N—Mg—N angles inside the five-membered rings (N—Mg—N 80°). The Mg—C bond distances have normal values (Mg—C 2.176 and 2.200 Å), probably because the effects of the carbon sp hybridization (resulting in shortening of the bond) and of the higher coordination number cancel each other. The Mg—N coordinative bonds are, as expected, elongated (Mg—N 2.375 Å) as compared to those of tetrahedral complexes. Very recently, the crystal structure of (*tert*-butylethynyl)$_2$Mg·[TMEDA]$_2$ has been determined (**18a**) (*36*). The structure is analogous to that of **18** with only minor differences in its structural parameters.

C. Complexes of Organomagnesium Compounds with Polyethers

Fascinating structures can be obtained on complexation of organomagnesium compounds with polyether ligands. High coordination numbers and unusual coordination geometries may be expected, and the reactivity of the organomagnesium reagents will be affected. For this reason the reactivity of dialkylmagnesium reagents in the presence of crown ethers was studied. An unusually high reactivity of these species was found in the presence of crown ether ligands: some reactions showed accelerations of several orders of magnitude. At first sight, this may seem paradoxical: intuitively, one would expect that strong coordination will result in stabilized and therefore unreactive complexes. The explanation was found in the disproportionation of the dialkylmagnesium reagent into alkylmagnesium cations and magnesate anions.

This effect was dramatically demonstrated in two complexes characterized by Richey and co-workers (*37*). In the crystal structures of [EtMg(2,2,1-cryptand)]$_2$·Et$_6$Mg$_2$ (**19**) and [NpMg(2,1,1-cryptand)]·

19

20

Np$_3$Mg (**20**), the cryptand induces a separation of the dialkylmagnesium species into an alkylmagnesium cation and a magnesate anion pair. This disproportionation is facilitated by the efficient coordination of the alkylmagnesium cation by the cryptand: the magnesium atom of this species is completely surrounded by the heteroatoms of the cryptand. In **19**, the EtMg$^+$ fragment is hexacoordinated to five heteroatoms of the cryptand but is remote from the two oxygens of one of the longer cryptand bridges. In **20**, even a pentagonal bipyramidal heptacoordination occurs involving all cryptand heteroatoms. The magnesate anions have no interactions with heteroatoms because of the negative charge of the anion. They can be regarded as dialkylmagnesium/alkylanion adducts and are responsible for the high reactivity. The magnesiums in the centrosymmetric Et$_6$Mg$_2{}^{2-}$ dianion in **19** are essentially tetracoordinated and have relatively long Mg—C bonds (Mg—μ-C 2.359 and 2.364 Å, Mg—C$_{term}$ 2.238 and 2.209 Å). The four-membered ring Mg$_2$C$_2$ has a large C—Mg—C angle (106.8°) and a small Mg—C—Mg angle (73.2°), and therefore a relatively short Mg—Mg distance (2.816 Å). The Np$_3$Mg$^-$ anion in **20** is distorted trigonal planar: it has C—Mg—C angles of 116.9°, 117.6°, and 124.8°. Its bond lengths also show a large variation (2.125, 2.240, and 2.296 Å).

With **20**, an unusual ether cleavage reaction occurs on heating in toluene. The reactive trineopentylmagnesate anion selectively abstracts a proton of the cryptand complex. One of the short bridges between the two nitrogen atoms is cleaved into a pendant enamine (N—CH=CH$_2$) and an alcoholate (N—CH$_2$CH$_2$—O—MgNp) group. The reaction is quantitative and has a remarkable selectivity. The cleavage product has a centrosymmetric dimeric stucture with a central Mg$_2$O$_2$ ring (**21**) as shown (O—Mg—O 89.0°, Mg—O—Mg 99.2°) (*38*). Each magnesium is pentacoordinated: apart from bonds to the alcoholate function (1.979 and 2.010 Å) and the residual neopentyl group (2.181 Å), only two heteroatoms of the newly formed monocyclic crown ether ring are involved (Mg—O 2.142 Å, Mg—N 2.377 Å). The structure clearly illustrates the strength of a bridging magnesium alcoholate bond: the pendant crown ether rings are

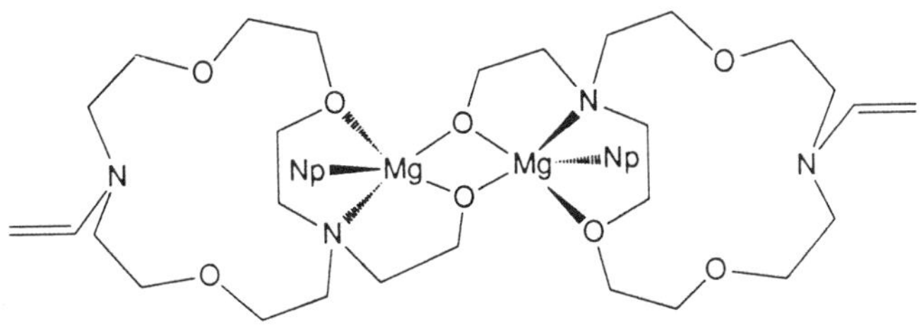

21

only weakly involved in the coordination and cannot prevent the dimerization to **21**.

Partial dissociation into an ion-separated complex is encountered in $[MeMg(15\text{-crown-}5) \cdot Me_5Mg_2]_n$ (**22**) (*39*), which is formed on complexation of dimethylmagnesium with 15-crown-5. In **22**, the magnesium atom

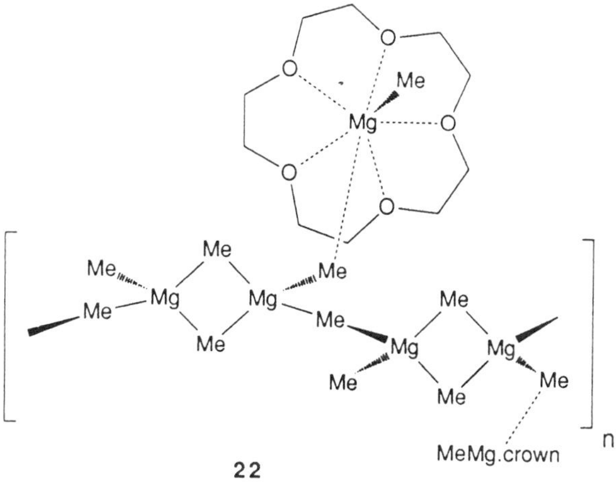

22

of the cation is equatorially surrounded by all crown ether oxygen atoms and lies 0.42 Å out of the mean plane. The apical positions are occupied by a single-bonded methyl group (Mg—C 2.140 Å) and a much weaker interaction with a terminal methyl group of the anionic fragment (Mg—C 3.28 Å) on the other side of the crown ether plane. The magnesium atoms in the anionic Me_5Mg_2 fragments are tetracoordinated, with a central four-membered Mg_2C_2 ring being formed by two bridging methyl groups (Mg—C 2.251–2.334 Å). Each magnesium atom has one terminal methyl group (Mg—C 2.170 Å) and shares a single μ_2-bridging methyl group with another anionic fragment (Mg—C 2.258 and 2.395 Å). Through these interactions, the anionic fragments are connected to form a continuous polymeric chain. The Mg—C bond lengths in **22** are slightly elongated as compared with polymeric dimethylmagnesium. This must be attributed to the negative charge on the $[Me_5Mg_2]_n$ chain.

Dissociation into ion pairs does not occur in the complex of diethylmagnesium with 18-crown-6. In $Et_2Mg \cdot (18\text{-crown-6})$, a linear diethylmagnesium fragment is equatorially surrounded by the crown either plane (**23**)

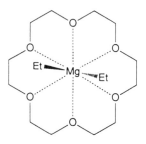

23

(*40*). The "threaded" geometry of **23** is hexagonal bipyramidal and very regular, since the magnesium atom is lying on an inversion center. The axial Mg—C bonds are perpendicular to the crown ether plane and are unexpectedly short (2.104 Å). The Mg—O coordinative bonds are very weak (2.767, 2.792, and 2.778 Å); **23** might even be regarded as a clathrate with a linear diethylmagnesium fragment encapsulated by the crown ether. The analogous diethylzinc complex was also characterized and proved to possess a comparable structure.

Complexes with a threaded structure were already known for two magnesium salts: $Mg(NCS)_2 \cdot (\text{benzo-15-crown-5})$ (**24**) (*41*) and $MgCl_2 \cdot (18\text{-crown-6})$ (**25**) (*42*). The structure of **24** has much in common with that of **23**, but has much shorter Mg—O bonds (2.171–2.205 Å); the relatively

24   **25**

small cavity of benzo-15-crown-5 is just large enough to enclose a magnesium, and the charge on magnesium is higher. The unsymmetrical structure of the crown ether in **24** causes a slight deviation of the N—Mg—N unit from linearity (177.9°). The structure of **25** has a remarkble feature: its crown ether cavity is contracted to facilitate the coordination of the (linear) Cl—Mg—Cl moiety by only five of the six oxygen atoms. Again, the Mg—O bond distances are much shorter (2.220–2.331 Å) than those of **23**, and the Mg—Cl bonds have normal lengths (2.429 Å). The strongly differing behavior of 18-crown-6 in **25** and **23** is also an indication of the higher charge on magnesium in $MgCl_2$ as compared to $MgEt_2$.

The crystal structure of $Ph_2Mg \cdot$ (1,3-xylyl-18-crown-5) (**26**) can also be

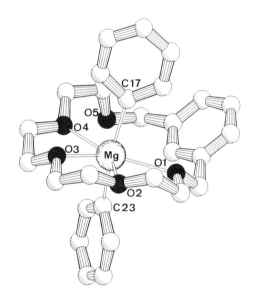

26 (from Ref. *43* with permission of the American Chemical Society)

regarded as threaded (*43*). Because the usual crown ether symmetry is broken by the presence of the 1,3-xylyl group, the coordination of magnesium in **26** is irregular. The magnesium atom has two short [2.204(3) and 2.222(4) Å] and two relatively long [2.516(4) and 2.520(4) Å] Mg—O interactions; one crown ether oxygen atom does not take part in the coordination. The diphenylmagnesium moiety deviates slightly from linear [163.8(2)°] and has rather normal Mg—C bond lengths [2.189(5) and 2.190(5) Å]. An interesting facet of **26** is its mode of formation: since the phenyl group of diphenylmagnesium is too large to penetrate the crown ether ring, **26** can be regarded as "rotaxane" complex in the sense that the inclusion of diphenylmagnesium is (also) of a "mechanical" nature. Dissociation into magnesate species, analogous to that in the cryptand complexes **19** and **20**, must occur to permit its formation (*43*).

Since the publication of **26**, we have determined the crystal structures of three additional diarylmagnesium–polyether complexes. Instead of diphenylmagnesium, its 4,4′-bis(*tert*-butyl) derivative was used in the complexation experiments since this species as a rule gives higher quality crystals. Complexation with 1,3-xylyl-18 crown-5 resulted in the formation of a rotaxane complex (**27**), too. Bond angles and distances, and even the

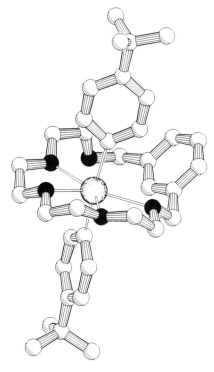

27

crown ether conformations of **26** and **27**, are closely similar. The structure of **27** is an additional argument for the above-mentioned mechanism of formation of a rotaxane complex: the sterically large *tert*-butylphenyl group absolutely precludes the formation of **27** by a direct threading mechanism.

With a much larger crown ether ligand, a completely different mode of complexation was found. In [(*p-t*-BuPh)$_2$Mg]$_2$·(1,3,16,18-dixylyl-30-

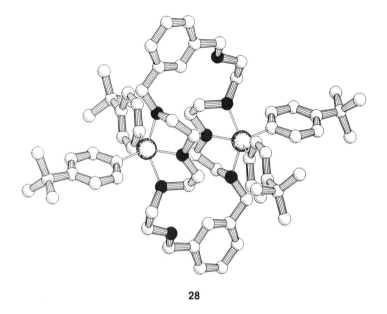

28

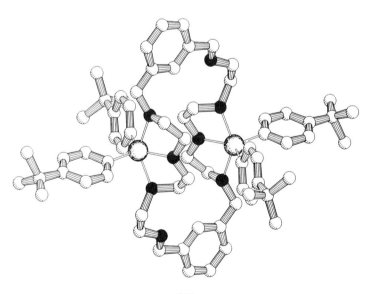

28'

crown-8) (**28**), the large crown ether ring coordinates two diarylmagnesium units at its periphery. This is facilitated by the relatively high flexibility of the 30-crown-8 ligand compared with that of 1,3-xylyl-18-crown-5. The two 1,3-xylyl units divide the macrocycle into two polyether regions, which are turned outside to facilitate the coordination of two diarylmagnesium molecules. Only three of the four oxygens of each region are involved in the coordination process. Compound **28** was crystallized from toluene and from benzene; both kinds of solvents yield similar structures (**28** and **28′**). Essentially the same coordination geometry is found in two complexes of $(p\text{-}t\text{-BuPh})_2\text{Mg}$ with glymes. Because of the high flexibility of these noncyclic polyether ligands, this geometry may be closer to a conformational optimum. In $(p\text{-}t\text{-BuPh})_2\text{Mg}\cdot$diglyme (**29**), a five-coordinated complex is

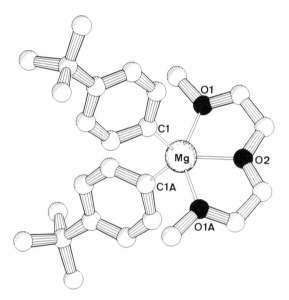

29

formed which has a twofold symmetry axis through the central Mg—O coordinative bond. The three oxygens and magnesium lie in one plane. The O—Mg—O angles inside the coordinative five-membered rings are relatively small [70.98(6)°]. The Mg—O bonds are slightly elongated [2.152(3) and 2.157(2) Å] relative to those in a type **5** complex. The C—Mg—C plane is arranged perpendicular to the MgO₃ plane; the C—Mg—C angle [112.6(1)°] and bond lengths [2.157(2) Å] are normal. In the crystal structure of $(p\text{-}t\text{-BuPh})_2\text{Mg}\cdot$tetraglyme (**30**), three chemically identical but conformationally different complexes (**30a**–**30c**) are present

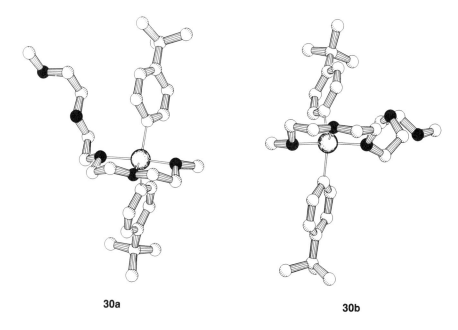

30a

30b

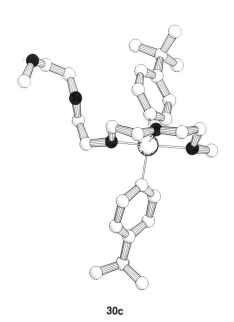

30c

in the unit cell. For each tetraglyme ligand, only three of its five oxygen atoms interact with a diarylmagnesium unit. The coordination geometry of the central magnesium atom is analogous to that in **29**. Two ether oxygen atoms remain "unused," suggesting that a coordinative saturation point has been reached.

D. *Intramolecular Coordination in Organomagnesium Compounds*

1. *Complexes with Simple Ligands*

In this section, organomagnesium compounds which carry electron-donating groups as part of the organic ligand are discussed. In the crystal structures, intramolecular coordination is usually found to be preferred over coordination by solvent molecules. In most cases, type **5** or **6** complexes are encountered with normal tetracoordinated magnesium atoms. Prominent examples are the ω,ω'-disubstituted dialkylmagnesium derivatives studied by Bogdanovic and co-workers. Intramolecular coordination occurs under formation of coordinative five- or six-membered rings. The crystal structure of the solvent-free $[MeO(CH_2)_4]_2Mg$ (**31**) shows a mono-

31

32: $R_1 = R_2 = Me$
33: $R_1 = Me$, $R_2 =$ cyclohexyl

meric type **5** complex (*44*). Its magnesium atom has a distorted tetrahedral coordination geometry. Analogous to the 1,7-dimagnesacyclododecane structure **9** in Section III,A,3, a large C—Mg—C angle (140.2°) and a small O—Mg—O angle (96.4°) were found. The relevant bond distances in **31** (Mg—C 2.144 Å, Mg—O 2.071 Å) are normal.

In the solvent-free complexes $[(N,N\text{-dimethyl-3-aminopropyl})MgEt]_2$ (**32**) and $[(N\text{-cyclohexyl-}N\text{-methyl-3-aminopropyl})MgEt]_2$ (**33**) (*45*), only one of the two alkyl substituents bears a coordinating group, the other being a normal ethyl group. This leads to the formation of dimers, since only one amino group per magnesium is available. The centrosymmetric complexes **32** and **33** have comparable structures, with a central four-membered Mg_2C_2 ring. The magnesiums are connected by amino-substituted carbon-bridging propyl groups, whereas the ethyl groups are terminally bonded to magnesium. Five-membered chelate rings are formed by the coordination of the amino groups to complete the coordination sphere of magnesium. The structures of **32** and **33** are the only known examples of type **6** complexes with bridging alkyl groups.

The centrosymmetric crystal structure of $[Me_2NC_2H_4N(Me)MgMe]_2$ (**34**) has some analogy with the cryptand cleavage product **21**, but has

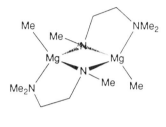

34

tetracoordinated magnesium atoms (*46*). It contains a central four-membered Mg_2N_2 ring in which both amido groups are involved (Mg—N 2.102 and 2.107 Å, N—Mg—N 91.7°, Mg—N—Mg 88.3°). Each magnesium atom is further coordinated by one amino group (under formation of a chelate ring, Mg—N 2.190 Å) and a terminally bonded methyl group (Mg—C 2.100 Å).

The solvent-free centrosymmetric complex $[(o\text{-pyridyl})(Me_3Si)_2C]_2Mg$ (**35**) was prepared from bis(trimethylsilyl)-2-pyridylmethane by metalla-

35

tion with *n*-Bu-*s*-BuMg in heptane (*47*). Its crystal structure shows a central tetracoordinated magnesium atom that has two Mg—C bonds (2.219 Å) and two Mg—N interactions (2.131 Å). The coordination geometry of **35** strongly deviates from ideal tetrahedral, since two four-membered chelate rings are involved [C—Mg—N (ring) 67.30°, C—Mg—N' 126.3°, C—Mg—C 157.0°, N—Mg—N 107.5°].

An unusual kind of hexacoordination is found in the crystal structure of 1-bromomagnesio-2-pivaloyl-tetrahydroisoquinoline·$(THF)_3$ (**36**) (*48*).

36

The metal atom in **36** is bonded to the organic group in a pseudoequatorial position (C—Mg 2.245 Å) and has an intramolecular interaction with the carbonyl oxygen in a planar five-membered chelate ring (Mg—O 2.049 Å, C—Mg—O 92.4°). The geometry is close to octahedral, with an enlarged C—Mg—Br angle (104.3°). Three THF molecules complete the coordination sphere (Mg—O 2.146, 2.136, and 2.236 Å). The unusually high coordination number might be explained by the high polarity of the Mg—C bond due to favorable charge stabilization in the organic group, analogous to the phenylacetylide complexes **17** and **18**. In addition, the chelate ring in **36** may help to induce a higher coordination geometry by its preference for a small C—Mg—O angle.

Finally, three unpublished structures from our group fall into this category. By investigating the crystal structure of $[(o\text{-anisyl})_2 Mg \cdot THF]_2$ (**37**), we intended to probe the possibility of a four-membered intramolecular chelate ring. A very unsymmetrical dimer was found on crystallization from THF/n-hexane. It contains one five-coordinated (Mg-1) and one tetrahedral (Mg-2) magnesium. The two magnesium atoms do not have the same number of bonds to carbon: Mg-1 has a full σ bond to one anisyl group [Mg—C 2.195(7) Å] but shares the other one [Mg—C 2.327(6) Å] in a three-center two-electron fashion with Mg-2 [Mg—C 2.302(6) Å]; Mg-2 is further σ bonded to its two anisyl groups. Thus, Mg-2 gains partial magnesate character, whereas, in turn, Mg-1 formally carries a positive charge, which is compensated by coordination to three oxygens. Incidentally, all methoxy groups are coplanar with their aryl rings and oriented such that the ether dipole has the best possible orientation toward a magnesium atom; only the methoxy group of the μ-bridging anisyl forms a chelate bond to Mg-1 [Mg-1—O-2 2.166(4) Å], but the resulting four-membered ring is not planar. One of the anisyl groups bridges in an η^2 fashion between the two magnesiums: it is σ bonded to Mg-2 by its carbon

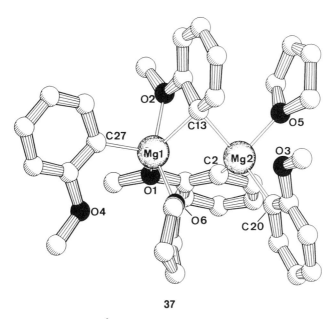

37

[Mg(2)—C(2) 2.195(7) Å] and to Mg-1 by its oxygen [Mg(1)—O(2) 2.052(5) Å].

In [2,6-bis(methoxymethyl)phenyl]$_2$Mg (**38**), the intramolecularly coordinating substituents force the complex into a distorted octahedral coordination mode. The complex is solvent free; its central magnesium atom is

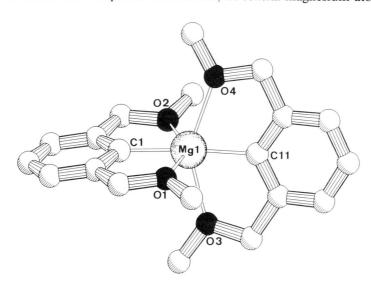

38

completely shielded by the methoxymethyl groups and therefore unaccessible for coordination by the solvent. The C—Mg—C moiety is almost linear [173.4(2)°] and has relatively short Mg—C bonds [2.093(4) and 2.105(4) Å]. The deviation from an ideal coordination geometry is mainly caused by the presence of four five-membered chelate rings with their small C—Mg—O angles (~73°). The Mg—O bonds are relatively long [2.282(3)–2.338(3) Å].

A combination of intra- and intermolecular coordination is present in the diphenylmagnesium derivative 1-oxa-6-magnesadibenzo(4,5,7,8)cyclodecadiene·(THF)$_2$ (**39**). This complex has a pseudo-trigonal bipyramidal (TBP) geometry, which is rare in organomagnesium chemistry (cf. **15**, **28**, and **29**). The two six-membered chelate rings in **39** must be responsible for

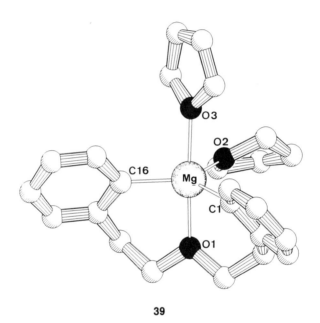

39

the induction of this geometry, as indicated by their tendency to form small intraannular C—Mg—O angles [88.7(2) and 88.2(2)°]. The conformational strain of chelate rings will be minimized in a TBP coordination geometry, which creates an extra position to bind a THF molecule.

2. Complexes Involving Polyether Side Chains and Crown Ethers

In the context of a broad investigation of coordination effects in organomagnesium compounds, we synthesized and studied a considerable number of aryl-Grignard reagents and diarylmagnesiums carrying a poly(ox-yethylene) substituent suitable for intramolecular complexation. Derivatives with open or with cyclic (crown ether) substituents have been obtained and several of them characterized by their crystal structures. Unusual coordination geometries were found, induced by the presence of several oxygen atoms in the close vicinity of the organomagnesium func-tion. The unusual structures often induce reactions that are atypical for normal organomagnesium compounds.

The crown ether-substituted aryl-Grignard 2-bromomagnesio-1,3-xylyl-15-crown-4 (**40**) crystallized as a solvent-free complex from its THF solu-

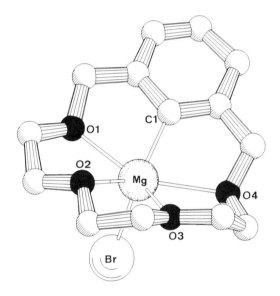

40 (from Ref. *49* with permission of the American Chemical Society)

tion (*49*). The two oxygen atoms farthest removed from the aromatic group show the strongest coordinative interaction [2.12(1) and 2.13(1) Å]. Since the crown ether ring has only a limited conformational freedom, the other oxygen atoms are also fixed in the vicinity of the Grignard function [2.33(1) and 2.49(1) Å] and must therefore weakly coordinate. Neverthe-less, the structure may in first approximation be considered to be distorted tetrahedral of type **5**; the C—Mg—Br angle [128.2(4)°] is slightly wi-dened. The Mg—C [2.10(1) Å] and Mg—Br [2.517(4) Å] bonds have normal lengths; there is no indication of an (incipient) ionic dissociation of the Mg—Br bond. A remarkably selective ether cleavage reaction occurred during the formation reaction of **40** from the corresponding aryl bromide and magnesium. This phenomenon could be related to the coordination in **40** and shed new light on the mechanism of the Grignard formation reaction (*49*).

The analogous ether cleavage reaction of the crown-5 analog of **40** rendered its synthesis from the aryl bromide and magnesium next to impossible: isolation of crystalline 2-bromomagnesio-1,3-xylyl-18-crown-5 (**41**), formed in 16% yield, from the crude reaction mixture was not feasible. Therefore, pure **41** had to be obtained by treatment of the aryl bromide with *n*-butyllithium, followed by the addition of 1 equiv magnesium bromide. The crystal structure of **41** contained two independent residues (**41a** and **41b**) and proved to be very irregular; the conformation of its crown

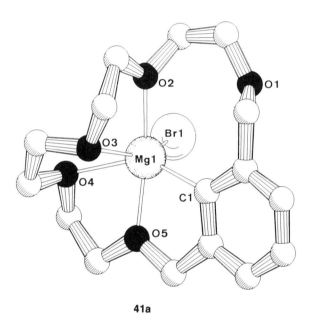

41a

ether ring appears to be strained. The central magnesium atom is coordinated by four of the five available oxygen atoms in a distorted octahedral fashion. The Mg—O bond distances (in **41a**) vary between 2.126(4) and 2.331(4) Å; the Mg—C [2.176(5) Å] and Mg—Br [2.597(2) Å] bonds in **41** are slightly elongated relative to those in **40**. The second residue (**41b**) has a very similar structure.

A remarkable halogen–metal exchange reaction was observed between diphenylmagnesium and 2-bromo-1,3-xylyl-15-crown-4 in diethyl ether, yielding the crystalline compound 2-phenylmagnesio-1,3-xylyl-15-crown-4 (**42**) (*50*). Reaction of diphenylmagnesium with 1,3-xylyl-15-crown-4 gave

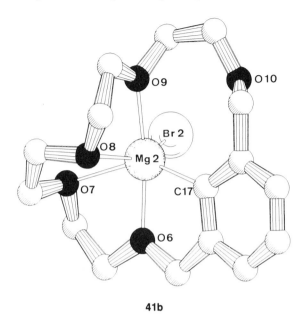

41b

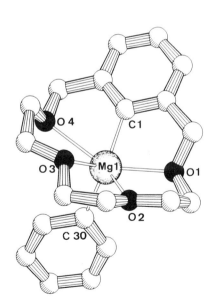

42 (from Ref. *50* with permission of VCH Verlagsgesellschaft, Weinheim)

the same product in a direct metallation of the intraannular aryl-H position. This type of reactivity is common in organolithium chemistry but quite unusual for organomagnesium compounds; it must be ascribed to the special coordination by the crown ether. Compound **42** is also remarkable as being a stable diarylmagnesium in which two different organic groups are attached to magnesium; usually, such unsymmetrical diorganylmagnesiums are in rapid equilibrium with the two corresponding symmetrical ones. The crystal structure of **42** is, *mutatis mutandis*, practically identical to that of **40**. By the halogen–metal exchange reaction between bis(*p-tert*-butylphenyl)magnesium and 2-bromo-1,3-xylyl-18-crown-5, the unsymmetrical diarylmagnesium 2-[(*p-t*-BuPh)Mg]-1,3-xylyl-18-crown-5 (**43**)

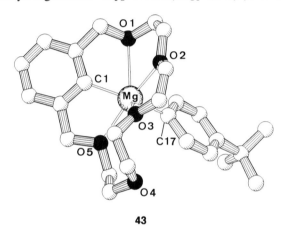

43

was obtained. The crystal structure of **43** differs remarkably from that of Grignard **41** and is rather irregular. The coordination geometry can be regarded as distorted trigonal bipyramidal, with the Mg—C bonds [2.127(4) and 2.155(4) Å] and a short Mg—O bond [2.146(4) Å] equatorial and two weakly bonded ether oxygen atoms [Mg—O 2.317(3) and 2.337(4) Å] axial. The remaining oxygen atom must be regarded nonbonding, both because of its distance [Mg—O(1) 2.713(4) Å] and because of the unfavorable orientation of the ether bisector toward magnesium; O(4) is completely outside of the coordination sphere of magnesium.

   To investigate the induction of higher coordination states by an intramolecular polyether substituent, a series of *ortho*-[CH$_2$(OCH$_2$CH$_2$)$_n$OMe]-substituted phenylmagnesium bromide derivatives ($n = 0$–3, **44**–**47**) were synthesized and characterized by their crystal structures. Already in the smallest member of this series, [1-bromomagnesio-2-methoxymethylbenzene·THF]$_2$ (**44**), differences with normal Grignard reagents become apparent. From the highly concentrated THF solution, **44** crystallized as a

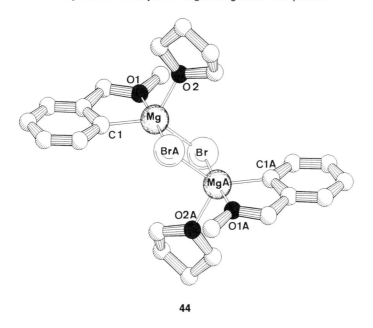

44

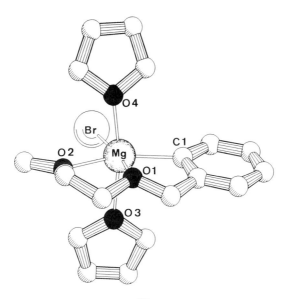

45

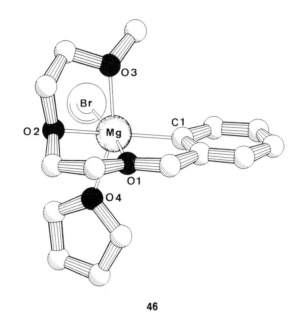

46

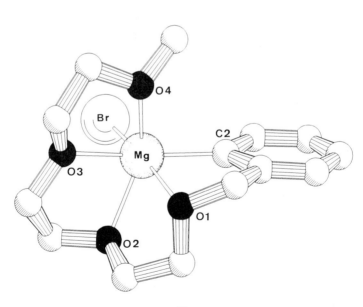

47

dimer. It differs from the simple dimeric Grignard **6b** ([EtMgBr · *i*-Pr$_2$O]$_2$, involving tetracoordinated magnesium) by having pentacoordinated magnesium in a distorted trigonal bipyramid; this comes from additional coordination to the methoxymethyl oxygens. The higher ether complexes from **45** on show the uncommon phenomenon of hexacoordination, with a distorted octahedral geometry (cf. Section III,B,2). Complexation of a poly(oxyethylene) chain to magnesium involves the formation of five-membered chelate rings which for reasons of geometrical constraint need to have small O—Mg—O angles (70–75°). Such angles are difficult to accommodate in a tetrahedral or trigonal bipyramidal coordination mode, where at least some of the preferred angles are larger (100–120°). In **47**, the molecule has four ether oxygens available, which is sufficient to complete the octahedral coordination. The tendency to attain this coordination mode is apparently so strong that it induces a regular pattern in the series **47, 46, 45**: as fewer intramolecular oxygens become available, hexacoordination is achieved by incorporating one or two THF molecules.

E. *Organomagnesium Compounds with Delocalized Organic Anions*

In this section, cyclopentadienyl, anthracenyl, and alkenyl derivatives are presented. Especially in the last two classes, considerable progress has been achieved recently. Structures containing η^1–σ bonds are included as counterparts of potentially feasible η^5–π complexes.

1. *Cyclopentadienylmagnesium Derivatives*

The structure of dicyclopentadienylmagnesium in the gas phase (structure **1**) has already been discussed in Section II; its crystal structure was also determined (**48**) (*51*). In contrast to the (probably) eclipsed orientation in **1**, the parallel rings of **48** have a staggered conformation in the crystalline state. The mean value for the Mg—C distances in the solid is shorter (2.304 Å, Mg—Cp centroid 1.98 Å). The crystal structure of [η^5-1,2,4-(Me$_3$Si)$_3$C$_5$H$_2$]$_2$Mg (**49**) (*52*) has much in common with that of **48**. Two molecules are present in the unit cell which differ in minor details only. For steric reasons, the bulky trimethylsilyl groups avoid interactions as much as possible. The cyclopentadienyl rings are not completely parallel (Cp—Mg—Cp 171°) because of repulsion between two trimethylsilyl groups which cannot fully avoid a gauche repulsion. Compared to **48**, the distance between magnesium and the cyclopentadienyl centroids is slightly greater (2.03 Å).

The crystal structures of Grignard reagents corresponding to **48** and **49**, CpMgBr · Et$_2$NCH$_2$CH$_2$NEt$_2$ (**50**) (*53*), and [η^5-1,2,4-

50

(Me$_3$Si)$_3$Cp]MgBr·TMEDA (**51**) (*52*) were also determined and found to be very similar. They show normal tetrahedral type **5** complexes, with the pentahapto cyclopentadienyl ligands instead of alkyl or aryl groups bound to magnesium. As expected for this kind of complex, large Cp—Mg—Br (125°) and small N—Mg—N angles (83°) are found. The relatively long Mg—Br bond (2.63 Å) in **50** might indicate incipient ionic dissociation. Remarkably, the corresponding bond is significantly shorter in **51** (2.52 Å). The coordination of the diamine ligands enlarges the Mg—Cp centroid distances (**50**: 2.21 Å; **51**: 2.167 Å) relative to that of **48** and **49**.

A much more complex structure was found for diindenylmagnesium (**52**) (*54*), from which single crystals could be obtained by slow high

52a   **52b**

vacuum sublimation. Association occurs via intermolecular contacts: each magnesium atom is coordinated to three indenyl moieties, one of which being bonded in a pentahapto fashion. The structure is polymeric and has both bridging and terminal indenyl groups. Two different types of magnesium atoms can be discerned; one type has contacts to the three associated indenyl groups in an $\eta^5 \sigma\sigma$ (or an $\eta^5\eta^1\eta^1$) fashion (52a), the other one in a $\eta^5\eta^2\eta^2$ fashion (52b). The magnesium atoms are bonded to the indenyl groups at distances varying over a broad range (η^5: 2.26–2.60 Å; η^2: 2.33–2.44 Å; σ: 2.26 and 2.32 Å) but with a clear tendency to form weaker bonds to the sterically less favorable C-4 and C-9 positions.

A crystal structure unique in organomagnesium chemistry was found for [CpMgOEt]$_4$ (53) (55): a tetrameric cluster (cf. 11) with a cubane skeleton,

53

the corners of which are occupied by magnesium and oxygen. The formation of this cluster must be facilitated by the μ^3-bridging capacity of the ethoxide groups. The central skeleton is quite regular, with the Mg—O distances varying from 2.059 to 2.075 Å. A terminal cyclopentadienyl group is bonded to each magnesium atom (Mg—Cp centroid 2.103 and 2.106 Å).

Recently some papers have appeared which deal with the related structures of the heavier alkali earth metal organometallic compounds. Cyclopentadienyl derivatives of calcium, strontium, and barium proved to possess sufficient stability to permit their synthesis and subsequent crystallization. These results are briefly included in this review (see Table III).

2. Magnesium Anthracene Derivatives

Reaction of magnesium metal with anthracene or its derivatives in THF yields magnesium anthracene reagents, which have versatile applications. In synthetic reactions, these compounds have an ambivalent character: they can react both as a soluble form of magnesium or as bifunctional

TABLE III

SELECTED DATA FOR HEAVY ALKALI EARTH METAL CYCLOPENTADIENYL DERIVATIVES

Compound	Cp—M—Cp (°)	Characteristic bonds (Å)	Ref.
Ba[Me$_5$C$_5$]$_2$	131	Ba—C, 2.987(2) av.	a
Ca[η^5-1,3-(Me$_3$Si)$_2$C$_5$H$_3$]$_2$·THF	154(3)	Ca—C, 2.66(1)–2.718(9)	b,c
		Ca—O, 2.310(9)	
Sr[η^5-1,3-(Me$_3$Si)$_2$C$_5$H$_3$]$_2$·THF	149(3)	Sr—C, 2.78(4)–2.84(5)	b
		Sr—O, 2.49 (3)	
[Me$_5$C$_5$CaI·(THF)$_2$]$_2$	—	Ca—I, 3.128(2)–3.274(2)	d
		Ca—C, 2.624(7)–2.722(6)	
		Ca—O, 2.373(5)–2.441(4)	
Ca[Me$_5$C$_5$]$_2$	154(3)	Ca—C, 2.609(6) av.	e

a R. A. Williams, T. P. Hanusa, and J. C. Huffman, *J. Chem. Soc., Chem. Commun.*, 1045 (1988).

b L. M. Engelhardt, P. C. Junk, C. L. Raston, and A. H. White, *J. Chem. Soc., Chem. Commun.*, 1500 (1988).

c P. Jutzi, W. Leffers, G. Müller, and G. Huber, *Chem. Ber.* **122**, 879 (1989).

d M. J. McCormick, C. S. Sockwell, C. E. H. Davies, T. P. Hanusa, and J. C. Huffman, *Organometallics* **8**, 2044 (1989).

e Gas-phase electron diffraction, R. A. Andersen, R. Blom, J. M. Boncella, C. J. Burns, and H. J. Volden, *Acta Chem. Scand. A* **41**, 24 (1987).

organomagnesium compounds derived from a 9,10-dihydroanthracene 9,10-dianion. An application based on the first property is the synthesis of Grignard reagents which are otherwise difficult to obtain from the corresponding bromides; in this regard they resemble the well-known radical anion reagents like lithium naphthalene. With other substrates, magnesium anthracenes behave as reagents containing two (benzylic) magnesium–carbon σ bonds involving carbon atoms 9 and 10 and yield dihydroanthracene products derivatized at these positions.

The crystallization of magnesium anthracene itself proved to be difficult, owing to its low solubility even in THF. Therefore, derivatives with substituents at the anthracene moiety were the first to be structurally characterized. Crystal structures were obtained for [9,10-bis(trimethylsilyl)anthracene]magnesium·(THF)$_2$ **(54)** *(56)* and a 1 : 1 adduct of [9,10-bis(trimethylsilyl)anthracene]magnesium·(THF)$_2$ and [9,10-bis(trimethylsilyl)anthracene]magnesium·TMEDA **(55)** *(57)*. Three comparable structures were found, differing only in details. The pseudotetrahedral magnesium atom is bridging the 9- and 10-anthracenyl positions, and it coordinates to two solvent heteroatoms. Typical bond lengths and angles lie in a narrow range (Mg—C 2.21–2.23 Å, C—Mg—C 78–79°). The Mg—O (2.01 Å) and Mg—N (2.12 and 2.14 Å) coordinative bonds are normal.

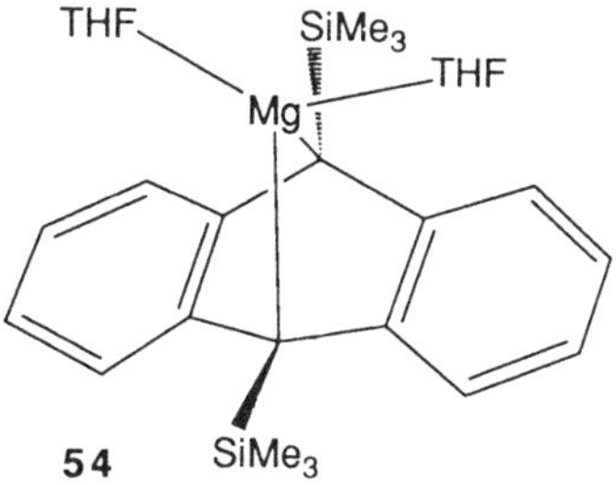

54

In (1,4-dimethylanthracene)Mg·(THF)$_3$ (**56**) (*58*), the coordination number of the bridging magnesium atom has increased to five; three THF molecules are coordinated. At first sight, the absence of steric hindrance by the large Me$_3$Si groups on the carbon atoms bonded to magnesium seems to be responsible. However, the Mg—C bond in **56** is enlarged (2.32 Å) compared with those in **54** and **55**. Together with a difference in color between **56** (orange) and **54** or **55** (yellow), this indicates a different character of the magnesium–hydrocarbon interaction. The absence of a Me$_3$Si group probably facilitates a better delocalization of negative charge in the anthracene unit; as a result, the structure of the anthracene dianion is approached, and thus the Mg—C bonds are weakened. Thereby magnesium becomes more electrophilic and can bind one extra THF molecule.

During the investigation of magnesium anthracene, a remarkable complex of a radical anion was isolated. In [(THF)$_3$Mg(μ-Cl)$_3$-Mg(THF)$_3$]$^+$·[anthracenyl]$^{\bar{}}$ (**57**) (*55*), an adduct of MgCl$^+$ with magne-

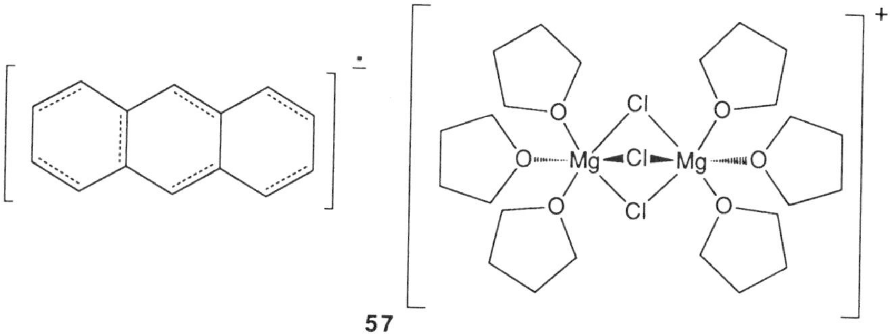

57

sium chloride is separated from the anthracenyl anion by six coordinating THF molecules. The μ-trichlorodimagnesium·[THF]$_6$ cation had already been described before as counterion of the [TiCl$_5$]·[THF] anion (*104*). Complex **57** allowed the first characterization of the anthracene radical anion by means of a crystal structure. The carbon skeleton of the anthracene anion clearly shows the presence of an electron in the LUMO of anthracene by changes of C—C bond lengths; this also follows from an electron density deformation map.

Recently, the crystal structure of unsubstituted magnesium anthracene was obtained. The structure of magnesium(anthracene)·$(THF)_3$ **(58)** is analogous to that of **56**, including the relatively long Mg—C bond (2.30 Å) (*59*). The short Mg—O (2.03–2.09 Å) bond length in **58** again signals a strongly electrophilic character of the magnesium atom.

Magnesium anthracene derivatives undergo an insertion reaction with ethylene; bicyclic products are formed in which a $MgCH_2CH_2$ chain connects the 9,10 positions of dihydroanthracene. The reaction is possibly facilitated by ring strain present in the magnesium bridge of the educt as no indications were found for the insertion of a second ethylene molecule. Two compounds with analogous structures, [9,10-(1′-magnesapropano)-9,10-dihydroanthracene]·$(THF)_2$ **(59)** and [10-methyl-9,10-(1′-magnesa-

Me   **59**

propano)-9,10-dihydroanthracene]·$(THF)_2$ **(60)** were obtained and structurally characterized (*60*). In **59** and **60**, the magnesium atom has a pseudotetrahedral coordination geometry of type **5**. In addition to the Mg–anthracenyl and Mg–alkyl bonds (average 2.113 Å), two THF molecules are coordinating.

Bimetallic adducts are formed on reaction of magnesium anthracene with organoaluminum compounds (AlR_2X, R = alkyl, R′ = alcoholate, hydride) which insert into one of the two Mg—C bonds. A bimetallic Mg—μ-R′—Al bridge is formed between the 9 and 10 positions of the anthracene moiety. In 9-(Et_2Al)-10-[Mg(μ-H)$(THF)_3$]-9,10-dihydroanthracene **(61)** (*61*), obtained from the reaction of magnesium anthracene with diethylaluminum hydride, the aluminum atom has a pseudotetrahedral coordination. The magnesium atom is solvated by three THF molecules (mean Mg—O distance 2.07 Å) and has a pseudo-trigonal bipyramidal coordination. Compound **61** can be regarded as a magnesium cation complex (RMg^+) and the anthracene unit as cis-substituted by a

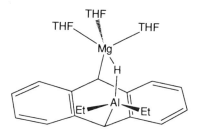

6 1

diethylhydridoaluminate anion and a positively charged Mg·[THF]$_3$ group. The hydrido atom transfers some negative charge to magnesium by weak bridging between both metals (Al—H 1.62 Å, Mg—H 1.96 Å, Al—H—Mg 139°). For 9-(Et$_2$Al)-10-[Mg(μ-OEt)(THF)$_2$]-9,10-dihydroanthracene (**62**) (*62*), a crystal structure analogous to that of **61** was found, with an OEt group in the bridging position between the two metal atoms. Owing to disorder, structural details could not be determined.

3. *Miscellaneous Organomagnesium Compounds with Unsaturated (Arene, Alkene) Ligands*

A few organomagnesium compounds with allyl, alkenyl, or polynuclear aromatic organic groups were characterized by means of their crystal structure. Magnesium here seems to have a strong tendency for η^1 coordination: no clear-cut example of interaction between the magnesium atom and the π system of the organyl group was found in any of the structures. This behavior is in clear contrast to that of the corresponding organolithium compounds, in which the lithium atoms as a rule are bonded to several carbon atoms. For magnesium, this type of bonding seems to be limited to compounds discussed previously (Sections III,E,1 and III,E,2) which involve (derivatives of) the cyclopentadienyl anion or anthracene.

A centrosymmetric dimeric structure was found for (η^1-allyl)MgCl·TMEDA (**63**) (*63*). The magnesiums are connected by highly unsymmetrically bridging chlorines (Mg—Cl 2.400 and 2.694 Å); therefore, **63** might be regarded as a weak associate of two monomers. Both magnesium atoms carry an allyl group and a bidentate TMEDA molecule as terminal ligands and are thus pentacoordinated. The Mg—C$_\alpha$ bond length is normal (2.179 Å); no interaction exists with the other carbon atoms. The coordination geometry around the magnesium atoms is irregular, and one of the Mg—N bonds is relatively weak (Mg—N 2.211 and 2.285 Å). Besides the allylic Grignard **63**, the allylic diorganylmagnesium

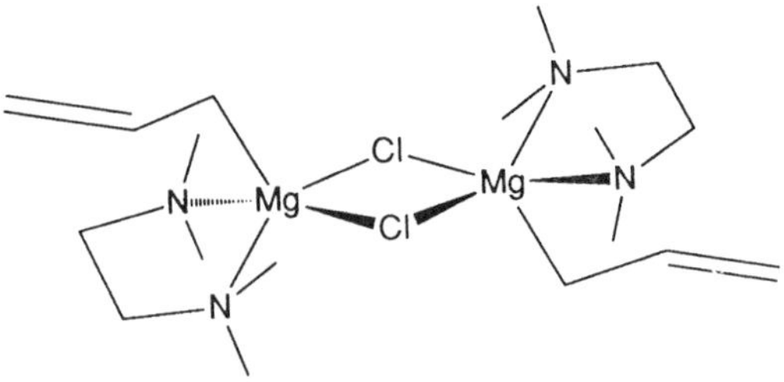

63

compound (2,4-dimethyl-2,4-pentadienyl)$_2$Mg·TMEDA (**64**) was characterized by its crystal structure (*64*). A normal type **5** complex was found,
with a tetracoordinated magnesium atom and normal bond lengths around
magnesium (Mg—C 2.179 Å, Mg—N 2.202 Å). Again, the organyl groups
are bonded to magnesium in a η^1 fashion, with no interaction with the
other carbon atoms being observed.

The magnesium-1,3-diene complex (s-*cis*-PhCH=CHCH=CHPh)Mg·
(THF)$_3$ (**65**) was synthesized from 1,4-diphenylbutadiene and activated

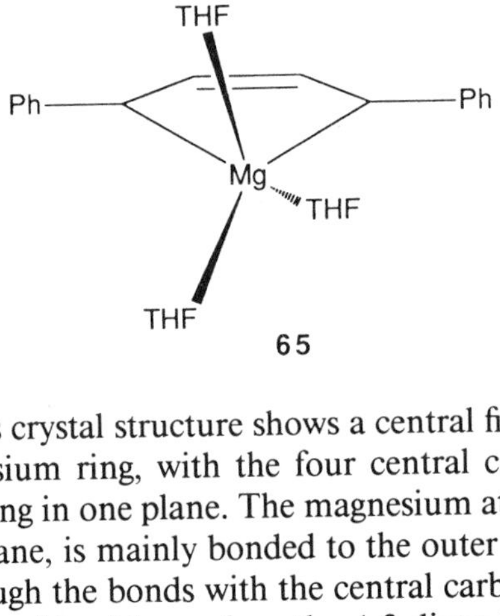

65

magnesium (*65*). Its crystal structure shows a central five-membered 2-bu-
tene-1,4-diylmagnesium ring, with the four central carbon atoms of the
butadiene ligand lying in one plane. The magnesium atom, at a distance of
1.71 Å from this plane, is mainly bonded to the outer carbon atoms (2.26
and 2.32 Å). Although the bonds with the central carbon atoms (2.52 and
2.56 Å), are still considerably weaker, the 1,3-diene ligand might be regarded as tending slightly toward η^4 coordination. An approximate
pseudo-trigonal bipyramidal coordination geometry is reached by three
THF molecules coordinating to magnesium (Mg—O 2.18, 2.06, and 2.12
Å). Again, this may reflect the high electrophilicity of magnesium owing to
weak covalent bonding to the delocalized organic dianion.

From the reaction of triphenylmethyl bromide with magnesium in
diethyl ether/benzene and subsequent crystallization, crystals of

66

Ph_3C—$MgBr \cdot (Et_2O)_2$ (**66**) were obtained (*59*). A normal type **5** structure is found with a pseudotetrahedrally coordinated magnesium atom, bonded only to the benzylic carbon of the carbanion (Mg—C 2.25 Å). An indication for some delocalization of negative charge in the organyl group may be found in the elongated Mg—C bond and in the large Ph—C_α—Ph angles (113–117°); on the other hand, **66** is colorless (in contrast to Ph_3CLi) and thus clearly does not possess a free triphenylmethyl anion. The Mg—Br bond length is normal (2.465 Å), and two diethyl ether molecules (Mg—O 2.02 and 2.04 Å) serve to complete the coordination sphere of the metal atom.

F. *Mixed-Metal Organomagnesium Complexes*

1. *Magnesate Complexes with Main Group Metals*

Organomagnesate complexes are obtained by addition of organolithium or -sodium reagents to diorganylmagnesium compounds. Structural characterization became possible after complexation with a strongly coordinating solvent; in particular, the chelating ligands TMEDA and PMDTA $[(Me_2NCH_2CH_2)_2NMe]$ proved to be a success. In this research area, much work has been performed by E. Weiss and co-workers. The structures of the organometallate complexes show clusters of metal atoms which are connected by bridging organyl ligands. Since the doubly positively charged magnesium atoms have the strongest complexing interactions with the organyl groups, these structures can be understood as associates of negatively charged magnesate units and positive alkali metal ions solvated by chelating Lewis bases. Bridging by the organyl groups connects the alkali metal cations to the clusters. So far, only one structure was found in which an alkali metal cation is completely separated from the magnesate counterion by solvation. In magnesate complexes with alkyl or aryl ligands, both lithium and magnesium prefer a normal pseudotetrahedral coordination geometry.

67

In the crystal structure of [TMEDA·Li]$_2$(μ-Me)$_4$Mg (67) (66), two TMEDA-coordinated lithium cations are connected with a central tetrahedral tetramethylmagnesate unit (C—Mg—C 108.4–109.9°, Mg—C 2.23–2.29 Å) by bridging of the methyl groups (Li—C 2.26–2.30 Å). Both the Mg—C and Li—C distances have normal values for three-center two-electron bonds. The Mg—C—Li angles are small (69.8–70.6°), which is usual for organometallic compounds with bridging alkyl or aryl groups. As a result, lithium and magnesium make a very short contact of 2.615 Å (note that the sum of the covalent radii of Mg and Li is about 2.7 Å!).

The crystal structure of [TMEDA·Li]$_2$(μ-Ph)$_2$Mg(μ-Ph)$_2$Mg(μ-Ph)$_2$ (68)

68

contains equimolar amounts of phenyllithium, diphenylmagnesium, and TMEDA (67). The center of the structure is formed by a dimeric triphenylmagnesate unit, isostructural to (Ph$_3$Al)$_2$. The structure of 68 is related to that of 67, since the terminal TMEDA-coordinated lithium cations are attached in the same fashion. The Ph$_6$Mg$_2$ magnesate center is different, however, and might be regarded as containing a section of the polymeric [Ph$_2$Mg]$_n$ chain of 7d. The bridging of the phenyl groups in the central four-membered Mg$_2$C$_2$ ring is unsymmetrical (Mg—C 2.29 and 2.33 Å), analogous to that in the dimeric structure of bis(p-tolyl)magnesium 6d.

The alkali metal sodium, having a larger ionic radius than lithium, prefers a pentacoordinated state. This can be accomplished by using the tridentate ligand PMDTA instead of TMEDA in the crystallization experiments. Thus, the adduct [PMDTA·Na]$_2$Mg(μ-Ph)$_4$ (69) was isolated (36),

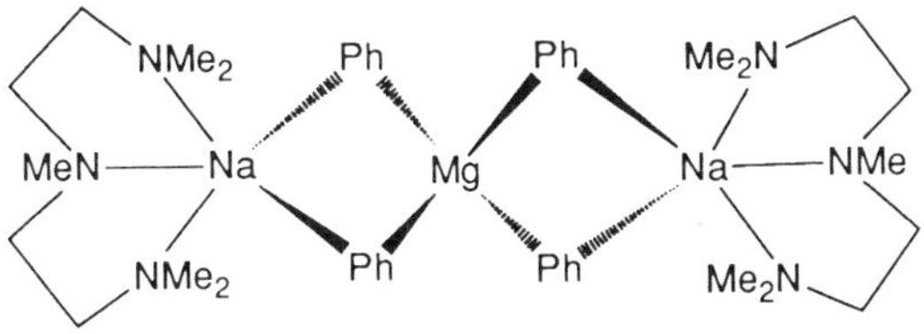

69

which contains phenylsodium and diphenylmagnesium in a 2 : 1 stoichiometry. Analogous to **67**, the central magnesium atom has close to perfect tetrahedral coordination geometry (C—Mg—C 107–115°, Mg—C 2.27–2.31 Å). A remarkable aspect of **69** is the unsymmetrical bridging of the phenyl groups: the Mg—C σ bonds lie almost in the planes of their phenyl groups, whereas the Na—C interactions are perpendicular to these planes. It must be concluded that the sodium ions are mainly bonded to the π electrons at the ipso carbon atoms (C—Na 2.73–2.79 Å).

The crystal structure of [Li·(TMEDA)$_2$]·[(benzyl)$_4$Mg·Li(TMEDA)] (**70**) (*68*) differs from that of **67**, although it contains organolithium and

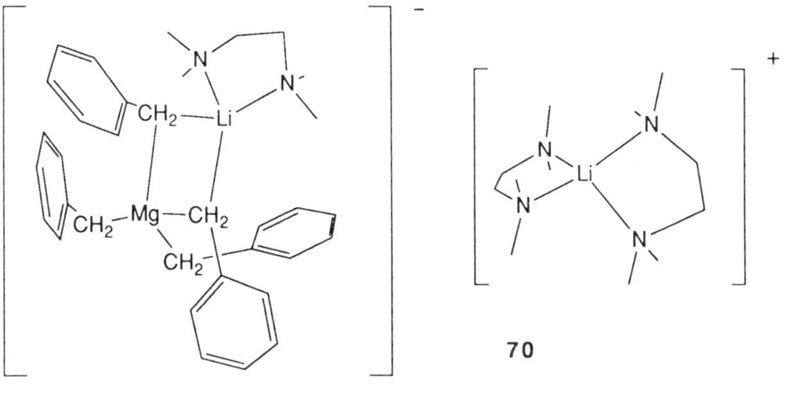

70

diorganylmagnesium reagents in the same stoichiometry. Only one lithium cation is connected to the tetrabenzylmagnesate counterion in a normal fashion; the other one is completely separated by solvation with two TMEDA ligands. As a result, the magnesium atom carries two bridging (Mg—C 2.32 and 2.31 Å) and two terminal (Mg—C 2.22 and 2.26 Å) benzyl groups in a distorted tetrahedral coordination geometry (C—Mg—C 105–114°). The benzyl groups are bonding to the metal atoms only with their C$_\alpha$ atoms, which according to the C$_{Ph}$—C$_\alpha$—Mg angles (103–111°) are sp^3 hybridized.

In the magnesate complexes **67–70**, all magnesium atoms are tetracoordinated by their alkyl or aryl ligands. Alkynyl ligands proved to facilitate higher coordination geometries, analogous to the bis(alkynyl)magnesium complexes (**17–18a**). This is probably related to the lower steric requirements and the higher electronegativity of these groups. In [(PhC≡C)$_3$Mg·Li·(TMEDA)]$_2$ (**71**) (*68*), the magnesium atoms have a

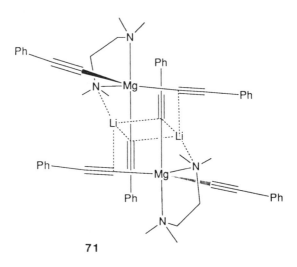

71

pseudo-trigonal bipyramidal coordination, with two phenylethynyl ligands in equatorial positions (Mg—C 2.18 Å) and one in an axial position (Mg—C 2.31 Å). In contrast with all previous magnesate complexes, the TMEDA is coordinating to magnesium and not to lithium, occupying the remaining two positions (Mg—N$_{ax}$ 2.35 Å, Mg—N$_{eq}$ 2.20 Å). This indicates the presence of a net positive charge on the (magnesate!) magnesium atoms. Two lithium atoms connect the R$_3$L$_2$Mg centers via contacts with four ethynyl C$_\alpha$ atoms. A three-dimensional electrostatic array is formed of alternating positive (metal) and negative (ethynyl C$_\alpha$) charges. Coplanarity of the three four-membered M$_2$C$_2$ rings in this structure is prevented by the spatial requirements of the ethynyl groups: the Mg—C≡C—Ph units are almost linear, whereas the lithium atoms interact mainly with the π electrons of the ethynyl groups.

Many characteristics of **71** are also found in the almost identical crystal structures of [TMEDA·(*tert*-butylethynyl)$_3$Mg·Na]$_2$ (**72**) (*36*) and [PMDTA·(*tert*-butylethynyl)$_3$Mg·Na]$_2$ (**73**) (*36*). The magnesium atoms are pseudo-trigonal bipyramidally coordinated, but in a different fashion. Two ethynyl groups occupy the axial, a third one an equatorial position.

72: L_2 = TMEDA
73: L_2 = PMDTA

The remaining coordination sites are occupied by the TMEDA or PMDTA ligands; only two nitrogen atoms of the PMDTA ligand in **73** are involved in the chelation process. As in **71**, the two magnesate units are connected by two bridging alkali metal ions via contacts with the π electrons of the ethynyl groups; each sodium atom has four contacts with ethynyl C_α atoms. Because of the different arrangement of ligands around the magnesium centra, a new three-dimensional structure results in which five four-membered M_2C_2 rings can be recognized.

The crystal structure of $[(Me_3Si)_3C](THF)Mg(\mu\text{-}Br)_2Li(THF)_2$ (**74**)

74

might be regarded as an adduct of the Grignard reagent $(Me_3Si)_3CMgBr$ with lithium bromide (*69*); the bridging bromine atoms of the central four-membered $LiBr_2Mg$ ring serve to connect the components. Until now, **74** represents the only example of an alkylhalomagnesate complex characterized by X-ray diffraction. The large size of the tris(trimethylsilyl)methyl group does not result in a lower coordination number for magnesium. Both metal atoms in **74** are tetracoordinated, and additional THF molecules serve to complete their coordination spheres. The metal–bromine bonds are relatively short (Mg—Br 2.551 and 2.515 Å, Li—Br 2.46 and 2.49 Å); the other bond distances in **74** have rather normal values. As is usual for halogen bridges between two positive metal atoms, the angles at the metal atoms (Mg 98°, Li 101°) are larger than those at bromine (80°).

A complex which has some analogy to **67** is formed on combination of trimethylaluminum with dimethylmagnesium in a 2:1 stoichiometry. In the crystal structure of $[Me_4Al]_2Mg$ (**75**) *(70)*, the almost linearly arranged

75

metal atoms (178.5°) are connected by bridging methyl groups, under formation of two almost perpendicular four-membered M_2C_2 rings. As expected (see **7a** and **7b**), the metal–carbon–metal bridge angles are small (77.7°); the metal atoms approach each other to 2.70 Å. In accordance with the differences in electronegativity between magnesium and aluminum, **75** might be regarded as a complex with a central positively charged Mg^{2+} atom interacting with two tetramethylaluminate anions.

2. *Compounds Containing Magnesium–Transition Metal Bonds*

Some organometallic compounds containing both magnesium and transition metal atoms have been synthesized and structurally characterized. In their crystal structures, the different metal atoms are connected both by direct Mg—M′ bonds and via bridging organyl groups. Formation of the first type of bonds is facilitated by large differences in electronegativities between magnesium and most (late) transition metals. For the sake of completeness, some compounds without any Mg—C bond are included in this section.

From the reaction of cuprous bromide and diphenylmagnesium, a complex with the composition $Cu_4Mg(\mu\text{-Ph})_6$ (**76**) was obtained *(71)*. Structural characterization of **76** revealed the presence of five metal atoms arranged in a trigonal bipyramidal fashion, with the magnesium atom at one of the axial positions. There are six phenyl groups bridging between the

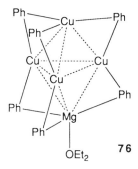

76

two axial metal atoms and each of the three equatorial metal atoms. The axial–equatorial Cu—Cu distances are relatively short (2.427 Å), indicating the presence of some direct metal–metal bonding. Complex **76** can be regarded as involving positive magnesium, analogous to complexes in Section III,F,1, since the Mg—Cu interactions are weak (2.754 Å). If these contacts are not regarded as real bonds, the magnesium atom is tetracoordinated; one Et_2O molecule serves to complete its coordination sphere (Mg—O 2.046 Å). The bridging of the phenyl groups between copper and magnesium is quite unsymmetrical (Cu—C 1.95 Å, Mg—C 2.35 Å), with the Cu—C bonds almost in the plane of the phenyl rings.

The presence of a direct Mg—M' bond is unambiguously established in the organocobalt complexes $CpCo(C_2H_4)(\mu\text{-Ph})MgBr \cdot TMEDA$ (**77**, Mg—Co 2.56 Å) (*72*) and $CpCo(allyl)MgBr \cdot (THF)_2$ (**78**) (*72*). The phenyl group in **77** is weakly bridging toward magnesium (Mg—C 2.56 Å, Co—C

77

1.98 Å), increasing the coordination number of magnesium to five. No Mg—C bond is present in **78** (this structure is included for completeness). The magnesium in **78** has a normal tetrahedral coordination. The crystal structure of the organonickel complex $(C_2H_4)_2Ni(\mu\text{-Me})MgMe \cdot TMEDA$ (**79**) (*73*) is to some extent analogous to that of **77**. Again, the metal atoms are connected by a direct bond (Mg—Ni 2.615 Å) and a bridging organyl

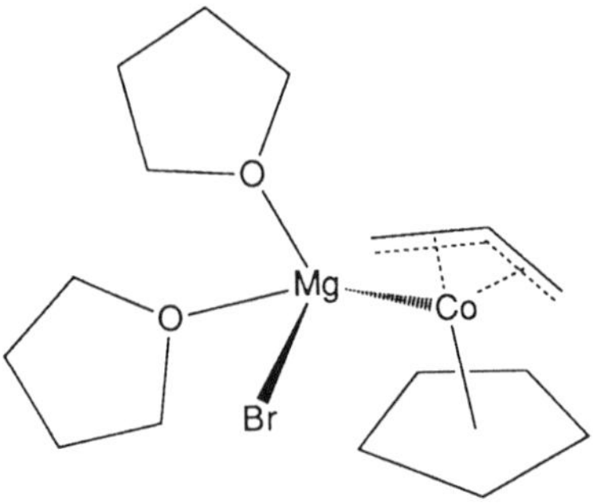

78

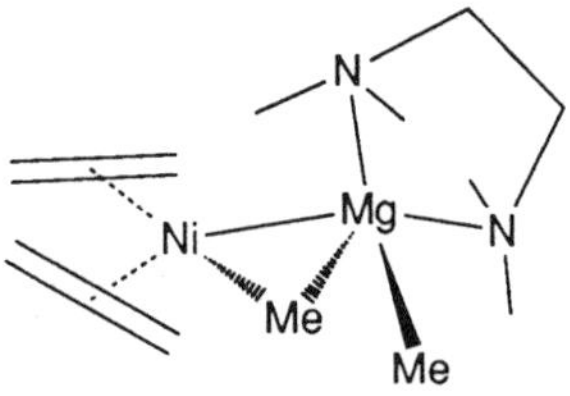

79

group (Mg—C 2.295 Å, Ni—C 2.031 Å). In addition to the TMEDA ligand, the magnesium atom carries one terminal methyl group (Mg—C 2.150 Å) and is therefore pentacoordinated.

A remarkable organometallic cluster with two molybdenum and four magnesium atoms is formed in $[(Cp_2MoH)\cdot\{RMg(\mu\text{-}Br)_2Mg(Et_2O)\}]_2$ (**80**) (*74*). Two four-membered Mg_2Br_2 rings with normal structural fea-

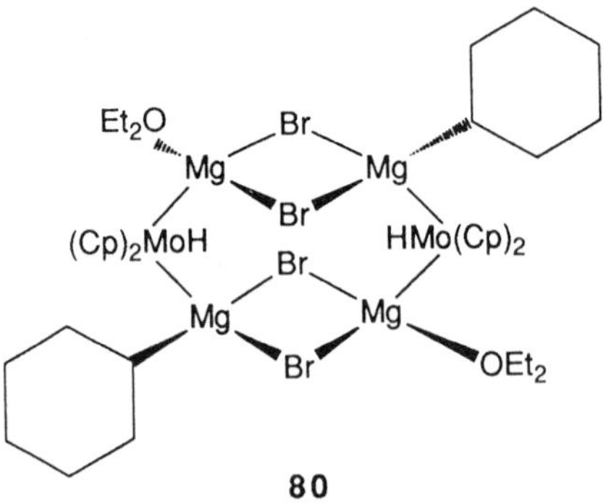

80

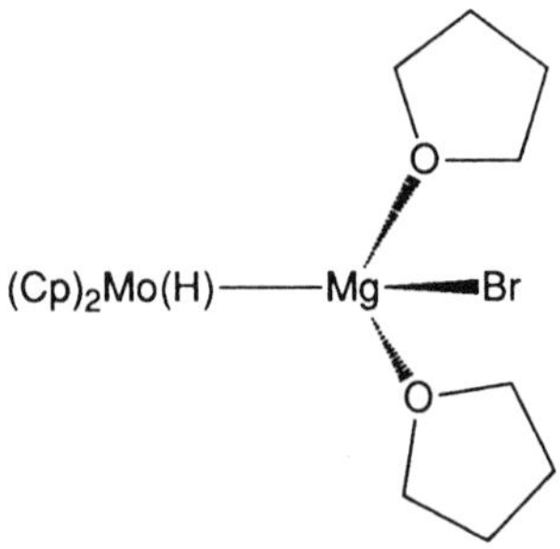

81

tures are connected by $(Cp)_2MoH$ residues via Mg—Mo—Mg bridges (Mg—Mo—Mg 108.8°). Two different types of magnesiums are present in **80**, both with a pseudotetrahedral coordination geometry. One of them carries a diethyl ether molecule and is formally positively charged; it has the shortest bond to molybdenum (Mg—Mo 2.74 Å). The magnesium carrying a cyclohexyl group has a weaker Mg—Mo bond (Mg—Mo 2.85 Å). The other bonds around the magnesium centra have normal lengths. Complex **80** might be regarded as an adduct of $Cp_2Mo(H)MgBr$ with cyclohexylmagnesium bromide (and additional solvent molecules). The crystal structure of $Cp_2Mo(H)MgBr$ was also determined separately as its bis(tetrahydrofuranate) complex **81** (*75*).

IV

COMPOUNDS WITHOUT A MAGNESIUM–CARBON BOND

In this section, the structures of some magnesium compounds which do not involve Mg—C contacts are reviewed. Although not "organometallic" in the proper sense, they are of interest because they contain structural elements which are also present in organometallic compounds and may serve for comparison purposes. The structures are divided into several classes, with the relevant data being summarized in tabular form. The listing is not intended to be complete. With the exception of Section IV,A, structural data of magnesium salts are not included.

A. Magnesium Dihalide Complexes

Magnesium dihalide complexes (Table IV) are included to serve as possible reference material for Grignard structures, as both contain magnesium–halogen bonds. Generally, magnesium dihalides can form complexes both with two and four solvent molecules. Compared with organomagnesium compounds, the tendency for formation of the six-coordinated state is high. This must be explained by the more polarized character of magnesium dihalides, giving stronger ion-induced dipole interactions, and by steric factors. The gas-phase and solution structures of magnesium bromide and iodide have already been mentioned in Section II.

TABLE IV

STRUCTURAL FEATURES OF MAGNESIUM DIHALIDE COMPLEXES

Compound	Bond lengths (averaged, Å)		Ref.
	Mg—X	Mg—O/N	
MgBr$_2 \cdot$(THF)$_4$ (discrete octahedral molecules, Br trans	2.625	2.16	76

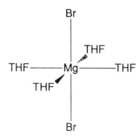

Compound	Mg—X	Mg—O/N	Ref.
MgBr$_2 \cdot$(THF)$_2$ [polymeric, distorted octahedral, two trans THF, two cis Br, two cis Br (bridging to next Mg)]	2.633 2.799	2.126	77

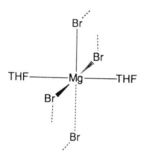

Compound	Mg—X	Mg—O/N	Ref.
[Mg$\cdot$(THF)$_4 \cdot$(H$_2$O)$_2$]$\cdot$Br$_2$ (octahedral complex, with trans H$_2$O ligands and out sphere Br atoms)	4.651	2.042 (aq) 2.14 (THF)	77
MgCl$_2 \cdot$(THF)$_2$ {incomplete data from powder diffraction study: [Mg(THF)$_2$(μ-Cl)$_2$]$_n$}	—	—	78
MgCl$_2 \cdot$(THF)$_4$ [incomplete data from powder diffraction study: analogous to MgBr$_2 \cdot$(THF)$_4$]	—	—	78
MgBr$_2 \cdot$(Et$_2$O)$_2$ (incomplete data: tetrahedral Mg)	—	—	79
MgCl$_2 \cdot$(pyridine)$_4$ (discrete octahedral complexes, Cl trans)	2.463 2.483	2.26 2.28	80
MgBr$_2 \cdot$(pyridine)$_6$ (discrete octahedral complexes, trans Br, two pyridines not coordinating)	2.767	2.17	81
MgBr$_2 \cdot$(5,7-dihydrodibenz[c,e]oxepin)$_4$ (octahedral complex, trans Br)	2.623	2.127 2.152 2.183	82
MgCl$_2 \cdot$[(−)-sparteine] (distorted tetrahedral)	2.269 2.279	2.150 2.160	17
MgBr$_2 \cdot$[(+)-6-benzylsparteine] (distorted tetrahedral)	2.45 2.47	2.08 2.17	19

TABLE V

STRUCTURAL FEATURES OF R_2Mg COMPLEXES WITH Mg—M' BONDS INVOLVING GROUPS 13, 14 OR 15

Compound	Geometry	Bond lengths (averaged, Å)		Bond angle M'—Mg—M' (°)	Ref.
		Mg—M'	Mg—O/N		
$(Me_3Si)_2Mg \cdot glyme$ 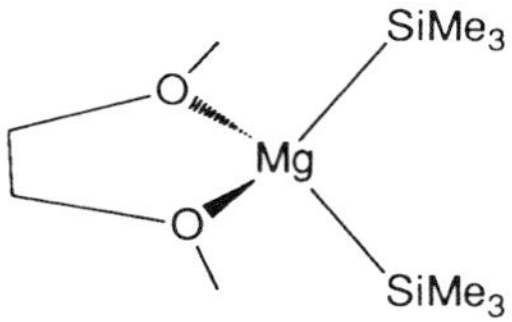	Tetrahedral	2.630	2.124	125.2	83
$(Me_3Si)_2Mg \cdot TMEDA$	Tetrahedral	2.628	2.192		84
$(Me_3Ge)_2Mg \cdot (glyme)_2$	Octahedral	2.719	2.22		85
$(B_6H_9)_2Mg \cdot (THF)_2$	Tetrahedral	2.592	2.306	122.5	86
	Tetrahedral	2.38^a	2.019	a	87
		2.48^a			

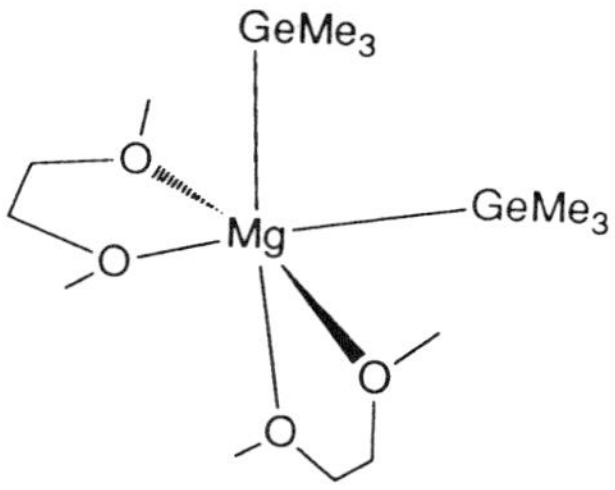

^{a} Three-center B—Mg—Br bond to each B_6H_9 ligand.

213

B. *R_2Mg Complexes with Magnesium–Element Bonds Involving Groups 13, 14, or 15*

Most R_2Mg complexes involving bonds to elements of groups 13, 14, or 15 have a pseudotetrahedral coordination, with a close resemblance to structures of type **5** (Section III,A). They are listed in Table V, where M′ is an element from group 13, 14, or 15.

C. *Complexes with Magnesium–Oxygen or Magnesium–Nitrogen Bonds*

In this section, magnesium compounds such as alcoholates, enolates, and amides are presented (Table VI). Most complexes were obtained by deprotonation reactions. Alcoholates can also be obtained from addition reactions of organomagnesium reagents to carbonyl substrates. In many cases, oligomeric structures were found owing to the high bridging capacity of the anionic oxygen or nitrogen ligands.

TABLE VI

STRUCTURAL FEATURES OF COMPLEXES WITH Mg—O OR Mg—N BONDS

Compound	Bond lengths (averaged, Å)		Ref.
	Mg—X	Mg—O/N	
$[\{t\text{-Bu}-C(O)=CHCH_3)MgBr\cdot(Et_2O)]_2$ (type **6** dimer with bridging enolate group)	2.417	2.049 (Et$_2$O) 1.953	*88*

| $[t\text{-BuOMgBr}\cdot Et_2O]_2$ (type **6** dimer with bridging alcoholate group) | — | 2.01 (Et$_2$O) 1.91 (RO$^-$) | *89* |
| $(THF)_2Mg_2Br_2(\mu\text{-B}=CPh_2)_2(\mu\text{-THF})$ (type **6** dimer with a μ_2 bridging THF molecule) | 2.474 | 2.078 (N) 2.066 (O$_{term}$) 2.453 (O$_{bridge}$) | *90* |

TABLE VI (*continued*)

Compound	Bond lengths (averaged, Å)		Ref.
	Mg—X	Mg—O/N	

| [(*N,N'*-Ethylenebis(acetylacetoneimine enolate)·Mg]$_2$ (dimeric polydentate magnesium complex) 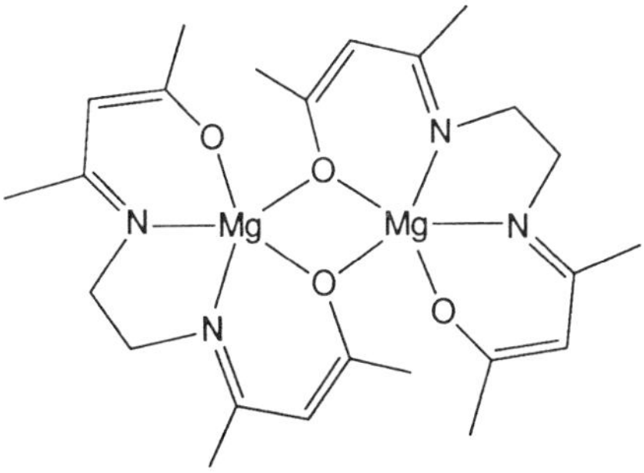 | — | 2.12 (N) 2.00 (O) | 91 |

| (6,8,15,17-Tetramethyldibenzo[*b,i*]-[1,4,8,11]-N$_4$-cyclotetradecine)Mg·(THF) (tetradentate macrocyclic magnesium complex) 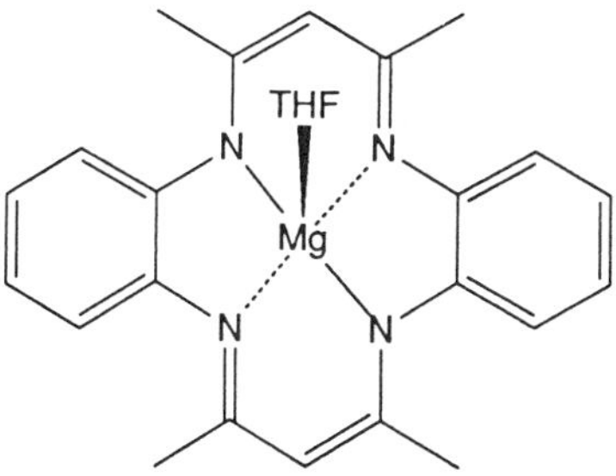 | — | 2.060 (N) 2.041 (O) | 91 |

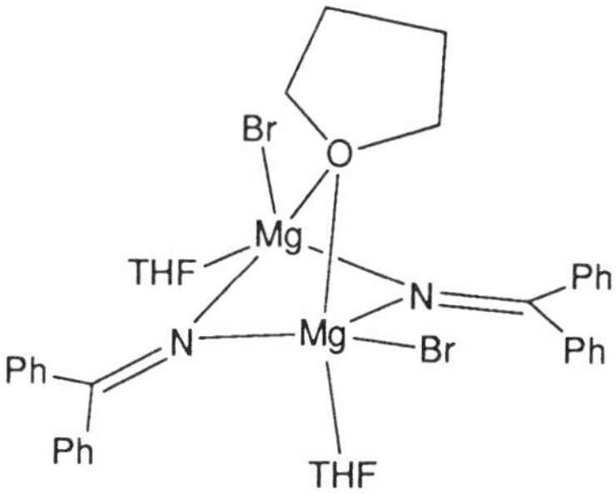

(*continued*)

TABLE VI (*continued*)

| | Bond lengths (averaged, Å) | | |
Compound	Mg—X	Mg—O/N	Ref.
$NMg_6(NH{-}t\text{-}Bu)_9$ cluster	—	2.148 (N^{3-})	92
(irregular cage cluster with central N^{3-} atom)		2.093 (N_{bridge})	
$[Me_2Si(t\text{-}BuN)_2Mg \cdot THF]_2$	—	2.049 (O)	93
(dimeric structure with a central Mg_2N_2 ring		2.15 (Mg_2N_2)	
and terminal amide/THF ligands)		1.993 (N)	

$[(Me_2Al \cdot Me_2Si(t\text{-}BuN)_2] \cdot Mg(\mu\text{-}I)]_2$	2.83	2.28 (N)	93
(dimeric structure with a central Mg_2I_2 ring)			

$[t\text{-}BuN{-}Si(\mu\text{-}N{-}t\text{-}Bu)Si{-}N{-}t\text{-}Bu]Mg \cdot (THF)_2$	—	2.12 (O)	94
(cage structure with tetrahedral Mg atom)	—	2.05 (N)	

TABLE VI (*continued*)

Compound	Bond lengths (averaged, Å)		Ref.
	Mg—X	Mg—O/N	
[{o-(NSiMe$_3$)$_2$C$_6$H$_4$)Mg·Et$_2$O]$_2$	—	2.041 (O)	95
(dimer with two bridging and two terminal	—	1.997 (N$_{term}$)	95
nitrogen atoms)		2.083 (N$_{bridge}$)	

[Phenazine-*N*,*N'*-(MgBr·(THF)$_3$)$_2$]·[MgBr$_2$·(THF)$_4$]	2.513	2.12 (THF)	96
(monomeric bisbromomagnesiodihydrophenazine		2.052 (N)	
with five-coordinate magnesium atoms)			

D. *Miscellaneous Magnesate Complexes*

Table VII gives selected bond lengths found in miscellaneous magnesate complexes.

V

APPENDIX

The investigation of organomagnesium structures is an active area of research; there is a steady stream of papers dedicated to this subject. After

TABLE VII

STRUCTURAL FEATURES OF MISCELLANEOUS MAGNESATE COMPLEXES

Compound	Bond lengths (averaged, Å)		Ref.
	Mg—X	Mg—O/N	
$[Cp^*U{=}O \cdot [CH_2P(Ph)_2CH_2]_2 \cdot [MgCl_2] \cdot [CH_2{=}PPh_2Me]_2$ 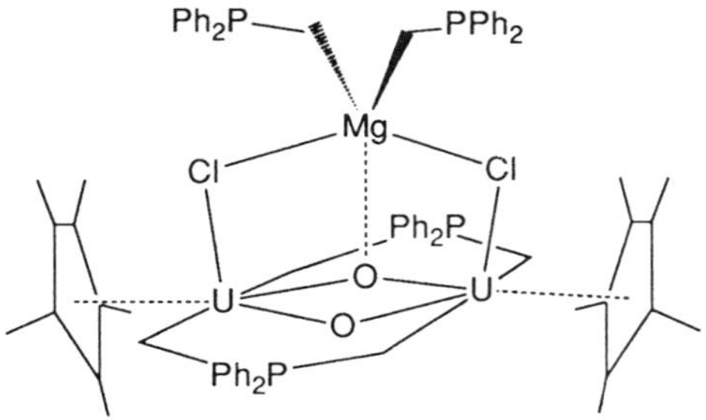	2.79	1.94 (0) 2.23 (C)	97
$[Me_2Al(\mu\text{-}OMe)_2]_2Mg \cdot (p\text{-}dioxane)$	—	2.05 (O_{bridge}) 2.24 (O_{diox})	98

$P = PPh_2$

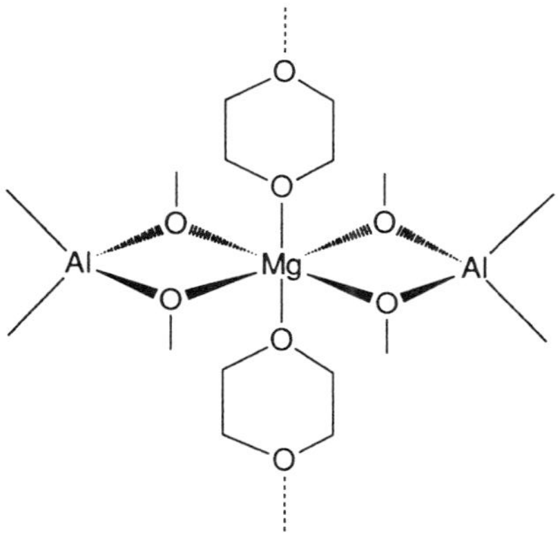

99

2.09 (O)

2.06 (H)

$(BH_4)_2Mg \cdot (THF)_3$

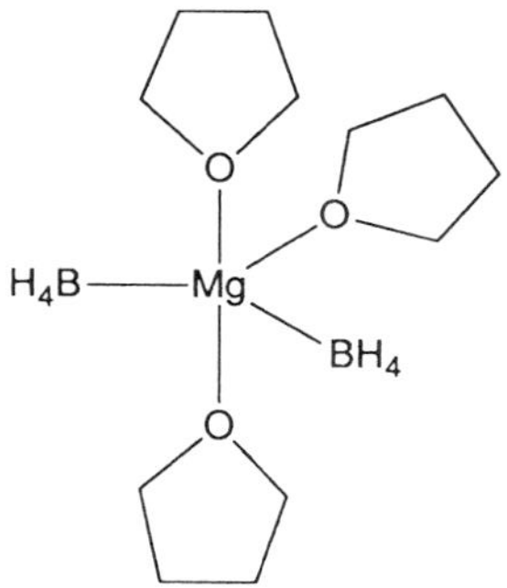

(continued)

TABLE VII (continued)

Compound	Bond lengths (averaged, Å)		Ref.
	Mg—X	Mg—O/N	
[(o-Xylidene)$_2$W=O]$_2$Mg·(THF)$_4$	—	2.07 (=O) 2.10 (THF)	100
(Me$_4$ReO)$_2$Mg·(THF)$_4$	—	2.029 (=O) 2.09 (THF)	101
[(Me$_3$Si)$_4$ReO]$_2$Mg·(THF)$_4$	—	1.923 (=O) 2.021 (THF)	101
[MgCl(THF)$_5$]·[FeCl$_4$]	2.384	2.13	102
[MgCl(THF)$_5$]·[AlCl$_4$]	2.396	2.13	102

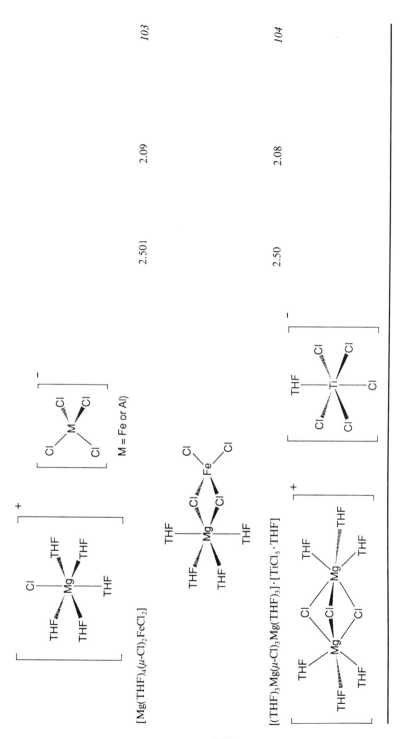

[Mg(THF)₄(μ-Cl)₂FeCl₂]

[(THF)₃Mg(μ-Cl)₃Mg(THF)₃] · [TiCl₅ · THF]

M = Fe or Al)

103

104

2.09

2.08

2.501

2.50

completion of the manuscript, the structure of HB·[3-(tl-Bu)pyrazo-lyl]$_3$MgCH$_3$ has been published. Relevant data are as follows: Mg—C 2.118 Å, Mg—N (average) 2.13 Å (*105*).

ACKNOWLEDGMENTS

This work was supported in part (P.R.M., W.J.J.S., A.L.S.) by the Netherlands Foundation for Chemical Research (SON) with financial aid from the Netherlands Organization for the Advancement of Research (NWO). X-Ray data for many of our new crystal structures were collected by A. J. M. Duisenberg.

REFERENCES

1. P. Chakrabarti and J. D. Dunitz, *Helv. Chim. Acta* **65**, 1482 (1982).
2. A. Haaland, J. Lusztyk, J. Brunvoll, and K. B. Starowieyski, *J. Organomet. Chem.* **85**, 279 (1975).
3. R. A. Andersen, R. Blom, J. M. Boncella, C. J. Burns, and H. V. Volden, *Acta Chim. Scand., Ser. A* **41**, 24 (1987).
4. E. C. Ashby, L. Fernholt, A. Haaland, R. Seip, and R. Scott Smith, *Acta Chim. Scand., Ser. A* **34**, 213 (1980).
5. R. A. Andersen, R. Blom, A. Haaland, B. E. R. Schilling, and H. V. Volden, *Acta Chim. Scand., Ser. A* **39**, 563 (1985).
6. A. Ericson and I. Persson, *J. Organomet. Chem.* **326**, 151 (1987).
7. P. A. Akishin, V. P. Spiridonov, G. A. Sobolev, and V. A. Naumov, *Z. Fiz. Khim. SSSR* **31**, 461 (1957); *Chem. Abstr.* **51**, 15212b (1957).
8. P. A. Akishin and V. P. Spiridonov, *Z. Fiz. Khim. SSSR* **32**, 1682 (1958); *Chem Abstr.* **53**, 804g (1958).
9. F. A. Schröder, *Chem. Ber.* **102**, 2035 (1969).
10. G. Stucky and R. E. Rundle, *J. Am. Chem. Soc.* **80**, 4825 (1964).
11. L. J. Guggenberger and R. E. Rundle, *J. Am. Chem. Soc.* **86**, 5344 (1964); L. J. Guggenberger and R. E. Rundle, *J. Am. Chem. Soc.* **90**, 5375 (1968).
12. J. Toney and G. D. Stucky, *J. Organomet. Chem.* **22**, 241 (1970).
13. T. Greiser, J. Kopf, D. Thoennes, and E. Weiss, *J. Organomet. Chem.* **191**, 1 (1980).
14. D. Thoennes and E. Weiss, *Chem. Ber.* **111**, 3381 (1978).
15. W. Clegg, D. A. Brown, S. J. Bryan, and K. Wade, *J. Organomet. Chem.* **325**, 39 (1987).
16. H. Kageyama, K. Miki, N. Tanaka, N. Kasai, Y. Okamoto, and H. Yuki, *Bull. Chem. Soc. Jpn.* **56**, 1319 (1983).
17. H. Kageyama, K. Miki, Y. Kai, N. Kasai, Y. Okamoto, and H. Yuki, *Bull. Chem. Soc. Jpn.* **56**, 2411 (1983).

18. H. Kageyama, K. Miki, Y. Kai, N. Kasai, Y. Okamoto, and H. Yuki, *Acta Crystallogr. Sect. B* **38**, 2264 (1982).
19. H. Kageyama, K. Miki, Y. Kai, N. Kasai, Y. Okamoto, and H. Yuki, *Bull. Chem. Soc. Jpn.* **57**, 1189 (1984).
20. J. Toney and G. D. Stucky, *J. Chem. Soc., Chem. Commun.,* 1168 (1967).
21. A. L. Spek, P. Voorbergen, G. Schat, C. Blomberg, and F. Bickelhaupt, *J. Organomet. Chem.* **77**, 147 (1974).
22. H. F. Klein, H. König, S. Koppert, K. Ellrich, and J. Riede, *Organometallics* **6**, 1341 (1987).
22a. P. R. Markies, G. Schat, O. S. Akkerman, F. Bickelhaupt, W. J. J. Smeets, P. van der Sluis, and A. L. Spek, *J. Organomet. Chem.* **393**, 315 (1990).
23. E. Weiss, *J. Organomet. Chem.* **2**, 314 (1964); E. Weiss, *J. Organomet. Chem.* **4**, 101 (1965).
24. P. R. Markies, G. Schat, O. S. Akkerman, F. Bickelhaupt, W. J. J. Smeets, A. J. M. Duisenberg, and A. L. Spek, *J. Organomet. Chem.* **375**, 11 (1989).
25. M. Vallino, Ph.D. Thesis, University of Paris VI, 1972.
26. A. L. Spek, G. Schat, H. C. Holtkamp, C. Blomberg, and F. Bickelhaupt, *J. Organomet. Chem.* **131**, 331 (1977).
27. M. F. Lappert, T. R. Martin, C. L. Raston, B. W. Skelton, and A. H. White, *J. Chem. Soc., Dalton Trans.,* 1959 (1982).
28. M.A.G.M. Tinga, G. Schat, O. S. Akkerman, F. Bickelhaupt, W.J.J. Smeets, A.L. Spek, *J. Am. Chem. Soc.* **113**, 3604 (1991) mentions another Mg_4 cluster: $[o\text{-}C_6H_4Mg]_4$; compound **11** and $[PhC(Mg)=CPh]_4$ are described in M.A.G.M. Tinga, Ph.D. Thesis, Vrije Universiteit te Amsterdam, 1991.
29. S. S. Al-Juaid, C. Eaborn, P. B. Hitchcock, C. A. McGeary, and J. D. Smith, *J. Chem. Soc., Chem. Commun.,* 273 (1989).
30. R. Shakir and J. L. Atwood, *Am. Crystallogr. Assoc., Ser.* **6**, 11 (1978).
31. L. M. Engelhardt, B. S. Jolly, P. C. Junk, C. L. Raston, B. W. Skelton, and A. H. White, *Aust. J. Chem.* **39**, 1337 (1986).
32. M. Vallino, *J. Organomet. Chem.* **20**, 1 (1969).
33. J. Toney and G. D. Stucky, *J. Organomet. Chem.* **28**, 5 (1971).
34. M. C. Perucaud, J. Ducom, and M. Vallino, *C. R. Seances Acad. Sci., Ser. C* **264**, 571 (1967).
35. B. Schubert, U. Behrens, and E. Weiss, *Chem. Ber.* **114**, 2640 (1981).
36. M. Geissler, J. Kopf, and E. Weiss, *Chem. Ber.* **122**, 1395 (1989).
37. E. P. Squiller, R. R. Whittle, and H. G. Richey, Jr., *J. Am. Chem. Soc.* **107**, 432 (1985).
38. E. P. Squiller, R. R. Whittle, and H. G. Richey, Jr., *Organometallics* **4**, 1154 (1985).
39. A. D. Pajerski, M. Parvez, and H. G. Richey, Jr., *J. Am. Chem. Soc.* **110**, 2660 (1988).
40. A. D. Pajerski, G. L. Bergstresser, M. Parvez, and H. G. Richey, Jr., *J. Am. Chem. Soc.* **110**, 4844 (1988).
41. J. D. Owen, *J. Chem. Soc., Dalton Trans.,* 1418 (1978).
42. N. R. Strel'tsova, L. V. Ivakina, P. A. Storozhenko, B. Bulychev, and V. K. Bel'skii, *Sov. Phys. Dokl. (Engl. Transl.)* **31**, 943 (1986).
43. P. R. Markies, T. Nomoto, O. S. Akkerman, F. Bickelhaupt, W. J. J. Smeets, and A. L. Spek, *J. Am. Chem. Soc.* **110**, 4845 (1988).
44. K. Angermund, B. Bogdanovic, G. Koppetsch, C. Krüger, R. Mynott, M. Schwickardi, and Y. H. Tsay, *Z. Naturforsch., B: Anorg. Chem., Org. Chem.* **41**, 455 (1986).
45. B. Bogdanovic, G. Koppetsch, C. Krüger, and R. Mynott, *Z. Naturforsch., B: Anorg. Chem., Org. Chem.* **41**, 617 (1986).
46. V. R. Magnuson and G. D. Stucky, *Inorg. Chem.* **8**, 1427 (1969).

47. M. J. Henderson, R. I. Papasergio, C. L. Raston, A. H. White, and M. F. Lappert, *J. Chem. Soc., Chem. Commun.*, 672 (1986).
48. D. Seebach, J. Hansen, P. Seiler, and J. M. Gromek, *J. Organomet. Chem.* **285**, 1 (1985).
49. P. R. Markies, O. S. Akkerman, F. Bickelhaupt, W. J. J. Smeets, and A. L. Spek, *J. Am. Chem. Soc.* **110**, 4284 (1988).
50. P. R. Markies, O. S. Akkerman, F. Bickelhaupt, W. J. J. Smeets, and A. L. Spek, *Angew. Chem.* **100**, 1143 (1988).
51. W. Bünder and E. Weiss, *J. Organomet. Chem.* **92**, 1 (1975).
52. C. P. Morley, P. Jutzi, C. Krüger, and J. M. Wallis, *Organometallics* **6**, 1084 (1987).
53. C. Johnson, J. Toney, and G. D. Stucky, *J. Organomet. Chem.* **40**, C11 (1972).
54. J. L. Atwood and K. D. Smith, *J. Am. Chem. Soc.* **96**, 994 (1974).
55. H. Lehmkuhl, K. Mehler, R. Benn, A. Rufinska, and C. Krüger, *Chem. Ber.* **119**, 1054 (1986).
56. H. Lehmkuhl, A. Shakoor, K. Mehler, C. Krüger, K. Angermund, and Y. H. Tsay, *Chem. Ber.* **118**, 4239 (1985).
57. T. Alonso, S. Harvey, P. C. Junk, C. L. Raston, B. W. Skelton, and A. H. White, *Organometallics* **6**, 2110 (1987).
58. B. Bogdanovic, N. Janke, C. Krüger, R. Mynott, D. Schlichte, and U. Westeppe, *Angew. Chem.* **97**, 972 (1985).
59. L. M. Engelhardt, S. Harvey, C. L. Raston, and A. H. White, *J. Organomet. Chem.* **341**, 39 (1988).
60. B. Bogdanovic, N. Janke, C. Krüger, K. Schlichte, and J. Treber, *Angew. Chem.* **99**, 1046 (1987).
61. H. Lehmkuhl, K. Mehler, R. Benn, A. Rufinska, G. Schroth, and C. Krüger, *Chem. Ber.* **117**, 389 (1984).
62. H. Lehmkuhl, K. Mehler, A. Shakoor, C. Krüger, Y. H. Tsay, R. Benn, A. Rufinska, and G. Schrott, *Chem. Ber.* **118**, 4248 (1985).
63. M. Marsch, K. Harms, W. Massa, and G. Boche, *Angew. Chem.* **99**, 706 (1987).
64. H. Yasuda, M. Yamauchi, A. Nakamura, T. Sei, Y. Kai, N. Yasuoka, and N. Kasai, *Bull. Chem. Soc. Jpn.* **53**, 1089 (1980).
65. Y. Kai, N. Kanehisa, K. Miki, N. Kasai, K. Mashima, H. Yasuda, and A. Nakamura, *Chem. Lett.*, 1277 (1982).
66. T. Greiser, J. Kopf, D. Thoennes, and E. Weiss, *Chem. Ber.* **114**, 209 (1981).
67. D. Thoennes and E. Weiss, *Chem. Ber.* **111**, 3726 (1978).
68. B. Schubert and E. Weiss, *Chem. Ber.* **117**, 366 (1984).
69. N. H. Buttrus, C. Eaborn, M. N. A. El-Kheli, P. B. Hitchcock, J. D. Smith, A. C. Sullivan, and K. Tavakkoli, *J. Chem. Soc., Dalton Trans.*, 381 (1988).
70. J. L. Atwood and G. D. Stucky, *J. Am. Chem. Soc.* **91**, 2538 (1969).
71. S. I. Khan, P. G. Edwards, H. S. H. Yuan, and R. Bau, *J. Am. Chem. Soc.* **107**, 1682 (1985).
72. K. Jonas, G. Koepe, and C. Krüger, *Angew. Chem.* **98**, 901 (1986).
73. W. Kaschube, K. R. Pörschke, K. Angermund, C. Krüger, and G. Wilke, *Chem. Ber.* **121**, 1921 (1988).
74. M. L. H. Green, G. A. Moser, I. Packer, F. Petit, R. A. Forder, and K. Prout, *J. Chem. Soc., Chem. Commun.*, 839 (1974); K. Prout and R. A. Forder, *Acta Crystallogr.* **31**, 852 (1975).
75. S. G. Davies and M. L. H. Green, *J. Chem. Soc., Dalton Trans.*, 1510 (1978).
76. M. C. Perucaud and M. T. Le Bihan, *Acta Crystallogr., Sect. B* **24**, 1502 (1968).
77. R. Sarma, F. Ramirez, B. McKeever, Y. Fen Chaw, J. F. Marecek, D. Nierman, and T. M. McCaffrey, *J. Am. Chem. Soc.* **99**, 5289 (1977).

78. K. Handlir, J. Holecek, and L. Benes, *Collect. Czech. Chem. Commun.* **50**, 2422 (1985).
79. H. Schibilla and M. T. Le Bihan, *Acta Crystallogr.* **23**, 332 (1967).
80. S. Halut-Desportes and C. Bois, *Acta Crystallogr., Sect. B* **35**, 2205 (1979).
81. S. Halut-Desportes, *Acta Crystallogr., Sect. B* **33**, 599 (1977).
82. L. M. Engelhardt, W. P. Leung, C. L. Raston, and A. H. White, *Aust. J. Chem.* **38**, 977 (1985).
83. A. R. Claggett, W. H. Ilsley, T. J. Anderson, M. D. Glick, and J. P. Oliver, *J. Am. Chem. Soc.* **99**, 1797 (1977).
84. D. W. Goebel, J. L. Hencher, and J. P. Oliver, *Organometallics* **2**, 746 (1983).
85. L. Rosch, C. Kruger, and A. P. Chiang, *Z. Naturforsch., B: Anorg. Chem., Org. Chem.* **39**, 855 (1984).
86. E. Hey, L. M. Engelhardt, C. L. Raston, and A. H. White, *Angew. Chem.* **99**, 61 (1987).
87. D. L. Denton, W. R. Clayton, M. Mangion, S. G. Shore, and E. A. Meyers, *Inorg. Chem.* **15**, 541 (1976).
88. P. G. Williard and J. M. Salvino, *J. Chem. Soc., Chem. Commun.*, 153 (1986).
89. P. T. Moseley and H. M. M. Shearer, *J. Chem. Soc., Chem. Commun.*, 279 (1968).
90. K. Manning, E. A. Petch, H. M. M. Shearer, K. Wade, and G. Whitehead, *J. Chem. Soc. Chem. Commun.*, 107 (1976).
91. F. Corazza, C. Floriani, A. Chiesi-Villa, C. Guastini, and S. Ciurli, *J. Chem. Soc., Dalton Trans.*, 2341 (1988).
92. G. Dozzi, G. Del Piero, M. Cesari, and S. Cucinella, *J. Organomet. Chem.* **190**, 229 (1980).
93. M. Veith, W. Frank, F. Töllner, and H. Lange, *J. Organomet. Chem.* **326**, 315 (1987).
94. M. Veith, F. Goffing, and V. Huch, *Z. Naturforsch., B: Anorg. Chem., Org. Chem.* **43**, 846 (1988).
95. A. W. Duff, P. B. Hitchcock, M. F. Lappert, R. G. Taylor, and J. A. Segal, *J. Organomet. Chem.* **293**, 271 (1985).
96. P. C. Junk, C. L. Raston, B. W. Skelton, and A. H. White, *J. Chem. Soc., Chem. Commun.*, 1162 (1987).
97. R. E. Cramer, M. A. Bruck, and J. W. Gilje, *Organometallics* **7**, 1465 (1988).
98. J. L. Atwood and G. D. Stucky, *J. Organomet. Chem.* **13**, 53 (1968).
99. H. Nöth, *Z. Naturforsch., B: Anorg. Chem., Org. Chem.* **37**, 1499 (1982).
100. M. F. Lappert, C. L. Raston, G. L. Rowbottom, and A. H. White, *J. Chem. Soc., Chem. Commun.*, 6 (1981).
101. P. Stavropoulos, P. G. Edwards, G. Wilkinson, M. Motevalli, K. M. A. Malik, and M. B. Hursthouse, *J. Chem. Soc., Dalton Trans.*, 2167 (1985).
102. P. Sobota, T. Pluzinski, J. Utko, and T. Lis, *Inorg. Chem.* **28**, 2217 (1989).
103. P. Sobota, T. Pluzinski, and T. Lis, *Polyhedron* **3**, 45 (1984).
104. P. Sobota, J. Utko, and T. Lis, *J. Chem. Soc., Dalton Trans.*, 2077 (1984).
105. P. Han, A. Looney, and G. Parkin, *J. Am. Chem. Soc.* **111**, 7276 (1989).

RELATED REVIEWS

N. S. Poonia and A. V. Bajaj, "Coordination Chemistry of Alkaline and Alkaline Earth Cations," *Chem. Rev.* **79**, 389 (1979).
W. E. Lindsell, "Magnesium, Calcium, Strontium and Barium," in *Comprehensive Organometallic Chemistry* (G. Wilkinson, F. G. A. Stone, and E. W. Abel, eds.), Vol. 1, p. 155. Pergamon, Oxford, 1982.
W. N. Setzer, and P. von Rague Schleyer, "X-Ray Structural Analyses of Organolithium Compounds," *Adv. Organomet. Chem.* **24**, 353 (1985).
C. Schade and P. von Rague Schleyer, "Sodium, Potassium, Rubidium, and Cesium: X-Ray

Structural Analysis of Their Organic Compounds," *Adv. Organomet. Chem.* **27,** 169 (1988).

P. Jutzi, "π-Bonding to Main-Group Elements," *Adv. Organomet. Chem.* **26,** 217 (1986).

D. E. Fenton, "Alkali Metals and Group IIA Metals," *in Comprehensive Coordination Chemistry* (G. Wilkinson, ed.), Chap. 23. Pergamon, Oxford, 1987.

P. Hubberstey, "Elements of Group 2," *Coord. Chem. Rev.* **75,** 100 (1986).

P. Hubberstey, "Elements of Group 2," *Coord. Chem. Rev.* **85,** 86 (1988).

N. S. Poonia, P. Bagdi, and K. S. Sidhu, "Structural Aspects of Crown Complexes with Alkali and Alkaline Earth Cations. Benzo-15-crown-5 as a Discriminating Macrocycle," *J. Incl. Phen.* **4,** 43 (1986).

D. R. Russell, "Structures of Organometallic Compounds Determined by Diffraction Methods," *Organomet. Chem.* **15,** 422 (1987).

J. L. Wardell, "Group II: The Alkaline Earths and Zinc and Its Congeners," *Organomet. Chem.* **15,** 14 (1987).

J. L. Wardell, "Group II: The Alkaline Earths and Zinc and Its Congeners," *Organomet. Chem.* **17,** 11 (1989).

A. V. Bajaj and N. S. Poonia, *Coord. Chem. Rev.* **87,** 55 (1988).

A. Haaland, "Kovalente und dative Bindungen zu Hauptgruppenmetallen, eine Nützliche Unterscheidung," *Angew. Chem.* **101,** 1017 (1989).

ADVANCES IN ORGANOMETALLIC CHEMISTRY, VOL. 32

Recent Advances in the Chemistry of Metal – Carbon Triple Bonds

ANDREAS MAYR and HANS HOFFMEISTER

Department of Chemistry
State University of New York at Stony Brook
Stony Brook, New York 11794-3400

I

INTRODUCTION

The chemistry of metal – carbon triple bonds has developed at an increasing pace since the discovery (*1*) of the first carbyne metal complexes by E. O. Fischer and co-workers in 1973 (*2–4*). Two extensive reviews on the field by Kim and Angelici (*5*) and by Fischer, Hofmann, Kreissl, Schrock, Schubert, and Weiss (*6*) have been published in recent years. In addition several articles focusing on the research areas of the respective authors have been published (*7–19*). The chemistry of metal – carbon triple bonds is also covered within the context of more general yearly literature surveys (*20*).

This article covers the literature from the beginning of 1986 to early 1990. For earlier developments in the chemistry of metal – carbon triple bonds we refer to the previous article in this series (5), which covered the literature up to early 1986, and to the more recent monograph (6). The scope of this review is confined to the chemistry of complexes containing terminal alkylidyne ligands. Reactions of metal alkylidynes with other transition metal complexes leading to polynuclear complexes with bridging alkylidyne ligands are not included. This aspect of the chemistry of metal–carbon triple bonds was pioneered and developed by Stone and now represents an important bridge between the chemistry of mononuclear and polynuclear alkylidyne metal complexes (14,15).

II

PHYSICAL AND THEORETICAL STUDIES OF ALKYLIDYNE METAL COMPLEXES

A. Structural Studies

General aspects concerning the solid-state structures of metal alkylidyne complexes have been previously discussed (6,21). Table I lists metal–carbon triple bond lengths and M—C—R bonding angles for the structures reported since 1986. Electron deformation density maps of $Cr(CPh)Cl(CO)_4$ were studied by Dao et al. using X-ray and neutron diffraction methods (22). The results suggest the presence of a partial negative charge on the alkylidyne carbon atom.

B. Spectroscopic Properties

^{13}C NMR resonances of the alkylidyne carbon atom and selected IR absorptions for new carbyne complexes are given in Table II. In other studies, the [95]Mo and [183]W NMR spectra of molybdenum and tungsten alkylidyne complexes were determined by Enemark and co-workers (23). The metal atoms in the alkylidyne complexes were found to be less shielded than those in nitrido complexes but more shielded than those in compounds containing metal–metal multiple bonds.

Bocarsly et al. reported the electronic absorption and emission spectra of a series of tungsten alkylidyne complexes of the type $W(CR)X(CO)_2L_2$ (R = Ph, CMe_3, and 2-naphthyl; X = halide; L_2 = tmeda, 2py, dppe) (24). For example, the absorptions at 450 and 327 nm for the complex $W(CPh)Br(CO)_2(tmeda)$ were assigned to $d \rightarrow \pi_{(MC)}{}^*$ and $\pi_{(MC)} \rightarrow \pi_{(MC)}{}^*$ transitions. Emission from the excited state occurs at 630 nm. The lifetime of the electronically excited state was measured to be 0.180 μseconds. Emission was observed only for the aryl-substituted carbyne complexes. (The photoreactivity of these compounds is discussed in Section IV, B.)

TABLE I

STRUCTURAL DATA FOR ALKYLIDYNE METAL COMPLEXES

Compound	M≡C bond distance (pm)	M—C—R bond angle (°)	Ref.
Ta[COSi(^{i}Pr)$_3$](CO)(dmpe)$_2$	185.0(1)	173.0(1)	73
[Cr{CNC(C$_{12}$H$_8$O)}(CO)$_5$][BF$_4$]	189.5(6)	179.0(5)	43
Mo(CCH$_2$^{t}Bu)(η^5-C$_5$H$_5$)(η^2-C$_6$N$_4$)[P(OMe)$_3$]	177.4(5)	177.1(5)	90
Mo(CCH$_2$^{t}Bu)(η^5-C$_5$H$_5$)[P(OMe)$_3$]$_2$	179.6(2)	176.7(2)	66
Mo(CPh)Cl[P(OMe)$_3$]$_4$	179.3(8)	^{a}	105
[Mo(CCH$_2$^{t}Bu)H(η^5-C$_5$H$_5$){P(OMe)$_3$}$_2$][BF$_4$]	179.8(2)	170.0(1)	90
W(CPh)Cl(C$_4$H$_2$O$_3$)(CO)(py)	180.1(6)	173.0(4)	103
W(CPh)Br(CN^{t}Bu)$_4$	176.4(6)	172.9(4)	175
[PPh$_4$][W(CTol)(η^5-C$_2$B$_9$H$_9$Me$_2$)(CO)$_2$]	182.6(7)	178.4(4)	213
W(CTol)(η^5-C$_5$H$_5$)(CO)(PMe$_3$)	180.0(2)	177.1(12)	169
W(CTol)[B(pz)$_4$](CO)$_2$	182.1(7)	164.0(6)	122
[NEt$_4$][W(CC$_6$H$_3$Me$_2$-2,6)(η^5-C$_2$B$_{10}$H$_{10}$Me$_2$)(CO)$_2$]	184.0(1)	175.6(9)	130
W(CNEt$_2$)(CO)$_2$(bpy)[PPh$_2$Cr(CO)$_5$]	187.7(8)	171.5(7)	114
W(CNCPh$_2$)Br(CO)$_4$	187.8(5)	169.6(5)	42
W(CPPh$_3$)Cl$_2$(CO)(PMePh$_2$)$_2$	182.3(7)	171.2(5)	172
W(CSMe)[HB(pz)$_3$](SMe)$_2$	178.8(7)	171.6(5)	143
[W(CEt)(CH$_2$^{t}Bu)(O^{i}Pr)$_2$]$_2$	176.5(7)	178.0(6)	78
[W(CEt)(NHMe$_2$)(O^{i}Pr)$_2$(μ-O^{i}Pr)]$_2$	177.2(11)	^{a}	77
W[CRu(η^5-C$_5$H$_5$)(CO)$_2$](O^{t}Bu)$_3$	175.0(2)	177.0(2)	200
[Re(CCH$_2$^{t}Bu)F(dppe)$_2$][BF$_4$]	177.2(7)	167.9(7)	58
Re(CPh)(η^5-C$_5$H$_5$)[C(O)C$_2$B$_{10}$H$_{11}$](CO)	176.0(3)	168.0(3)	158
[Re(CNH$_2$)Cl(dppe)$_2$][BF$_4$]	180.2(4)	171.9(3)	68
[Re(CNH^{t}Bu)Cl(CN^{t}Bu)$_2$(PMePh$_2$)][SbF$_6$]	182.0(1)	175.7(9)	73
Re(C^{t}Bu)(η^5-C$_5$Me$_5$)Br$_2$	167.0(1)	173.0(1)	54
Re(C^{t}Bu)(η^5-C$_5$Me$_5$)Br$_3$	175.5(6)	179.1(5)	55
Re(CSiMe$_3$)(CH$_2$SiMe$_3$)$_3$Cl	172.6(11)	178.8(6)	53
Os(CTol)Cl(CO)(PPh$_3$)$_2$	178.0(2)	165.0(2)	56

^{a} Not reported.

C. *Gas-Phase, Matrix Isolation, and Surface Studies*

Aristov and Armentrout (*25*) and Freiser *et al.* (*26,27*) derived the bond energies of metal–carbon triple bonds from gas-phase experiments. The values are listed in Table III. Table III also lists bond energies for metal–carbon double and single bonds.

The cocondensation and irradiation of Zn atoms and diazomethane in an argon matrix were studied by Billups and co-workers (*28*). Irradiation of the cocondensate with light beyond 400 nm produces a species which is formulated as ZnCH$_2$, whereas further irradiation with light of wavelengths between 360 and 280 nm gives rise to HZnCH. The assignments are based on IR stretches at v equals 513 cm^{-1} for Zn=CH$_2$ and at v

TABLE II

New Alkylidyne Metal Complexes:
^{13}C NMR Data for the Alkylidyne Carbon and Characteristic IR Metal Absorptions

Compound[a]	^{13}C NMR[b] $\delta M\underline{C}R$ (J_{CW}, Hz)	IR[c] (cm^{-1})	Ref.
Nb[CSi(^{i}Pr)$_3$](CO)(dmpe)$_2$	253.2^b		73
Ta[COSi(^{i}Pr)$_3$](CO)(dmpe)$_2$	297.6^a	1994, 1912	73
Cr(CPh)Cl(CO)$_2$(tmeda)	302.5	1993, 1933	44
Cr(CPh)Br(CO)$_2$(PMe$_3$)$_2$	296.0^a	1954	97
Cr(CPh)Br(CO)[P(OMe)$_3$]$_3$	326.7	1997, 1935^c	105
Cr(CTol)(η^5-C$_5$H$_5$)(CO)$_2$	310.0	1997, 1913^c	121
Cr(CTol)[HB(pz)$_3$](CO)$_2$	329.7^a	1992, 1924	121
Cr(CC$_6$H$_3$Me$_2$-2,6)(η^5-C$_5$H$_5$)(CO)$_2$		2140, 2062, 2024, 1850	35
[Cr(CNC^{t}Bu$_2$)(CO)$_5$][BF$_4$]	200.4	2103, 2031, 2008, 1888	43
[Cr(CNCPh$_2$)(CO)$_5$][BF$_4$]		2104, 2032, 2005, 1890	43
[Cr(CNC(C$_6$H$_2$Me$_3$-2,4,6)$_2$)(CO)$_5$][BF$_4$]		2102, 1993	43
cis-Cr[CNC(C$_6$H$_2$Me$_3$-2,4,6)$_2$](BF$_4$)(CO)$_4$		2104, 2051	96
trans-Cr[CNC(C$_6$H$_2$Me$_3$-2,4,6)$_2$](BF$_4$)(CO)$_4$		2090, 2039^b	96
trans-Cr[CNC(C$_6$H$_2$Me$_3$-2,4,6)$_2$]Br(CO)$_4$		2107, 2005, 1994, 1926	43
[Cr(CNC(C$_{12}$H$_8$O))(CO)$_5$][BF$_4$]		2103, 2000, 1991, 1932	43
[Cr(CNC(C$_6$H$_4$OMe-4)$_2$)(CO)$_5$][BF$_4$]		2106, 2038, 2009, 1882	43
[Cr(CNC(C$_6$H$_4$Br-4)$_2$)(CO)$_5$][BF$_4$]		2153, 2070, 2062, 2130	43
[Cr(CNCCl$_2$)(CO)$_5$][AlCl$_4$]			43
Mo(CMe)(η^5-C$_9$H$_7$)[P(OMe)$_3$]$_2$	295.8^b		66
Mo(CMe)[HB(dmpz)$_3$](CO)$_2$	304.0	1982, 1889^a	48
Mo(CMe)[HB(dmpz)$_3$](CO)(PMe$_3$)	288.1	1841^a	48
Mo(CEt)[HB(dmpz)$_3$](CO)$_2$	310.6	1980, 1885^a	48
Mo(CPr)[HB(dmpz)$_3$](CO)$_2$	310.5	1977, 1882^a	48
Mo(CCH$_2$^{i}Pr)(η^5-C$_5$H$_5$)[P(OMe)$_3$]$_2$	302.5		66
Mo(CCH$_2$^{t}Bu)(η^5-C$_5$H$_5$)(CO)$_2$	332.8^b	2002, 1931^b	91
Mo(CCH$_2$^{t}Bu)(η^5-C$_5$H$_5$)(CO)[P(OMe)$_3$]	316.3^b	1915^b	91
Mo(CCH$_2$^{t}Bu)(η^5-C$_5$H$_5$)(η^2-C$_2$(CN)$_4$)[P(OMe)$_3$]	338.9^a	2222	90

Compound			Ref.
Mo(CCH$_2$'Bu)(η^5-C$_5$H$_5$)[P(OMe)$_3$]$_2$	299.8[b]		66
Mo(CCH$_2$'Bu)(η^5-C$_9$H$_7$)[P(OMe)$_3$]$_2$	302.2[b]		66
Mo(CCH$_2$Ph)(η^5-C$_5$H$_5$)(PEt$_3$)[P(OMe)$_3$]	294.1[b]		64
Mo(CCH$_2$Ph)(η^5-C$_5$H$_5$)[P(OMe)$_3$]$_2$	292.7[b]		64
Mo(CCH$_2$Ph)(η^5-C$_9$H$_7$)[P(OMe)$_3$]$_2$			66
Mo(CCH$_2$SiMe$_3$)(η^5-C$_9$H$_7$)[P(OMe)$_3$]$_2$	297.3[c]		66
Mo(C'Pr)[HB(dmpz)$_3$](CO)$_2$	314.4	1980, 1885[a]	48
Mo(CCHPh$_2$)(η^5-C$_5$H$_5$)[P(OMe)$_3$]$_2$			64
Mo[CCH(Ph)(SiMe$_3$)](η^5-C$_9$H$_7$)[P(OMe)$_3$]$_2$	297.1[b]		66
Mo(C'Bu)[H$_2$B(pz)$_2$](CO)[P(OMe)$_3$]$_2$	312.7*	1912[d]	124
Mo(CCMe$_2$Ph)(η^5-C$_5$H$_5$)[P(OMe)$_3$]$_2$	311.3[b]		64
Mo(CC(CN)(CO$_2$Et)(N$_2$C$_6$H$_4$NMe$_2$-4)][HB(dmpz)$_3$](CO)$_2$		2202, 2015, 1940[a]	165
Mo[CC(CN)$_2$Cu][HB(dmpz)$_3$](CO)$_2$		2202, 2180, 1975, 1892[c]	165
Mo[CC(CN)(CO$_2$Et)(Cu)][HB(dmpz)$_3$](CO)$_2$		2210, 1989, 1885[e]	165
Mo[CC(CN)(CO$_2$Et)(HgCl)][HB(dmpz)$_3$](CO)$_2$		2190, 2011, 1940[a]	165
Mo(CPh)(η^5-C$_5$H$_5$)[P(OMe)$_3$]$_2$	286.3[b]		98
Mo(CPh)[HB(dmpz)$_3$](CO)$_2$	288.8	1979, 1890[a]	48
Mo(CPh)Cl(CO)$_2$(tmeda)	274.3[a]	1997, 1912	44
trans,trans-Mo(CPh)Cl(CO)$_2$(PMe$_3$)$_2$	280.9	2006, 1946	97
Mo(CPh)Cl(CO)[P(OMe)$_3$]$_3$	272.8[a]	1964	105
Mo(CPh)Cl[P(OMe)$_3$]$_4$	265.8[a]		105
Mo(CPh)Br(CO)$_2$(py)$_2$	276.2[a]	2003, 1924	44
Mo(CPh)Br(PMe$_3$)$_4$	265.0[d]		82
[NEt$_4$]{Mo(CTol)(η^5-C$_2$B$_9$H$_9$Me$_2$)(CO)[P(OMe)$_3$]}	293.8	1877	129
Mo(CTol)(η^5-C$_5$H$_5$)(CO)$_2$			169
Mo(CTol)(η^5-C$_5$H$_5$)(CO)(PMe$_3$)	294.0	1900[c]	170
Mo(CTol)[HB(pz)$_3$](CO)$_2$	293.1	1998, 1921[c]	121
Mo(CTol)(CO)$_2${Co(η^5-C$_5$H$_5$)[P(O)(OMe)$_2$]$_3$}	288.2[a]	1977, 1881	125
Mo(CTol)Cl(CO)[P(OMe)$_3$]			129
Mo(CC$_6$H$_4$Me-2)(η^5-C$_5$H$_5$)(CO)$_2$	309.5[a]	1995, 1920	35
Mo(CC$_6$H$_3$Me$_2$-2,6)(η^5-C$_5$H$_5$)(CO)$_2$	310.5[a]	1992, 1919	35
Mo(CC$_6$H$_3$Me$_2$-2,6)(η^5-C$_5$H$_5$)(CO)(PPh$_3$)	303.7	1889	35
Mo(CC$_6$H$_4$OMe-2)(η^5-C$_5$H$_5$)(CO)$_2$	305.9[a]	1995, 1919[d]	35

(continued)

231

TABLE II (continued)

Compound[a]	^{13}C NMR[b] $\delta\underline{\text{M}}\text{CR}$ ($\underline{J}_{\text{CW}}$, Hz)	IR[c] (cm^{-1})	Ref.
Mo(CC$_6$H$_4$OMe-2)[H$_2$B(pz)$_2$](CO)[P(OMe)$_3$]$_2$	281*	1919[d]	124
Mo(CC$_6$H$_4$OMe-2)Cl(CO)[P(OMe)$_3$]$_3$			124
MoCr($\mu-\sigma$:η^6-CC$_6$H$_4$OMe-2)(η^5-C$_5$H$_5$)(CO)$_5$	295.6[a]	2009, 1967, 1932, 1903[d]	47
MoCr($\mu-\sigma$:η^6-CC$_6$H$_4$OMe-2)[HB(pz)$_3$](CO)$_5$	278.1[a]	2011, 1966, 1929, 1900[d]	47
MoCr($\mu-\sigma$:η^6-CC$_6$H$_4$OMe-2)(O$_2$CCF$_3$)(CO)$_5$(tmeda)	268.7[a]	2011, 1961, 1927, 1890[f]	47
MoCr($\mu-\sigma$:η^6-CC$_6$H$_4$OMe-2)(O$_2$CCF$_3$)(CO)$_5$(bpy)	274.4[a]	2009, 1960, 1928, 1894[f]	47
Mo$_2$Fe($\mu-\sigma,\sigma'$:η^5-CC$_5$H$_4$)[HB(pz)$_3$]$_2$(CO)$_4$	293.2[a]	1987, 1904	46
Mo$_2$Fe($\mu-\sigma,\sigma'$:η^5-CC$_5$H$_4$)$_2$(O$_2$CCF$_3$)$_2$(CO)$_4$(tmeda)	286.9[a]	1992, 1909, 1735[f]	46
Mo(CC$_6$H$_4$NMe$_2$-4)(η^5-C$_5$H$_5$)(CO)$_2$	312.7[a]	1991, 1924	35
Mo(CCC^{t}Bu)(η^5-C$_5$H$_5$)(CO)$_2$	274.9	2006, 1940[c]	45
Mo(CCC^{t}Bu)[HB(pz)$_3$](CO)$_2$	259	2001, 1919	45
Mo(CCC^{t}Bu)[HB(dmpz)$_3$](CO)$_2$	256.7	1993, 1908	45
Mo(CCC^{t}Bu)(O$_2$CCF$_3$)(CO)$_2$(tmeda)	252.3	2007, 1923, 1705	45
Mo(CCC^{t}Bu)(O$_2$CCF$_3$)(CO)$_2$(bpy)		2007, 1923, 1705	45
Mo(μ-^{t}BuC$_2$C)[HB(pz)$_3$](CO)$_2$Co$_2$(CO)$_6$	263.2	2089, 2053, 2035, 2029, 1978, 1902[c]	45
Mo(μ-^{t}BuC$_2$C)[HB(dmpz)$_3$](CO)$_2$Co$_2$(CO)$_6$	269.1[a]	2090, 2055, 2032, 1975, 1989	45
Mo(CNH$_2$)Cl(dppe)$_2$	225.9*		70
Mo(CNEt$_2$)(η^5-C$_5$H$_5$)(CO)$_2$	273.0	1952, 1866, 1558	106
Mo(CC^{i}Pr)(SC$_6$H$_2$^{i}Pr$_3$-2,4,6)$_3$	333.6		133
Mo{[C(CH$_2$)$_2$C]Mo(O^{t}Bu)$_3$}(O^{t}Bu)$_3$	281.4[b]		75
[Mo(CCH$_2$^{t}Bu)H(η^5-C$_5$H$_5$){P(OMe)$_3$}$_2$][BF$_4$]	346.4[a]		90
[Mo(CCH$_2$Ph)(η^5-C$_5$H$_5$)Br{P(OMe)$_3$}$_2$][BF$_4$]	384.7[a]		64
Mo(CC^{t}Bu)[NH(C$_6$H$_3$^{i}Pr$_2$-2,6)][OCMe(CF$_3$)$_2$]$_2$(dme)$_{0.5}$	316.3		181
Mo(CC^{t}Bu)[NH(C$_6$H$_3$^{i}Pr$_2$-2,6)][OC$_6$H$_3$^{i}Pr$_2$-2,6]$_2$	317		181
Mo(CC^{t}Bu)[NH(C$_6$H$_3$^{i}Pr$_2$-2,6)]Cl$_2$(dme)	321.2		154
Mo(CC^{t}Bu)(O$_2$CMe)$_3$	312.4[b]		152
Mo(CC^{t}Bu)(O$_2$C^{i}Pr)$_3$	311.2[b]		152
Mo(CC^{t}Bu)(O$_2$C^{t}Bu)$_3$	311.3[b]		152
Mo(CC^{t}Bu)(O$_2$CCF$_3$)$_3$(dme)	341.3[b]		152

232

Compound	Value	IR bands	Ref.
[Mo(C'Bu)(SC₆H₂Me₃-2,4,6)₃]₂	336.0		133
Mo(C'Bu)(SC₆H₂Me₃-2,4,6)₃	341.5		133
[NEt₃H][Mo(C'Bu)(SC₆H₂Me₃-2,4,6)₃Cl]	316.3		133
[Mo(CPh)(η⁵-C₅H₅)Cl(PMe₃)₂]Cl	354.6ᵃ		98
Mo(CPh)Br₃(PMe₃)₃	387ᵃ		82
Mo(CPh)Br₃(dme)	329.2		81
Mo(CNEt₂(η⁵-C₅H₅)Br₂(CO)	314.7	2034, 1612	72
Mo(CNEt₂)(η⁵-C₅H₅)I₂(CO)	309.3	2022, 1614	72
[NEt₄][W(CMe)(η⁵-C₂B₉H₉Me₂)(CO)₂]	309.7	1955, 1871	126
W(CMe)(η⁵-C₅H₅)(CO)₂	313.3ᵃ (208)	1979, 1900ᵈ	121
W(CMe)[H₂B(pz)₂](CO)₃	293.9 (189)	1976, 1881	116
W(CMe)[HB(pz)₃](CO)₂	295.2 (189)	1983, 1899ᵇ	122
W(CMe)Cl(CO)₂(py)₂	273.9ᵃ (197.5)	1982, 1889	44
W(CMe)Cl(CO)₂(dppe)	279.3ᵃ (200)	2001, 1928	44
[NEt₄][W(CMe)Cl(CO)₂(py)]	277.6	1961, 1857	103
[NEt₄][W(CMe)Cl₂(C₄H-N₂)(CO)(py)]	285.1, 281.2ᵃ·ᶜ	2037	103
[NEt₄][W(CMe)Cl₂(C₄H₂O₃)(CO)(py)]	285.8ᵃ (211.8)	2037	103
WCo(CMe)(CO)₈	299.4 (184)	2050, 2034, 1954ᶜ	112
W(CCH₂'Bu)(η⁵-C₅H₅)[P(OMe)₃]₂	284.5ᵇ		66
W(CCH₂'Bu)(η⁵-C₉H₇)[P(OMe)₃]₂	286.2ᵇ		66
W(CPr)(η⁵-C₅H₅)(CO)₂	327.2	1993, 1921	168
W(CPr)(η⁵-C₅H₅)(CO)[P(OMe)₃]	301.2ᵇ	1883ᵍ	99
W(CCH₂CO₂Me)F(dppe)₂	291		60
[W(CPh)(CN'Bu)₅][PF₆]	281.9	2191, 2154, 2102, 2060	106
[W(CPh)(CN'Bu)₅]I	279.9	2190, 2155, 2101, 2060	106
[NEt₄][W(CPh)(η⁵-C₂B₉H₉Me₂)(CO)₂]	297.3 (199)	1964, 1880	127
Li[W(CPh){C(O)Me}Br(CO)₂(PPh₃)]	267.2ⁱ	1986, 1916, 1522ᶠ	182
W(CPh)(CN)(CN'Bu)₄	269.3	2155, 2096, 2060	106
W(CPh)[HB(pz)₃](CO)₂		1987, 1905	123
W(CPh)(O₂CCF₃)(CO)₂(tmeda)	274.1ᵃ (194.2)	1987, 1895	44
W(CPh){Co(η⁵-C₅H₅)[PO(OMe)₃]₂]₃}(CO)₂	278.5ᵃ	1963, 1862	125
W(CPh){Co(η⁵-C₅H₅)[PO(O'Pr)₂]₃}(CO)₂	276.8ᵃ	1958, 1855	125
W(CPh)Cl(CO)₂(tmeda)	262.5ᵃ (198)	1985, 1892	44

(continued)

233

TABLE II (continued)

Compound[a]	^{13}C NMR[b] δMCR ($\underline{J}_{CW}$, Hz)	IR[c] (cm^{-1})	Ref.
W(CPh)Cl(CO)$_2$(py)$_2$	262.7[a] (198)	1985, 1897	44
W(CPh)Cl(CO)$_2$(bpy)	265.8[a]	1986, 1899	44
W(CPh)Cl(CN^{t}Bu)(CO)(PMe$_3$)$_2$		1909	82
W(CPh)Cl(CO)$_2$(PMe$_3$)$_2$	266.0[a] (194)	2000, 1926	44
trans,trans-W(CPh)Cl(CO)$_2$(PMe$_3$)$_2$	270.5[a] (183.1)	2018, 1928	97
W(CPh)Cl(CO)$_2$(dppe)	267.3[a]	2003, 1934	44
W(CPh)Cl(C$_4$H$_2$N$_2$)(CO)(py)$_2$	264.7, 264.3[a,e]	2218, 2049	103
W(CPh)Cl(C$_4$H$_2$N$_2$)(CO)(tmeda)	266.9[a]	2218, 2210, 2030	103
W(CPh)Cl(C$_4$H$_2$O$_3$)(CO)(py)$_2$	268.0[a] (208.2)	2044, 1806, 1738	103
W(CPh)Cl(C$_4$H$_2$O$_3$)(CO)(tmeda)	269.7[a] (208.2)	2032, 1806, 1736	103
W(CPh)Cl(CO)(py)(PMe$_3$)$_2$	255.6[a] (201.1)	1870	82
W(CPh)Cl(CO)(PMe$_3$)$_3$	261.0[a] (201.1)	1896	82
W(CPh)Cl(CO)(PMe$_3$)$_2$[P(OMe)$_3$]		1916	82
W(CPh)Cl(CO)[P(OMe)$_3$]$_3$	257.9[a] (161.6)	1946	105
W(CPh)Cl(dppe)$_2$	250.4[a]		105
W(CPh)Cl(dppe)[P(OMe)$_3$]$_2$	253.6[a] (186.7)		105
W(CPh)Cl[P(OMe)$_3$]$_4$	251.1[a]		105
W(CPh)Br(CNEt)$_3$(CN^{t}Bu)	260.2	2164, 2116, 2099	176
W(CPh)Br(CN^{t}Bu)$_4$	260.5	2156, 2097, 2064	107
W(CPh)Br(CN^{t}Bu)$_3$(CO)	263.9	2173, 2124, 1942	107
W(CPh)Br(CN^{t}Bu)$_2$(CO)$_2$	263.1	1988, 1900	107
W(CPh)Br(CO)$_2$(pic)$_2$	263.1	1988, 1900	107
W(CPh)Br(CO)$_2$(tmeda)	262.8[a] (198)	1983, 1892	44
trans,trans-W(CPh)Br(CO)$_2$(PPh$_3$)$_2$	271.4[a]	2029, 1937	97
W(CPh)Br(PMe$_3$)$_4$			82
W(CPh)I(CN^{t}Bu)$_4$	262.2	2153, 2097, 2064	106
W(CPh)I(CN^{t}Bu)$_2$(CO)$_2$	267.9	2184, 2167, 2015, 1960	106
W(CPh)I(CO)$_2$(pic)$_2$		1988, 1901	106
[NEt$_4$][W(CPh)Cl$_2$(CO)$_2$(py)]	267.2	1965, 1867	103

234

Compound	δ (J)	IR (ν_{CO})	Ref.
[NEt₄][W(CPh)Cl₂(C₄H₂N₂)(CO)(py)]	271.6, 270.4^{a,c} (208)	2210, 2046	103
[PPN][W(CPh)Cl₂(C₄H₂N₂)(CO)(py)]	269.8, 269.1^{a,c}	2210, 2046	103
[NEt₄][W(CPh)Cl₂(C₄H₂O₃)(CO)(py)]	273.8^a	2041, 1796, 1728	103
[PPN][W(CPh)Cl₂(C₄H₂O₃)(CO)(py)]	274.1	2046, 1796, 1727	103
[(CH₂)₄CN(CH₂)₄][W(CPh)Cl₂(CO)(PMe₃)₂]	256.3 (194)	1833	82
[N(PPh₃)₂][W(CTol)(η⁵-C₂B₅H₁₁)(CO)₂]	293.9 (200)	2525, 1965, 1880	126
[N(PPh₃)₂][W(CTol)(η⁵-C₂B₉H₉Me₂)(CO)₂]	298.3 (198)	2516, 1956, 1874	126
[P(CH₂Ph)Ph₃][W(CTol)(η⁵-C₂B₉H₉Me₂)(CO)₂]		1960, 1876	127
[PPh₄][W(CTol)(η⁵-C₂B₉H₉Me₂)(CO)₂]		1959, 1874	127
[Rh(dppe)₂][W(CTol)(η⁵-C₂B₉H₉Me₂)(CO)₂]	297.9	2517, 1956, 1874	126
[NEt₄][W(CTol)(η⁶-C₂B₁₀Me₂)(CO)₂]	299.8	1990, 1930	130
W(CTol)(η⁵-C₅Me₅)(CO)₂	301.3 (211)	1981, 1910^b	120
W(CTol)(η⁵-C₅H₅)(CO)(PMe₃)	285.1	1880	169
W(CTol)[H₂B(pz)₂](CO)₃	284.5 (189)	1980, 1888	116
W(CTol)[H₂B(pz)₂](CO)₂(PMe₃)	298.3	1989, 1899	116
W(CTol)[H₂B(pz)₂](CO)₂(PPh₃)	285.4	1991, 1905	116
W(CTol)[HB(pz)₃](CO)₂	284.8 (189)	1986, 1903	122
W(CTol)[B(pz)₄](CO)₂	286.4 (190)	1986, 1903	122
W(CTol)(CO)₂{Co(η⁵-C₅H₅)[PO(OMe)₃]₂}₃	279.1^a	1961, 1859	125
W(CTol)(CO)₂{Co(η⁵-C₅H₅)[PO(OPr)₂]₂}₃	277.5^a	1955, 1853	125
W(CTol)Cl(CO)(PMe₃)₃	261.9^a	1893	82
[NEt₄][W(CC₆H₄Me-2)(η⁵-C₂B₉H₉Me₂)(CO)₂]	298.1 (198)	1959, 1858^f	128
[NEt₄][W(CC₆H₃Me₂-2,6)(η⁵-C₂B₉H₉Me₂)(CO)₂]	299.2 (199)	1958, 1877^f	128
[PPh₄][W(CC₆H₃Me₂-2,6)(η⁵-C₂B₉H₉Me₂)(CO)₂]		1957, 1875^f	128
[NEt₄][W(CC₆H₃Me₂-2,6)(η⁶-C₂B₁₀H₁₀Me₂)(CO)₂]	299.8	1989, 1930	130
W(CC₆H₃Me₃-2,6)(η⁵-C₅H₅)(CO)₂	297.4 (207)	1992, 1922^c	35
W(CC₆H₃Me₂-2,6)(η⁵-C₅H₅)(CO)(PMe₃)	285.2	1868	35
W(CC₆H₃Me₂-2,6)(η⁵-C₅H₅)(CO)(PPh₃)	289.6	1879	35
WCr(μ-σ:η⁶-CC₆H₄OMe-2)(η⁵-C₅H₅)(CO)₅	282.9^a	1996, 1964, 1912, 1900^d	47
WCr(μ-σ:η⁶-CC₆H₄OMe-2)(O₂CCF₃)(CO)₅(tmeda)	257.1	1995, 1957, 1907, 1888^f	47
WCr(μ-σ:η⁶-CC₆H₄OMe-2)(O₂CCF₃)(CO)₅(bpy)	263.3^a	1999, 1956, 1906, 1886^f	47
W(C-2-C₄H₃O)[HB(pz)₃](CO)₂		1990, 1915^b	123
W(CC^{t}Bu)(η⁵-C₅H₅)(CO)₂	266.7 (226)	1988, 1929^c	45

(continued)

235

TABLE II (continued)

Compound[a]	^{13}C NMR[b] $\delta\underline{M}CR$ ($J_{C\underline{W}}$, Hz)	IR[c] (cm^{-1})	Ref.
W(CCC^{t}Bu)[HB(pz)$_3$](CO)$_2$	253.5 (202	1985, 1896	45
W(CCC^{t}Bu)(O$_2$CCF$_3$)(CO)$_2$(tmeda)	246.8 (208)	1993, 1903	45
W(CCC^{t}Bu)(O$_2$CCF$_3$)(CO)$_2$(py)$_2$		1995, 1908^f	45
W[μ-CC$_2^t$BuMo$_2$((η^5-C$_5$H$_5$)$_2$)(CO)$_4$][HB(pz)$_3$](CO)$_2$	272.9	2001, 1954, 1932, 1874, 1840	45
W[μ-CC$_2^t$BuCo$_2$(μ-dppm)(CO)$_4$][HB(pz)$_3$](CO)$_2$	278.5^a	2023, 1997, 1971, 1955, 1873	45
[W(CNEt$_2$)(CN^{t}Bu)$_5$]I	244.1	2167, 2131, 2089, 2041	109
[W(CNEt$_2$)(CN^{t}Bu)$_4$(CO)]I	250.1	2179, 2151, 2120, 2064, 1898	108
[W(CNEt$_2$)(CO)$_2$(bpy)(PMe$_3$)]I	253.3	1964, 1884	115
[W(CNEt$_2$)(CO)$_2$(bpy)(PEt$_3$)]I	252.7	1962, 1882	115
[W(CNEt$_2$)(CO)$_2$(ophen)(PMe$_3$)]I	253.5	1964, 1884	115
[W(CNEt$_2$)(CO)$_2$(ophen)(PEt$_3$)]I	252.6	1962, 1883	115
[W(CNEt$_2$)(CO)(bpy)(PMe$_3$)$_2$]I	254.1	1824	83
[W(CNEt$_2$)(CO)(ophen)(PMe$_3$)$_2$]I	253.8	1825	83
W(CNEt$_2$)(η^5-C$_5$H$_5$)(CN^{t}Bu)(CO)	259.9^b	1925, 1829	88
W(CNEt$_2$)(η^5-C$_5$Me$_5$)(CO)$_2$	268.7	1930, 1841	87
W(CNEt$_2$)[HB(pz)$_3$](CO)$_2$		1930, 1830	123
cis-W(CNEt$_2$)[PPh$_2$Cr(CO)$_5$](CO)$_2$(bpy)	259.7	2042, 1962, 1945, 1918, 1892, 1859^f	114
trans-W(CNEt$_2$)[PPh$_2$Cr(CO)$_5$](CO)$_2$(bpy)		2042, 1962, 1945, 1918, 1892, 1859^f	114
cis-W(CNEt$_2$)[PPh$_2$W(CO)$_5$](CO)$_2$(bpy)		2052, 1965, 1943, 1918, 1890, 1857^f	114
trans-W(CNEt$_2$)[PPh$_2$W(CO)$_5$](CO)$_2$(bpy)		2052, 1965, 1943, 1918, 1890, 1857^f	114
cis-W(CNEt$_2$)[PPh$_2$Cr(CO)$_5$](CO)$_2$(ophen)	259.2	2041, 1960, 1943, 1920, 1892, 1858^f	114
trans-W(CNEt$_2$)[PPh$_2$Cr(CO)$_5$](CO)$_2$(ophen)		2041, 1960, 1943, 1920, 1892, 1958^f	114
W(CNEt$_2$)(AsPh$_2$)(CO)$_2$(bpy)	248.7^f	1932, 1847^f	113
cis-W(CNEt$_2$)[AsPh$_2$Cr(CO)$_5$](CO)$_2$(bpy)	258.7	2040, 1947, 1917, 1892, 1860^f	114
trans-W(CNEt$_2$)[AsPh$_2$Cr(CO)$_5$](CO)$_2$(bpy)	246.6	2040, 1947, 1917, 1892, 1860^f	114
cis-W(CNEt$_2$)[AsPh$_2$Cr(CO)$_5$](CO)$_2$(ophen)	258.2	2040, 1947, 1916, 1892, 1860^f	114
trans-W(CNEt$_2$)[AsPh$_2$Cr(CO)$_5$](CO)$_2$(ophen)	246.2	2040, 1947, 1916, 1892, 1860^f	114
cis-W(CNEt$_2$)[AsPh$_2$Mo(CO)$_5$](CO)$_2$(bpy)	259.6	2052, 1944, 1928, 1895, 1858^f	114
trans-W(CNEt$_2$)[AsPh$_2$Mo(CO)$_5$](CO)$_2$(bpy)	246.5	2052, 1944, 1928, 1895, 1858^f	114

cis-W(CNEt$_2$)[AsPh$_2$Mo(CO)$_5$](CO)$_2$(ophen)	259.2	2053, 1948, 1930, 1895, 1859[f]	114
trans-W(CNEt$_2$)[AsPh$_2$Mo(CO)$_5$](CO)$_2$(ophen)	246.6	2053, 1948, 1930, 1895, 1859[f]	114
cis-W(CNEt$_2$)[AsPh$_2$W(CO)$_5$](CO)$_2$(bpy)	259.1	2052, 1960, 1945, 1918, 1890, 1858[f]	114
trans-W(CNEt$_2$)[AsPh$_2$W(CO)$_5$](CO)$_2$(bpy)	246.6	2052, 1960, 1945, 1918, 1890, 1858[f]	114
cis-W(CNEt$_2$)[AsPh$_2$W(CO)$_5$](CO)$_2$(ophen)	258.7	2053, 1960, 1949, 1917, 1891, 1860[f]	114
trans-W(CNEt$_2$)[AsPh$_2$W(CO)$_5$](CO)$_2$(ophen)	246.6	2053, 1960, 1949, 1917, 1891, 1860[f]	114
cis-W(CNEt$_2$)[SbPh$_2$Cr(CO)$_5$](CO)$_2$(bpy)	257.7	2032, 1947, 1916, 1899, 1861[f]	114
trans-W(CNEt$_2$)[SbPh$_2$Cr(CO)$_5$](CO)$_2$(bpy)	247.7	2032, 1947, 1916, 1899, 1861[f]	114
cis-W(CNEt$_2$)[SbPh$_2$Cr(CO)$_5$](CO)$_2$(ophen)	257.6	2030, 1948, 1916, 1899, 1863[f]	114
trans-W(CNEt$_2$)[SbPh$_2$Cr(CO)$_5$](CO)$_2$(ophen)	247.9	2030, 1948, 1916, 1899, 1863[f]	114
W(CNEt$_2$)(OSO$_2$CF$_3$)(CN^{t}Bu)$_3$(CO)	247.4	2168, 2122, 2067, 1885	138
W(CNEt$_2$)(OSO$_2$CF$_3$)(CN^{t}Bu)$_2$(CO)$_2$	247.0	2182, 2157, 1989, 1918	138
W(CNEt$_2$)(OSO$_2$CF$_3$)(CO)$_2$(PMe$_3$)$_2$	247.5	1980, 1893	138
[NEt$_4$][W(CNEt$_2$)(S$_2$C$_4$N$_2$)(CO)$_2$]	298.2[f]	2196, 1903, 1821[h]	114
W(CNEt$_2$)Cl(CO)$_2$(bpy)	241.0[f]	1948, 1850	113
W(CNEt$_2$)I(CN^{t}Bu)$_4$	230.7[c]	2088, 2040	109
W(CNEt$_2$)I(CN^{t}Bu)$_3$(CO)	236.7	2151, 2112, 2068, 1882	108
W(CNEt$_2$)I(CNMe)$_2$(CO)$_2$	234.9	2199, 2175, 1985, 1915	85
W(CNEt$_2$)I(CN^{t}Bu)$_2$(CO)$_2$	236.0	2170, 2143, 1980, 1911	85
W(CNEt$_2$)I(CO)$_3$(PMe$_3$)	241.0	2044, 1969, 1932	84
W(CNEt$_2$)I(CO)$_2$(py)$_2$	237.8	1953, 1855	84
W(CNEt$_2$)I(CO)$_2$(pic)$_2$	237.9	1951, 1852	87
cis-W(CNEt$_2$)I(CO)$_2$(PMe$_3$)$_2$	236.1	1970, 1884	84
trans-W(CNEt$_2$)I(CO)$_2$(PMe$_3$)$_2$	246.3	1870	84
W(CNEt$_2$)I(CO)(PMe$_3$)$_3$	240.2	1836	84
[PPN][W(CN^{i}Pr$_2$)Cl$_3$(CO)$_3$]	246.2	2027, 1924	101
W[CN(C$_6$H$_{11}$)$_2$]Cl(CO)$_2$(PPh$_3$)$_2$	239.6	2049, 1970, 1930	101
[PPN]{W[CN(C$_6$H$_{11}$)$_2$]Cl$_3$(CO)$_3$}	247.2	2032, 1924	101
[C$_{15}$H$_{15}$N$_2$]{W[CN(C$_6$H$_{11}$)$_2$]Br$_2$(CO)$_3$}	248.2	2036, 1925	102
[NEt$_4$]{W[CN(C$_6$H$_{11}$)$_2$]Br$_2$(CO)$_3$}	248.8	2034, 1925	101
trans-W(CNPh$_2$)Br(CO)$_4$	225.0[d]	2113, 2038, 2002, 1987[b]	42
[W(CNC^{t}Bu$_2$)(CO)$_5$][BF$_4$]		2129, 2055, 2022, 1846	43
[W(CNCPh$_2$)(CO)$_5$][BF$_4$]		2117, 2024, 1999, 1880	43

(continued)

237

TABLE II (continued)

Compound[a]	¹³C NMR[b] $\delta M\underline{C}R$ (J_{CW}, Hz)	IR[c] (cm⁻¹)	Ref.
trans-W(CNCPh₂)Br(CO)₄	197.8[d]	2105, 2019, 1996, 1540[b]	42
[W(CNC(C₆H₂Me₃-2,4,6)₂)(CO)₅][BF₄]		2115, 2020, 1997, 1881	43
trans-W[CNC(C₆H₂Me₃-2,4,6)₂](BF₄)(CO)₄		2107, 2025	96
trans-W[CNC(C₆H₂Me₃-2,4,6)₂]Cl(CO)₄		2104, 2014	96
trans-W[CNC(C₆H₂Me₃-2,4,6)₂]Br(CO)₄		2103, 2021, 2010, 1515[b]	42
[W(CNC(C₁₂H₈O))(CO)₅][BF₄]		2117, 1998, 1988, 1919	43
trans-W[CNC(C₁₂H₈O)]Br(CO)₄		2093, 1995, 1982, 1575[b]	42
[W(CNC(C₆H₄OMe-4)₂)(CO)₅][BF₄]	197.6[d]	2113, 1993, 1984, 1918	43
trans-W[CNC(C₆H₄OMe-4)₂]Br(CO)₄		2099, 2022, 2011, 1995, 1560[b]	42
[W(CNC(C₆H₄Br-4)₂)(CO)₅][BF₄]		2117, 2028, 1999, 1872	43
trans-W[CNC(C₆H₄Br-4)₂]Br(CO)₄		2106, 2016, 1990, 1540[b]	42
[W(CPMe₃)(HB(dmpz)₃)(CO)₂][PF₆]	253.4[g]	2015, 1935	164
[W(CPEt₃)(HB(dmpz)₃)(CO)₂][PF₆]		2020, 1935	164
W(CPPh₂)Cl₂(CO)(PMePh₂)₂	218.7	1903[i]	172
W(CPPh₃)Cl₂(CO)(PMePh₂)₂	215.4[a]	1905[i]	172
[W(CSMe)(HC(pz)₃)(CO)₂][PF₆]	276.0[i]	1991, 1906	74
W(CMe)[HB(pz)₃](SMe)₂	303.9		143
W(CMe)(SC₆H₂ⁱPr₃-2,4,6)₃	312.1		133
W(CMe)(SC₆H₂ⁱPr₃-2,4,6)₃(py)	308.2		133
[NEt₃H][W(CMe)(SC₆H₂ⁱPr₃-2,4,6)Cl]	294.8		133
W(CMe)Br₃(dme)	337.5 (219)		81
[W(CEt)(CH₂ⁱBu)(OⁱPr)₂]₂	278.3[c] (292)		78
[W(CEt)(CH₂ⁱBu)(OⁱPr)₂(py)]₂	277.3[c]		78
[W(CEt)(NHMe₂)(OⁱPr)₂(μ-ⁱPr)]₂			77
W(CH₂ⁱBu)Br₃(dme)	354.4[a] (215)		81
W[[C(CH₂)₂C]W(OⁱBu)₃](OⁱBu)₃	259.5[b]		75
W[[C(CH₂)₄C]W(OⁱBu)₃](OⁱBu)₃	261.4[b]		75
W[[C(CH₂)₅C]W(OⁱBu)₃](OⁱBu)₃	261.6[b]		75
W[[C(CH₂)₆C]W(OⁱBu)₃](OⁱBu)₃	261.7[b]		75

Compound			Ref.
$W(C'Bu)(NHC_6H_3{}^iPr_2\text{-}2,6)[OCMe(CF_3)_2]_2(dme)$	291.6[b] (279)		155
$W(C'Bu)(NHC_6H_3{}^iPr_2\text{-}2,6)Cl_2(dme)$	304.5[b]		155
$W(C'Bu)(OC_6H_3{}^iPr_2\text{-}2,6)_3(dmf)$	292.4[b]		207
$W(C'Bu)(O_2CMe)_3$	286.5[b]		152
$W(C'Bu)(O_2C^iPr)_3$	298.2[b] (244)		152
$W(C'Bu)(O_2C'Bu)_3$	285.5[b]		152
$W(C'Bu)(O_2CCF_3)_3(dme)$	319.7		152
$[W(C'Bu)(SC_6H_2Me_3\text{-}2,4,6)_3]_2$	324.3		133
$W(C'Bu)(SC_6H_2{}^iPr_3\text{-}2,4,6)_3$	329.4		133
$[NEt_3H][W(C'Bu)(SC_6H_2Me_3\text{-}2,4,6)_3Cl]$	304.0 (216)		133
$[NEt_3H][W(C'Bu)(SC_6H_2{}^iPr_3\text{-}2,4,6)_3Cl]$	306.9		133
$W(C\text{-}cis\text{-}CHCHCCEt)[OCMe_2(CF_3)]_3(bpy)$	271.2[b]		199
$W[[C\text{-}cis,cis\text{-}(CH)_2CC(CH)_2C]W(O'Bu)_3](O'Bu)_3$	257.8[b]		199
$W[[C\text{-}trans\text{-}CHCHC]W(O'Bu)_3](O'Bu)_3$	255.5[b]		199
$W[[C\text{-}trans\text{-}CHCHC]W[OC(CF_3)Me_2]_3][OC(CF_3)Me_2]_3$	260.3[b]		199
$W(CPh)HBr_2(PMe_3)_3$	269.5		82
$W(CPh)[OCH(CF_3)_2](dme)$	281[b]		202
$W(CPh)Br_3(dme)$	331.7[b] (219)		81
$W[C(1,2\text{-}C_6H_4)CC(1,2\text{-}C_6H_4)CW(O'Bu)_3]('BuO)_3$	259.6[j]		199
$W[C(1,4\text{-}C_6H_4)CW(O'Bu)_3]('BuO)_3$	257.8[b]		199
$W[CCW(O'Bu)_3]('BuO)_3$	278.6[j]		75
$W(CNMe_2)(O^iPr)_3(py)_2$			185
$W(CNEt_2)(\eta^5\text{-}C_5Me_5)Br_2(CO)$	306.0	1987	87
$[W(CNEt_2)(\eta^5\text{-}C_5H_5)I(CN'Bu)_2]I$	302.3	2172	88
$[W(CNEt_2)(\eta^5\text{-}C_5H_5)I(CNEt)(CO)][PF_6]$	302.7	2231, 2053, 1636	71
$[W(CNEt_2)(\eta^5\text{-}C_5H_5)I(CN'Bu)(CO)]I$	302.3	2199, 2053	88
$[W(CNEt_2)(\eta^5\text{-}C_5Me_5)I(CNEt)(CO)][PF_6]$	298.2	2210, 2036, 1620	71
$W(CNEt_2)(\eta^5\text{-}C_5H_5)I_2(CN'Bu)$	305.5	2145	88
$W(CNEt_2)(\eta^5\text{-}C_5H_5)I_2(CO)$	303.2	2008	86
$W(CNEt_2)(\eta^5\text{-}C_5Me_5)I_2(CO)$	299.6	1984	87
$W(CNEt_2)(O^iPr)_3(py)_2$			185
$[W(CNEt_2)I_2(CNMe)_4]I$	250.4	2250, 2216	85
$[W(CNEt_2)I_2(CN'Bu)_4]I$	251.1	2216, 2203, 2183	85

(continued)

TABLE II (*continued*)

Compound[a]	^{13}C NMR[b] $\delta\underline{M}CR$ ($\underline{J}_{CW}$, Hz)	IR[c] (cm^{-1})	Ref.
[W(CNEt$_2$)I$_2$(CN^{t}Bu)$_2$(bpy)$_2$]I	265.3	2169	132
[W(CNEt$_2$)I$_2$(CN^{t}Bu)$_2$(ophen)$_2$]I	265.5	2169	132
W(CNEt$_2$)I$_3$(CNMe)$_2$(CO)		2226, 2041	85
W(CNEt$_2$)I$_3$(CN^{t}Bu)$_2$(CO)	257.4	2196, 2040	85
W(CNEt$_2$)I$_3$(CO)(bpy)		1982	83
W(CNEt$_2$)I$_3$(CO)(ophen)		1980	83
W(CNEt$_2$)I$_3$(CO)(PMe$_3$)$_2$	284.3		84
W(CSMe)[HB(pz)$_3$](SMe)$_2$	268.0	1912	143
W(CSEt)[HB(pz)$_3$](SMe)$_2$,			143
W(CSMe)[HB(pz)$_3$](SMe)(SEt)[d]	271.1		143
W(CSMe)[HB(pz)$_3$](SMe)(SPh)	270.8		143
W(CSMe)[HB(pz)$_3$](SMe)(STol)			143
W[CRu(η^5-C$_5$H$_5$)(CO)$_2$](O^{t}Bu)$_3$	237.3[b] (290.1)	2030, 1980[b]	200
[Re(CCH$_2^t$Bu)F(dppe)$_2$][BF$_4$]			58
[Re(CCH$_2^t$Bu)Cl(dppe)$_2$][BF$_4$]	280.2		58
[Re(CCH$_2$Ph)Cl(dppe)$_2$][BF$_4$]	269.9		58
[Re(C^{t}Bu)(η^5-C$_5$H$_5$)(PMe$_3$)$_2$]Cl	304.1[a]		54
[Re(C^{t}Bu)(η^5-C$_5$H$_5$)(PMe$_3$)$_2$]Br	304.7		55
Re(C^{t}Bu)Cl$_2$(PMe$_3$)$_3$	279.3[a]		54
Re(CPh)(η^5-C$_5$H$_5$)[C(O)C$_2$B$_{10}$H$_{11}$](CO)		1982	158
[Re(CNH$_2$)Cl(dppe)$_2$][BF$_4$]	222.4	1595*	68
[Re(CNHMe)Cl(CN^{t}Bu)$_2$(PMePh$_2$)$_2$][SbF$_6$]	228.5	2171, 1594	69
[Re(CNH^{t}Bu)Cl(CN^{t}Bu)$_2$(PMePh$_2$)$_2$][SbF$_6$]	227.5	2147, 1588[a]	73
[Re(CTol)(η^5-C$_5$H$_5$)(CO)$_2$][BPh$_4$]	316.9	2033, 2087	40
[Re(CTol)(η^5-C$_5$H$_5$)(CO)$_2$][BCl$_4$]	316.3	2033, 2087	40
[Re(CTol)(η^5-C$_9$H$_7$)(CO)$_2$][BF$_4$]	319.2	2086, 2034	41
[Re(CC$_6$H$_3$Me$_2$-2,6)(η^5-C$_9$H$_7$)(CO)$_2$][BF$_4$]	312.0	2083, 2033	41
Re(C^{t}Bu)(η^5-C$_5$Me$_5$)Cl$_3$			54
Re(C^{t}Bu)(η^5-C$_5$Me$_5$)Br$_2$			54

Compound				Ref.
Re(C'Bu)(η⁵-C₅Me₅)I₂				54
Re(C'Bu)(CHEt)[OCMe(CF₃)₂]₂	299.7, 303.2ᵉ			52
Re(C'Bu)(CH'Bu)[OCMe(CF₃)₂]₂	295.8ᵇ			52
[Re(C'Bu)(CH'Bu)Cl₂]ₓ	293.9ⁱ			52
Re(C'Bu)(CH'Bu)Cl₂[(NH₂)₂(C₆H₄-1,2)]	295.6			52
(Re(C'Bu)(CH'Bu)Cl₂[NH₂(C₆H₃'Pr₂-2,6)])₂	292.1ʲ			52
Re(C'Bu)(NC₆H₃'Pr₂-2,6)(O'Bu)₂		291.0		51
Re(C'Bu)(NC₆H₃'Pr₂-2,6)[OCMe₂(CF₃)]₂		297.6		51
Re(C'Bu)(NC₆H₃'Pr₂-2,6)[OCMe(CF₃)₂]₂		304.6		51
Re(C'Bu)(NC₆H₃'Pr₂-2,6)(OC₆H₃'Pr₂-2,6)₂		304.4		51
[Re(C'Bu)(η⁵-C₅Me₅)Cl₂][SbF₆]		338.0ᵃ		54
Re(C'Bu)(η⁵-C₅-Me₅)Cl₃		373.8ᵃ		55
Re(C'Bu)(η⁵-C₅Me₅)Cl₂I		368.4ᵃ		55
Re(C'Bu)(η⁵-C₅Me₅)Cl₂Br		373.6ᵃ		55
Re(C'Bu)(η⁵-C₅Me₅)ClBr₂		373.6ᵃ		55
Re(C'Bu)(η⁵-C₅Me₅)ClI₂		363.6ᵃ		55
Re(C'Bu)(η⁵-C₅Me₅)Br₃		371.8		55
Re(C'Bu)(η⁵-C₅Me₅)I₃				55
[NEt₄][Re(C'Bu)[NH(C₆H₃'Pr₂-2,6)]Cl₄}	315.9			51
Re(CSiMe₃)(CH₂SiMe₃)₃Cl				53
Os(CTol)Cl(CO)(PPh₃)₂			1864	56
[Ir(CMe)Cl(P'Pr₃)₂][BF₄]	261.5ⁱ			61
[Ir(CEt)Cl(P'Pr₃)₂][BF₄]	265.7ⁱ			61
[Ir(CCH₂Ph)Cl(P'Pr₃)₂][BF₄]				61

ᵃ List of abbreviations: bpy, 2,2'-bipyridine; dme, dimethoxyethane; dmf, dimethylformamide; dmpe, bisdimethylphosphinoethane; dmpz, bisdi-methylpyrazolyl; dppe, bisdiphenylphosphinoethane; dppm, bisdiphenylphosphinomethane; ophen, *ortho*-phenanthroline; PPN, bis(triphenylphos-phine)iminium; pic, 4-picoline; py, pyridine; pz, pyrazolyl; thf, tetrahydrofuran; tmeda, tetramethylethylenediamine; Tol, *p*-tolyl.

ᵇ ¹³C NMR data were determined in CD₂Cl₂ solution unless otherwise indicated as follows: (a) CDCl₃; (b) C₆D₆; (c) toluene-*d₆*; (d) acetone-*d₆*; (e) two diastereomers; (f) DMF; (g) CD₃CN; (h) CD₃NO₂; (i) thf-*d₈*; (i) pyridine-*d₅*.

ᶜ IR data were determined in CH₂Cl₂ solution unless otherwise indicated: (a) KBr; (b) hexane; (c) light petroleum ether; (d) ether; (e) nujol; (f) thf; (g) CD₃ CN; (h) acetone; (i) fluorolube mull.

ᵈ Inseparable mixture.

* Solvent not reported.

241

TABLE III

BOND DISSOCIATION ENERGIES D° (kJ/mol)

	La (27)	V (25)	Nb (27)	Fe (26)	Co (26)
$M\equiv CH$	523	477	607	423	418
$M=CH_2$	444	318	456	343	351
$M-CH_3$		209		272	238

equals 647 cm^{-1} for Zn$\equiv$CH. The cocondensation of iron atoms and diazomethane in an argon matrix gave directly FeCH$_2$ and N$_2$FeCH$_2$ (29). UV photolysis of the matrix resulted in conversion of FeCH$_2$ to HFeCH. Photolysis of the carbyne through a cutoff filter (with $\lambda < 400$ nm) results in the reformation of Fe(CH$_2$). An infrared absorption at 624 cm^{-1} was assigned to the Fe$=$CH$_2$ stretch, and an absorption at 674 cm^{-1} was assigned to the HFe$\equiv$CH stretch.

Metal alkylidyne fragments are frequently invoked as intermediates in the transformation of hydrocarbons on metal surfaces. These species are usually formulated as triply bridging alkylidynes; however, terminal surface alkylidynes may be considered as reactive surface intermediates (30). Evidence for metal carbyne intermediates on Pt—Co bimetallic surfaces was found in a study of the isomerization and hydrogenolysis of C$_6$ alkanes (31).

D. Bonding Description

Current bonding models for metal–carbon triple bonds were reviewed by Hofmann and co-workers (6). Metal–carbon triple bonds are generally described as consisting of one σ or two π bonds. The electronic structure of Cr(CH)Cl(CO)$_4$ was studied by Poblet et al. using ab initio methods at the SCF level (32). With the inclusion of electron correlation a highly covalent bonding model is obtained. The alkylidyne carbon was found to carry a partial negative charge. The bond energy of the chromium–carbon triple bond was calculated to be 481 kJ/mol, in good agreement with a values obtained from gas-phase studies (Table III).

The origin of inequivalent chromium–carbonyl bond lengths and the bending of the methylcarbyne ligand in W(CCH$_3$)Cl(CO)$_4$, as previously determined by X-ray crystallography (33), was investigated by Low and Hall (34). The bending of the carbyne ligand was found to be primarily a consequence of the interaction of a p orbital of the carbyne carbon with a C—H bond of the methyl group. This electronic feature weakens one of the two MC p bonds and causes the inequivalent bonding of the carbonyl ligands. The valence orbitals of $Mo(CCH_3)(\eta^5\text{-}C_5H_5)(CO)_2$ with a bent geometry on the carbyne carbon were reported by Stone and co-workers (*35*). The interaction of the $Mo\equiv C$ fragment with $Fe(CO)_3$ and $Fe(CO)_4$ was studied.

In a study by Brower, Templeton, and Mingos the electronic properties of bis-oxo, oxo-carbyne, and bis-carbyne metal complexes were compared (*36*). The question of π conflict between these ligands was investigated. Calculations for $[W(CH)_2(H)_4]^{4-}$ showed that both cis and trans bis-carbyne complexes will be destabilized. Based on calculations for $[W(CH)(O)(H)_4]^{3-}$, oxo-carbyne complexes are expected to be viable molecules, especially in the cis form.

Carvalho *et al.* reported molecular orbital calculations on the electron-rich rhenium isocyanide complex $ReCl(CNH)(H_2PCH_2CH_2PH_2)$ and the aminocarbyne complex $[ReCl(CNH_2)(H_2PCH_2CH_2PH_2)]^+$ (*37*). They found that in the isocyanide complex a negative charge is accumulated on the Cl and N atoms. These two sites may therefore compete for electrophiles; however, the product of N attack is thermodynamically favored.

The influence of hybridization effects on metal–ligand bond lengths was analyzed by Dobbs and Hehre (*38*). Principles similar to those guiding the variations of bond lengths for main group elements were found to be effective for transition metals. Metal–ligand bonds are considered to be made up of primarily nd and $(n + 1)$ s-type orbitals. Valence $(n + 1)$ p-type functions are assumed to play a lesser role. The d functions are less diffuse than the higher energy s- (or p-) type orbitals, and hybridization changes which lead to reduced d character result in an increase of metal–ligand σ bond lengths. (This trend is opposite to main group elements where an increase in s character leads to shorter bonds.) Thus, the M—H bond lengths increase slightly in the series H_3Ti—$CH_3 < H_2Ti$=$CH_2 <$ $HTi\equiv CH$, owing to the change of hybridization (sd^3, sd^2, sd) on the metal center.

III

SYNTHESIS OF METAL–CARBON TRIPLE BONDS

A. *Removal of Substituents from α-Carbon Atoms*

1. *Alkoxide Abstraction from Alkoxycarbene Ligands*

The first transition metal carbyne complexes, $[M(CR)X(CO)_4]$, were synthesized in Fischer's laboratory by treatment of alkoxycarbene pentacarbonylmetal complexes of chromium, molybdenum, and tungsten with boron trihalides (*39*). BBr_3 is reactive enough to generate the metal carbynes below their temperature of decomposition [Eq. (1)]. If pentane is

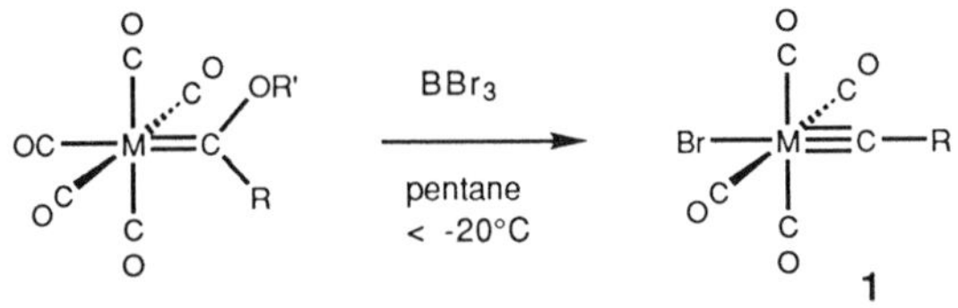

$$ (1) $$

M = Cr, Mo, W
R = alkyl, aryl, amino
R' = Me, Et

used as the solvent the *trans*-alkylidyne(bromo)tetracarbonyl complexes of chromium and tungsten, **1**, precipitate in pure form from the reaction solution. The corresponding molybdenum complexes are difficult to isolate in pure form owing to their high thermolability. The reaction shown in Eq.(1) is the method of choice if isolation of the *trans*-alkylidyne(bromo)tetracarbonylmetal complexes is desired. Several new cationic rhenium alkylidyne complexes were synthesized by reaction of alkoxy and siloxycarbene rhenium complexes with boron trihalides according to Eqs. (2) and (3) (*40,41*).

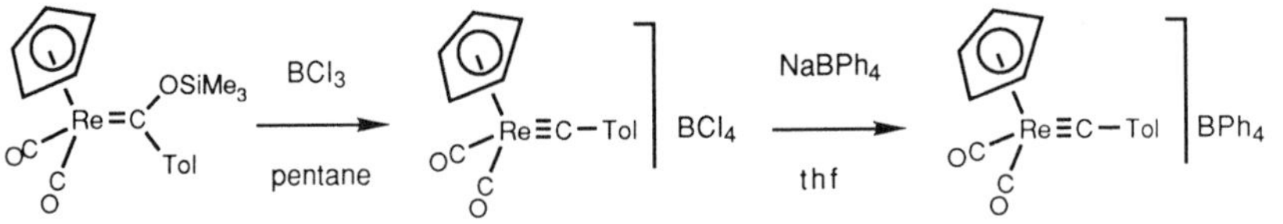

$$ (2) $$

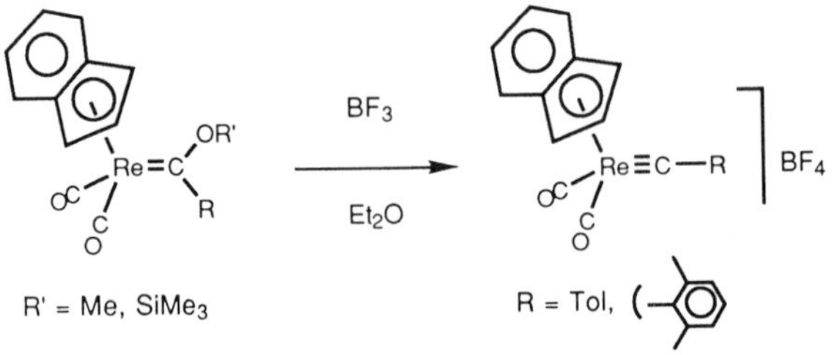

$$ (3) $$

R' = Me, SiMe₃   R = Tol, (─⬡)

H. Fischer and co-workers synthesized several 2-azaallenylidene metal complexes by the α-alkoxide abstraction reaction shown in Eq. (4) (*42,43*). 2-Azaallenylidene complexes are closely related to aminocarbyne complexes [see, e.g., Eq. (80), Section IV,C].

$$M = Cr; \ M' = B, \ X = F \qquad (4)$$
$$M = W; \ M' = Al, \ X = Br$$

$M = Cr, W; \ CR_2 = CPh_2, \ C(-\text{Ⓞ}-X)_2 \ (X = Br, OMe), \ C(-\text{Ⓞ}-)_2,$

$M = Cr; \ CR2 = C(t\text{-}Bu)_2$

2. *Oxide and Oxygen Abstraction from Acyl Ligands*

An efficient method for the formal abstraction of oxide from acyl ligands was developed in our laboratory (*44*). Oxalyl halides react directly with the pentacarbonylmetal acyl complexes of chromium, molybdenum, and tungsten to form the *trans*-alkylidyne(halo)tetracarbonyl complexes [Eq. (5)]. Other suitable Lewis acids are $COCl_2$, Cl_3COCl, $Cl_3COCOCl$, and

$$(5)$$

$M = Cr, Mo, W; \quad X = Cl, Br$
$R = alkyl, aryl; \quad L_2 = 2py, bpy, tmeda$

$(CF_3CO)_2O$. For the reaction of $[NMe_4][W\{C(O)(C_6H_4\text{-}OMe\text{-}4)\}(CO)_5]$ with $C_2O_2Br_2$ the (bromooxalato)carbene intermediate $W\{C(OCO-COBr)(C_6H_4\text{-}OMe\text{-}4)\}(CO)_5$ was characterized spectroscopically and demonstrated to transform cleanly into the carbyne complex $W(CC_6H_4\text{-}OMe\text{-}4)Br(CO)_4$. The tetracarbonylmetal carbyne complexes may be isolated from these reactions. However, the synthetic scheme is designed to include substitution of two carbonyl ligands by donor ligands [Eq. (5)]. With nitrogen-based donor ligands such as pyridine or tmeda the products are thermally more stable than the tetracarbonylmetal systems yet still retain a high degree of reactivity owing to the coordinative lability of the nitrogen-based donor ligands. Stone and co-workers further developed this method, using primarily trifluoroacetic anhydride as the Lewis acid, and prepared a large variety of new carbyne complexes [Eqs. (6)–(8)] (*35,45–47*). A similar synthetic procedure was employed by Templeton for the synthesis

$$M(CO)_6 \xrightarrow[\substack{thf}]{\substack{1.\ LiR \\ 2.\ (CF_3CO)_2O \\ -78\ to\ -10°C \\ 3.\ L_2}} \quad CF_3\!-\!C(=O)\!-\!O\!-\!M(L)(L)(CO)(CO)\!\equiv\!C\!-\!R \tag{6}$$

$$M = Mo,\ W;\ R = C\!\equiv\!C\text{-}t\text{-Bu},\ (\text{—MeO-C}_6H_4\text{-Cr(CO)}_3),\ (\text{—C}_5H_4\text{-Fe-C}_5H_5\text{—})\ ;\ L_2 = 2py,\ tmeda,\ bpy$$

$$CF_3\!-\!C(=O)\!-\!O\!-\!M(L)(L)(CO)(CO)\!\equiv\!C\!-\!R \xrightarrow{L_3^-} L_3\!-\!M(CO)(CO)\!\equiv\!C\!-\!R \tag{7}$$

$$M = Mo,\ W;\ R = C\!\equiv\!C\text{-}t\text{-Bu},\ (\text{—MeO-C}_6H_4\text{-Cr(CO)}_3)\ ;\ L_3 = C_5H_5,\ HB(\text{—N}_2C_3H_3)_3,\ (\text{—N}_2C_3HMe_2)_3$$

$$M(CO)_6 \xrightarrow[\substack{Et_2O}]{\substack{1.\ LiR,\ r.\ t. \\ 2.\ (CF_3CO)_2O \\ <\ -20°C \\ 3.\ Na(C_5H_5)^-\ dme,\ r.\ t.}} \quad (C_5H_5)(OC)(CO)M\!\equiv\!C\!-\!R \tag{8}$$

$$M = Cr,\ Mo;\ R = \text{(—2,6-Me}_2\text{C}_6H_3)$$

$$M = Mo;\ R = \text{(—2,6-Me}_2\text{C}_6H_3),\ (\text{—C}_6H_4\text{-X})\ (X = OMe,\ NMe_2)$$

of Mo(CR)(Tp′)(CO)$_2$ [R = Me, Ph; Tp′ = hydridotris(3,5-dimethyl-pyrazolyl)borate] (*184*).

Templeton and co-workers described a novel transformation of η^2-acyl ligands into alkylidyne ligands, which formally represents the abstraction of an oxygen atom [Eq. (9)] (*48*). The mechanism of this transformation is probably complex, and the yield of the alkylidyne complexes is relatively low. Nevertheless, this reaction deserves attention, since overall it repre-

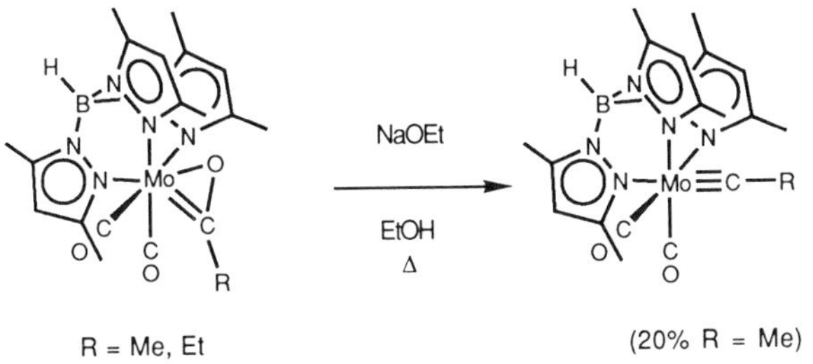

$$\text{(9)}$$

R = Me, Et   (20% R = Me)

sents a less "oxidizing" procedure than "oxide abstraction from acyl ligands" as described in the previous paragraph.

3. α-Hydrogen Elimination and Abstraction Reactions

The most general starting materials in the chemistry of high-valent alkylidyne complexes of molybdenum and tungsten are the dme-stabilized alkylidyne trichlorometal complexes **2** (7,8). The synthesis of these complexes is based on the reaction of MoO_2Cl_2 and $W(OMe)_3Cl_3$ with neopentylmagnesium chloride [Eq. (10)] (49,50). The reaction steps leading to the

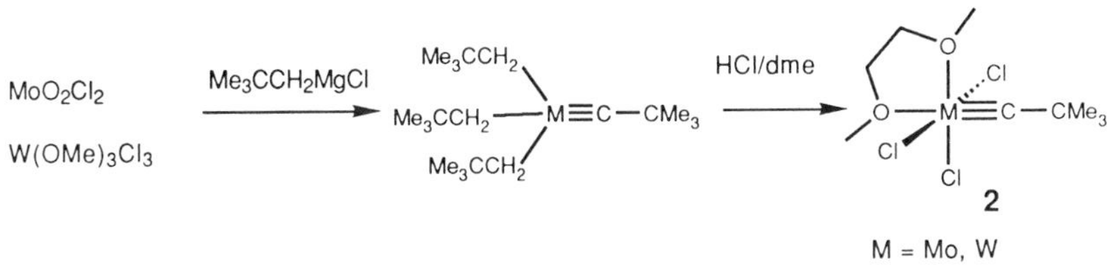

$$\text{(10)}$$

M = Mo, W

formation of the metal–carbon triple bonds are believed to involve α-hydrogen elimination steps.

Schrock described the synthesis of the rhenium neopentylidyne complexes **3** and **4** as shown in Eqs. (11) and (12) (51,52). Formation of these

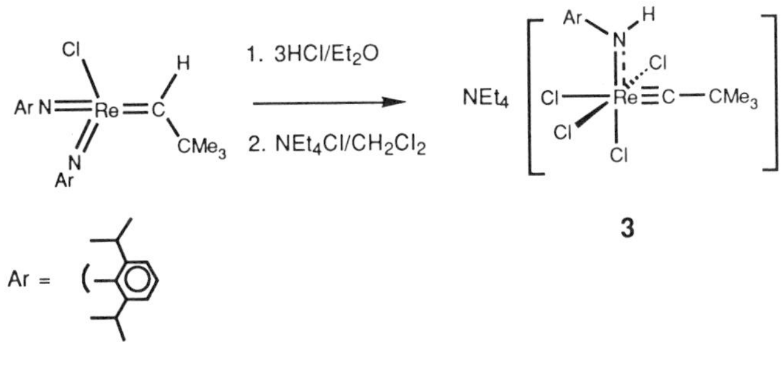

$$\text{(11)}$$

3

Ar =

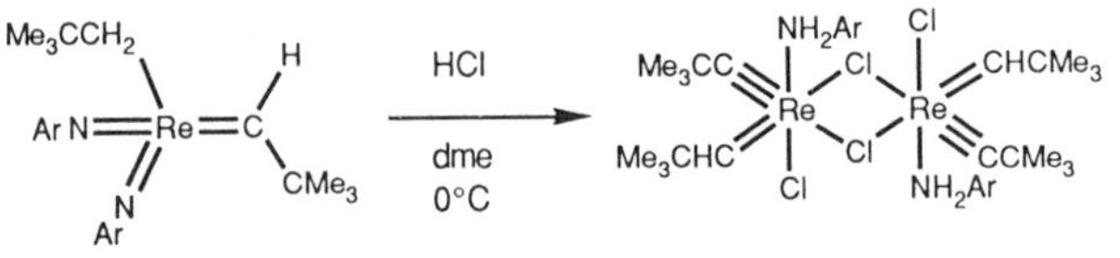

(12)

4

products is surprising since the imido nitrogen atom appears to be more basic than the alkylidyne carbon atom. Wilkinson and co-workers characterized $Re(CSiMe_3)(CH_2SiMe_3)_3Cl$ as a minor product of the reaction shown in Eq. (13) (*53*). Herrmann and co-workers found that the oxo(din-

$$ReCl_4(thf)_2 \xrightarrow[thf]{Me_3SiCH_2MgCl} Re(CSiMe_3)Cl(CH_2SiMe_3)_3\ (10\%) + [Re(CH_2SiMe_3)_4]_2N_2 + Re(\mu\text{-}CSiMe_3)_2(CH_2SiMe_3)_4 \qquad (13)$$

eopentyl)rhenium complex **5** can be transformed into the neopentylidyne complexes **6** by treatment with $Ti(\eta^5\text{-}C_5H_5)X_3$ [Eq. (14)] (*54,55*). ESR data

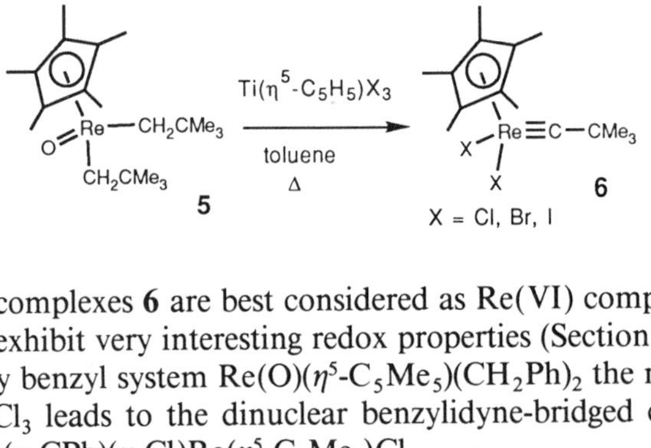

(14)

5 **6**

X = Cl, Br, I

suggest that complexes **6** are best considered as Re(VI) complexes. These compounds exhibit very interesting redox properties (Section IV,A). With the less bulky benzyl system $Re(O)(\eta^5\text{-}C_5Me_5)(CH_2Ph)_2$ the reaction with $Ti(\eta^5\text{-}C_5H_5)Cl_3$ leads to the dinuclear benzylidyne-bridged complex $(\eta^5\text{-}C_5Me_5)ClRe(\mu\text{-}CPh)(\mu\text{-}Cl)Re(\eta^5\text{-}C_5Me_5)Cl$.

4. *Other Reactions Based on Removal of α-Substituents*

Complex **7**, which is obtained by insertion of diphenylcyanamide into the metal–carbene bond in $W\{CPh(OMe)\}(CO)_5$, reacts with BBr_3 to form the diphenylaminocarbyne complex **8** [Eq. (15)] (*42*).

Roper and co-workers described the synthesis of the osmium carbyne complex **10** from the reaction of the dihalocarbene complex **9** with 2 equiv *p*-tolyllithium [Eq. (16)] (*56*).

Beck and co-workers achieved the synthesis of the μ-carbido complex **12** by reaction of the porphyrin iron dihalocarbene complex **11** with penta-

$$
\mathbf{7} \xrightarrow[\substack{CH_2Cl_2 \\ -10°C}]{BBr_3} \mathbf{8} \tag{15}
$$

$$
\mathbf{9} \xrightarrow[\substack{thf \\ -40°C}]{2 \ LiTol} \mathbf{10} \tag{16}
$$

carbonylrhenate anion [Eq. (17)] (*57*). Complex **12** is a rare example of a

$$
\underset{\mathbf{11}}{(TPP)Fe=CCl_2} \xrightarrow[thf]{Na[Re(CO)_5]} \underset{\mathbf{12}}{(TPP)Fe=C=Re(CO)_4Re(CO)_5} \tag{17}
$$

TPP = tetraphenylporphyrin

dinuclear complex bridged by a single carbon atom. [For another example, see Eq. (195), Section IV,G.]

B. *β-Addition of Substituents to Unsaturated Ligands*

1. *β-Addition to Vinylidene and Acetylide Ligands and Related Reactions of Alkyne and Vinyl Ligands*

The synthesis of alkylidyne complexes by β-addition of electrophiles to vinylidene and acetylide ligands is now well established (*5,6*). Pombeiro and co-workers synthesized several new rhenium alkylidyne complexes by protonation of the electron-rich vinylidene complexes **13** [Eq. (18)] (*58*). The mechanism of formation of the benzylcarbyne complex **14** (R = Ph)

$$
\mathbf{13} \xrightarrow[CH_2Cl_2]{HBF_4} \mathbf{14} \tag{18}
$$

X = Cl; R = CMe₃, Ph
X = F; R = CMe₃

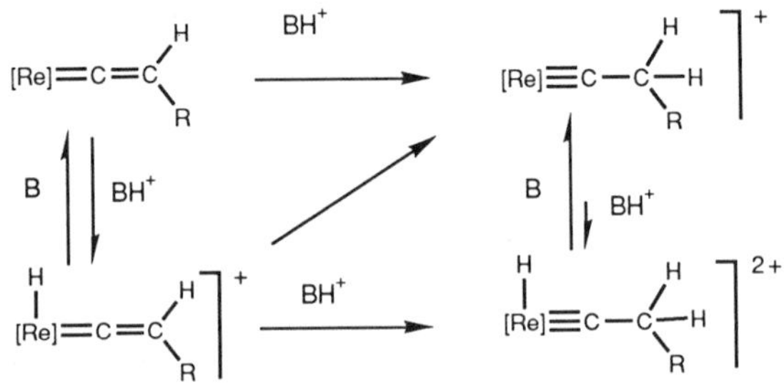

SCHEME 1

was investigated by kinetics methods (59). It was found that in addition to the direct protonation of the β-carbon atom there exists a second reaction pathway involving protonation at the metal center to give a vinylidene hydrido rhenium intermediate which rearranges to the alkylidyne complex (Scheme 1). Protonation of the bis(acetylide) complex 15 (R = CO_2Me) with HBF_4 affords carbyne complex 16 [Eq. (19)] (60). When R is Ph the

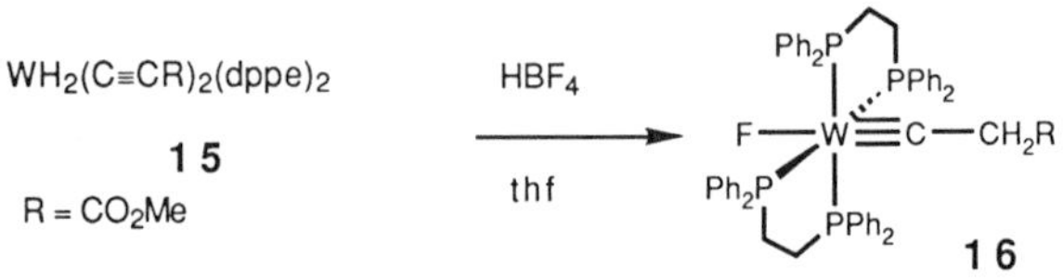

$$\tag{19}$$

alkylidene complex $[WF(=CHCH_2Ph)(dppe_2)][BF_4]$ is obtained, probably via the alkylidyne complex $[W(\equiv CCH_2Ph)F(dppe)_2]$ as an intermediate.

A rather significant result was reported by Höhn and Werner: the spectroscopic characterization of the first iridium alkylidyne complexes [Eq. (20)] (61,62). Protonation of the iridium vinylidene complexes 17 was found to occur initially at the metal center to afford the vinylidene hydrido

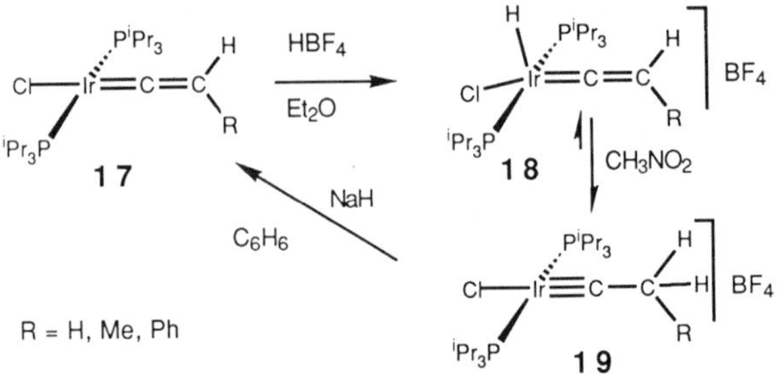

$$\tag{20}$$

complexes **18**. However, in a solution of nitromethane an equilibrium between the vinylidene hydrido complexes and the alkylidyne complexes **19** was found to exist, in which the alkylidyne complexes were favored. Unfortunately, on removal of the solvent, only the vinylidene hydrido complexes are obtained in the solid state. The prospects for a general synthesis of group 9 transition metal alkylidyne complexes by β-addition of electrophiles to vinylidene ligands are limited. Attack of electrophiles at the metal center appears to be the general case. Nevertheless, the observation of the novel iridium alkylidyne complexes, and to some extent the previous characterization of a nickel aminocarbyne complex by Fischer and Schneider (*63*), presage the viability of metal–carbon triple bonds involving late transition metals.

Protonation of vinylidene and acetylide ligands was also found to be useful for the synthesis of high-oxidation state molybdenum alkylidyne complexes. Green and co-workers demonstrated protonation of the vinylidene complex **20** as shown in Eq. (21) (*64*). Selegue and co-workers

$$(21)$$

accomplished double β-protonation of the acetylide ligand in **21** [Eq. (22)]

$$(22)$$

(*65*). The second protonation step is accompanied by loss of the carbonyl ligand, since the alkylidyne complex **22** is a d^0 system (considering the alkylidyne ligand as a trianion) and there are no available electrons for π backbonding to carbon monoxide.

The molybdenum carbyne complex **24** is the product of the unusual rearrangement, shown in Eq. (23) (*66*), involving formal α,β-migration of hydrogen in a vinyl ligand. Heating of the cyclopentadienyl complex **23** to 80°C in hexane results in loss of one trimethylphosphite ligand and

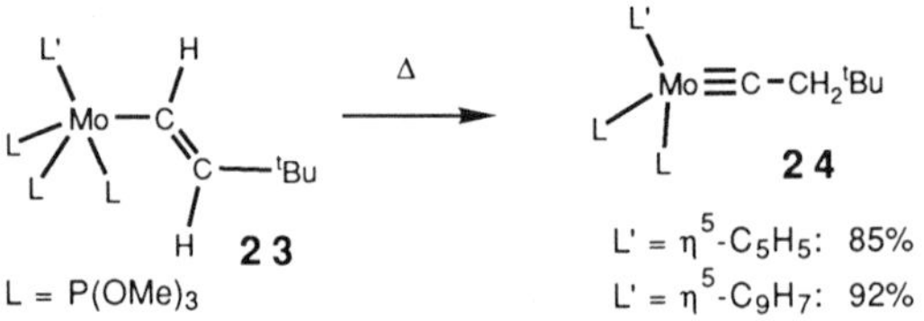

$$L' = \eta^5\text{-}C_5H_5: \ 85\%$$
$$L' = \eta^5\text{-}C_9H_7: \ 92\%$$

(23)

rearrangement of the *tert*-butylvinyl ligand to the alkylidyne ligand. External trimethylphosphite was found to slow down the reaction, indicating that the ligand rearrangement step is preceded by dissociation of phosphite. The analogous indenyl complex **23** undergoes the same rearrangement at 100°C. The σ-vinyl complexes **23** are prepared by hydride addition to the cationic alkyne complexes **25** in the presence of trimethylphosphite. Complexes **25** may be transformed into the metal alkylidynes **26** directly by reaction with hydride donor reagents in the absence of $P(OMe)_3$ [Eq. (24)].

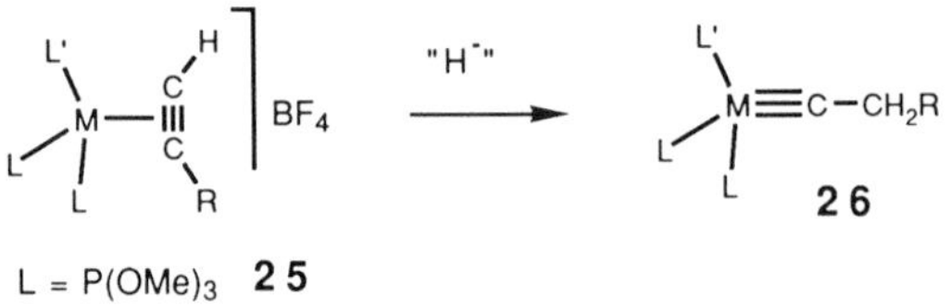

$L = P(OMe)_3$ **2 5**

(24)

M = Mo; $L' = \eta^5\text{-}C_5H_5$; R = t-Bu; "H$^-$" = KBH(s-Bu)$_3$
M = Mo; $L' = \eta^5\text{-}C_5H_5$; R = CHMe$_2$; "H$^-$" = NaBH$_4$
M = W; $L' = \eta^5\text{-}C_5H_5$; R = t-Bu; "H$^-$" = NaAlH$_2$(OCH$_2CH_2$OMe)$_2$
M = W; $L' = \eta^5\text{-}C_9H_7$; R = t-Bu; "H$^-$" = KBH(s-Bu)$_3$

Two mechanistic possibilities were discussed for the rearrangement reactions (Scheme 2). The 16-electron σ-vinyl intermediate, generated either by dissociation of phosphite from **23** or by hydride addition to **25**, may undergo α-hydride elimination (for R' = H) to generate a hydrido vinyli-

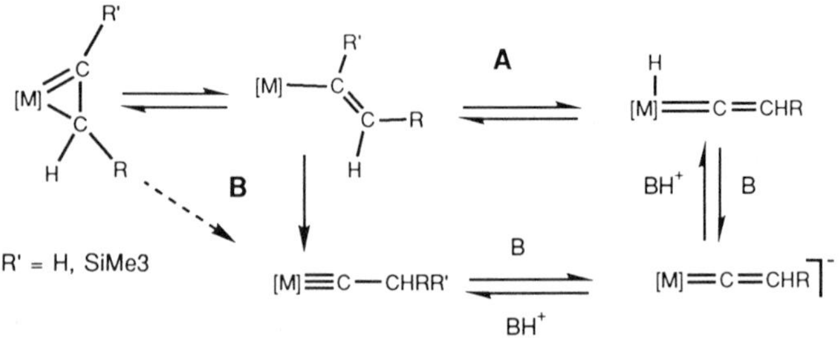

R' = H, SiMe$_3$

SCHEME 2

dene intermediate (path A). Abstraction of a proton from the metal hydride and readdition to the β-carbon of the vinylidene ligand would generate the carbyne complex. In the case of Eq. (23) the proton transfer may be mediated by free trimethylphosphite. Alternatively, the carbyne complexes may form via an intramolecular 1,2-hydrogen shift (path B). [In Eq. (24) no free phosphite is available to act as a base for the hydrogen transfer.]

Closely related reactions were observed for $[Mo(\eta^5\text{-}C_9H_7)\text{-}(Me_3SiC_2H)\{P(OMe)_3\}_2][BF_4]$ (66). This compound accepts hydride from $K[BH\text{-}s\text{-}Bu_3]$ at the nonsubstituted carbon of the alkyne ligand to afford the η^2-vinyl complex **27** (R = H), which may be isolated below 0°C. Complex **27** rearranges on warming to room temperature to the carbyne complex **28** [Eq. (25)]. For R equals Ph rearrangement occurs only on

$$\text{(25)}$$

L = P(OMe)₃

heating. Silyl migration is not observed in the diphenyl-substituted η^2-vinyl complex $Mo[C(SiMe_3)CPh_2](\eta^5\text{-}C_9H_7)[P(OMe)_3]_2$, presumably for steric reasons. The rearrangement of the silyl-substituted vinyl complexes has been proposed to occur by 1,2-silyl migration steps (path B of Scheme 2). Rearrangement via α-silyl elimination in analogy with path A in Scheme 2 is unlikely as no precedent for such a step is known.

In a related system the 1-bromo-*tert*-butylacetylene complex **29** reacts with $K[BH\text{-}s\text{-}Bu_3]$ to give the alkylidyne complex **30** (R = CMe₃) [Eq. (26)] (64). In the analogous reaction of the 1-bromophenylacetylene complex **29** with $K[BH\text{-}s\text{-}Bu_3]$ the alkylidyne complex **30** (R = Ph) is obtained only as the minor product. The major product is the vinylidene (bromo)

$$\text{(26)}$$

L = P(OMe)₃
R = CMe3, Ph; "Nu" = H (KBH(s-Bu)₃)
R = Ph; "Nu" = Me (Li[CuMe₂])

molybdendum complex **31** [Eq. (27)]. This compound reacts with several nucleophilic reagents to give carbyne complexes **32** and **33**. The addition

of nucleophiles to the β-carbon of vinylidene ligands is contrary to previously observed β-additions of electrophiles (67). It may be understood in terms of an S_N2' type mechanism. The typical nucleophilic behavior of the β-carbon of vinylidene ligands, however, is also observed for **31** [Eq. (21)]. Reaction of $Mo(\eta^5\text{-}C_5H_5)(Br)(HC_2Ph)(PEt_3)[P(OMe)_3]$ with $K[BH\text{-}s\text{-}Bu_3]$ provides the unsymmetrically substituted carbyne complex $Mo(CCH_2Ph)(\eta^5\text{-}C_5H_5)(PEt_3)[P(OMe)_3]$.

2. β-Addition to Isocyanide Ligands

Addition of electrophiles to electron-rich isocyanide complexes is a proven synthetic method for the synthesis of aminocarbyne complexes (5,6). Reaction of the trimethylsilyl isocyanide rhenium complex **34** with HBF_4 was reported by Pombeiro and co-workers (68). It leads to loss of the trimethylsilyl group and formation of the parent aminocarbyne ligand CNH_2 [Eq. (28)]. The CNH_2 ligand is easily interconverted to the CNH

ligand by removal and readdition of protons. In several cases isocyanide complexes in medium oxidation states were reduced before electrophile addition. Warner and Lippard reported reduction of complexes **35** with Zn or Al in the presence of water to afford the aminocarbyne complexes **36** [Eq. (29)] (*69*). The reaction conditions of Eq. (29) are the same as those

$$[ReCl_2(CNR)_3(PMePh_2)_2]Cl \xrightarrow[\text{thf}]{\substack{\text{Zn or Al} \\ \text{ZnCl}_2(\text{trace}) \\ \text{H}_2\text{O, KSbF}_6}} \left[\begin{array}{c} R \\ N \\ C \\ Cl-Re\equiv C-N \\ MePh_2P \\ C \\ N \\ R \end{array} \right] SbF_6 \quad (29)$$

R = Me, CMe$_3$ **35**   **36**

which led to the reductive coupling of isocyanide ligands in isoelectronic molybdenum complexes [Eq. (170), Section IV,F]. Complex **36** (R = Me) is easily deprotonated to give the neutral isocyanide complex $ReCl(CNMe)_3(PMePh_2)_2$. Electrochemical reduction of the cyano molybdenum complex **37** (−1.9 V versus ferrocinium/ferrocene) in the presence of phenol generates the aminocarbyne complex **38** [Eq. (30)] (*70*). This

$$(30)$$

37   **38**

result was proposed to provide chemical precedent for the possible role of ligating aminocarbyne in the biological reduction of aqueous cyanide. Very efficient syntheses of aminocarbyne tungsten and molybdenum complexes were developed by Filippou and co-workers [Eqs. (31) and (32)] (*71,72*). The key for the high-yield conversions of the reduced isocyanide complexes to the aminocarbyne complexes **39** and **40** is the use of the hard

R = H, Me

EtNC/Me$_3$NO → cis and trans-WI(C$_5$R$_5$)(CNEt)(CO)$_2$ 85-90%

$$(31)$$

Na

Na   [Et$_3$O][BF$_4$]

100%   73-85% **39**

$$(32)$$

alkylating agent $[Et_3O][BF_4]$. With MeI alkylation occurs at the metal center followed by isocyanide insertion to give iminoacyl complexes.

3. β-Addition to Carbonyl and Thiocarbonyl Ligands

Lippard and co-workers reported the synthesis of the siloxycarbyne complexes **41** by direct silylation of coordinated carbon monoxide [Eq. (33)] (73). This is the first example of this type of transformation. On the

$$(33)$$

other hand, electrophilic addition to thiocarbonyl ligands is a well-established method for the synthesis of thiocarbyne complexes (5). Doyle and Angelici synthesized the cationic thiocarbyne complex **42** according to Eq. (34) (74).

$$(34)$$

C. Metathesis Reactions

The metathesis of metal–metal triple bonds with carbon–carbon triple bonds directly connects the chemistry of inorganic, organometallic, and organic multiple bonds. This reaction is synthetically very useful. Krouse and Schrock prepared the series of μ-α,ω-alkylidene–bisalkylidyne complexes $(t\text{-BuO})_3W\equiv C-(CH_2)_n-C\equiv W(O\text{-}t\text{-Bu})_3$ ($n = 2, 4, 5,$ and 6), as well as the C_2-bridged complex $(t\text{-BuO})_3W\equiv C-C\equiv W(O\text{-}t\text{-Bu})_3$ [Eqs. (35) and (36)] (75).

$$W_2(O^tBu)_6 \longrightarrow \begin{cases} (^tBuO)_3W \equiv C\text{-}(CH_2)_6\text{-}C \equiv W(O^tBu)_3 \\ \\ (^tBuO)_3W \equiv C\text{-}(CH_2)_x\text{-}C \equiv W(O^tBu)_3 \\ \qquad\qquad x = 5,6 \\ \\ (^tBuO)_3W \equiv C\text{-}(CH_2)_y\text{-}C \equiv W(O^tBu)_3 \\ \qquad\qquad y = 2,4 \end{cases} \qquad (35)$$

$$W_2(O^tBu)_6 \xrightarrow{\quad EtC \equiv C \text{---} C \equiv CEt \quad} (^tBuO)_3W \equiv C\text{-}C \equiv W(O^tBu)_3 \qquad (36)$$

While the bulky *tert*-butoxide ligand is one of the best to promote the metathesis of metal–metal triple bonds, the reaction is also successful with less bulky phenoxide and alkoxide ligands [Eqs. (37) and (39)] (76,77), and

$$W_2(O\text{-} \bigcirc)_3 \xrightarrow[Et_2O]{RC \equiv CR/py} (\bigcirc \text{-}O)_3(py)_2W \equiv CR \qquad (37)$$
$$R = Me, Et$$

even in the presence of alkyl ligands [Eq. (40)] (78). Chisholm and co-workers established that dimetallatetrahedranes are involved as interme-diates in the metathesis reactions. There exists an equilibrium (K) between dimetallatetrahedranes—the adducts of metal–metal triple bonds and carbon–carbon triple bonds—and the corresponding alkylidyne com-plexes (79,80). The rates of formation of the trinuclear alkylidyne com-plexes **45** from $W \equiv W$ species **43** and dimetallatetrahedranes **44** increase with the size of the substituents R in the metallatetrahedranes [Eq. (38)].

$$W_2(OR)_6(\mu\text{-}C_2R'_2)(py)_n \underset{\xleftarrow{\hspace{1cm}}}{\overset{K}{\xrightarrow{\hspace{1cm}}}} 2\ (RO)_3W \equiv CR' + npy$$
$$\begin{array}{c} C\text{---}C \\ \diagdown\!\!\diagup\!\!\diagdown \\ M\text{------}M \end{array} \qquad\qquad M \equiv C$$

$$\begin{array}{c} 2\ W_2(OCH_2{}^tBu)_6(py)_2 \\ + \quad \textbf{4 3} \\ W_2(OCH_2{}^tBu)_6(\mu\text{-}C_2R'_2)(py)_n \\ \textbf{4 4} \end{array} \xrightarrow[\substack{25°C}]{hexane} \begin{array}{c} 2\ W_3(\mu_3\text{-}CR')(OCH_2{}^tBu)_9 \\ + \ (n{+}4)py \qquad \textbf{4 5} \end{array} \qquad (38)$$
$$k_{obs}\text{: Et} \approx \text{Ph} > \text{Me}$$

Fragmentation of the dimetallatetrahedranes **44** gives mononuclear alkylidyne complexes which combine with **43** to afford the trinuclear μ^3-alkylidyne complexes **45**. The observed rates of formation of complexes **45** probably reflect the equilibrium (K) of fragmentation of **44**.

Reaction of **46** with 3-hexyne at temperatures below $-20\,^{\circ}C$ results in clean metathesis [Eq. (39)] (77). At room temperature the di- and trinu-

$$W_2(O^iPr)_6(HNMe_2)_2 \xrightarrow[\substack{pentane \\ -20\,^{\circ}C}]{EtC\equiv CEt} [W(CEt)(O^iPr)_3(HNMe_2)]_2 \qquad (39)$$

$$\mathbf{46}$$

clear complexes $W_2(O\text{-}i\text{-}Pr)_6(\mu\text{-}C_4Et_4)(\eta^2\text{-}C_2Et_2)$ and $W_3(\mu^3\text{-}CEt)(\mu^2\text{-}O\text{-}i\text{-}Pr)_3(O\text{-}i\text{-}Pr)_6$ are obtained. The metathesis reactions shown in Eqs. (40) and (41) are noteworthy because of the presence of alkyl ligands on the metal

$$\qquad (40)$$

$$\qquad (41)$$

centers. The importance of steric encumbrance of the metal center for metathesis activity is evident for the system in Eq. (40). With the sterically less demanding CH_2SiMe_3 ligand, instead of CH_2CMe_3, the alkyne adduct **48** is obtained. The reaction in Eq. (40) is believed to proceed via such an alkyne adduct. The dimeric alkylidyne complex **47** was shown to react with pyridine to give $[W(CEt)(CH_2\text{-}t\text{-}Bu)(O\text{-}i\text{-}Pr)_2(py)]_2$ and the mononuclear complex $W(CEt)(CH_2\text{-}t\text{-}Bu)(O\text{-}i\text{-}Pr)_2(py)_2$. Metathesis and metathesis-like reactions of alkylidyne complexes are discussed in Section IV,G. Two novel methods for the synthesis of metal–carbon triple bonds based on ligand cleavage reactions are discussed in Section IV,F [Eqs. (160) and (169)].

IV

REACTIONS OF TRANSITION METAL ALKYLIDYNE COMPLEXES

A. Oxidation and Reduction Reactions

Metal alkylidyne complexes undergo a variety of oxidation and reduction reactions as well as redox-induced transformations of the alkylidyne ligands. A method for the direct transformation of Fischer-type carbyne complexes into Schrock-type alkylidyne complexes was developed in our laboratory. Bromine oxidation of the *trans*-carbyne bromo tetracarbonyl complexes **49** of molybdenum and tungsten in the presence of dimethoxyethane affords the dme-stabilized alkylidyne tribromo metal complexes **50** [Eq. (42)] (*81*). For alkyl-substituted complexes $(R = Me, CH_2CMe_3)$

$$\tag{42}$$

M = Mo; R = Ph
M = W; R = Me, CH₂CMe₃, Ph

the tetracarbonylmetal carbyne complexes need to be purified before the oxidation step, but the phenyl-substituted carbyne complexes could be used directly as obtained by reaction of the acyl complexes $[NMe_4][M(COPh)(CO)_5]$ with oxalyl bromide [Eq. (43)]. Since dme-stabi-

$$\tag{43}$$

M = Mo, W

lized alkylidyne trihalometal complexes such as **50** are among the most useful precursors in the chemistry of high-valent alkylidyne metal complexes (*7,8*), this oxidative conversion of carbyne complexes is potentially very useful. In the presence of trimethylphosphine the complexes **50** may be reduced to carbonyl-free Fischer-type carbyne complexes **51** [Eq. (44)]

$$ \tag{44} $$

(*82*). Green and co-workers reduced complex **52** to complex **53** using magnesium amalgam [Eq. (45)] (*64*).

$$ \tag{45} $$

A new class of aminocarbyne complexes was discovered by Filippou and Fischer. Oxidation of complexes **54** by iodine affords the triiodo monocarbonyl aminocarbyne complexes **55** [Eq. (46)] (*83–85*). Complexes **58** and

$$ \tag{46} $$

60 are oxidized by halogens to give complexes **59** and **61**, respectively [Eqs. (47) and (48)] (*72,86–88*). The oxidized products are aminocarbyne complexes at the same oxidation state as Schrock-type alkylidynes. The stability of these compounds is remarkable, especially in view of the observation that the reaction of related high-oxidation state alkylidyne and aminocarbyne complexes with CO led to various types of ligand coupling reactions (Section IV,F). The stability of these complexes may be rationalized in

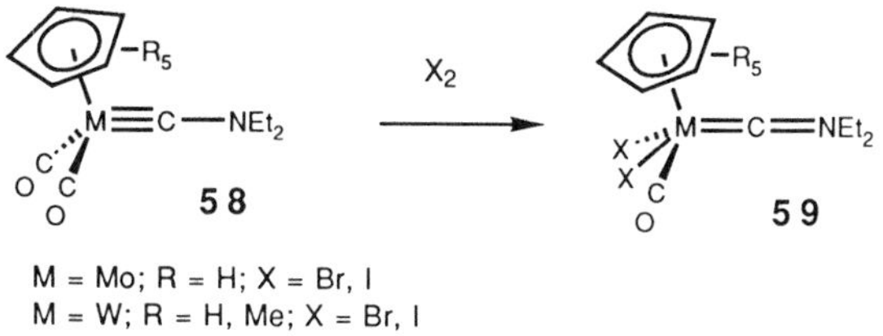

$$M = Mo; R = H; X = Br, I$$
$$M = W; R = H, Me; X = Br, I$$

(47)

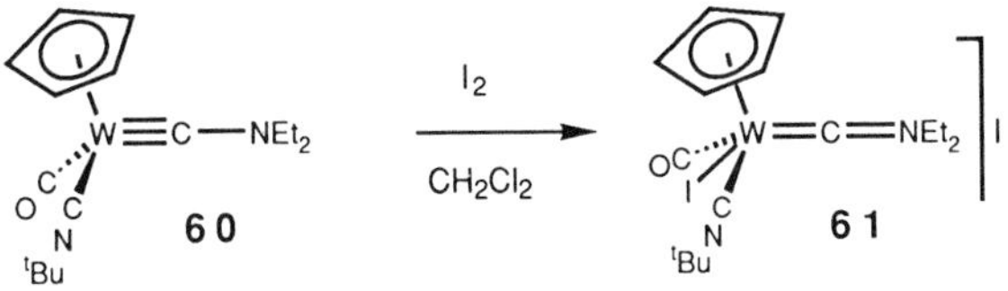

(48)

terms of a very strong contribution of the 2-azavinylidene resonance structure **B** to the bonding description (89). Since high-oxidation state carbyne

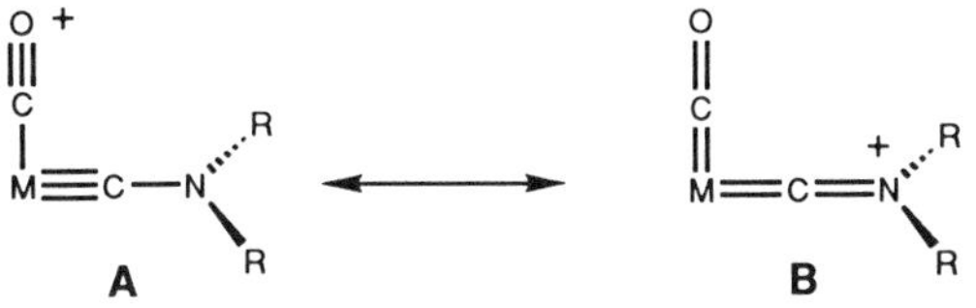

complexes have no metal d electrons other than those in the M≡C π bonds, the carbonyl and aminocarbyne ligands in **55, 59,** and **61** must be sharing two of these MC π electrons. Because the interaction of the nitrogen lone pair raises the energy of the adjacent MC π bond (6), the CO ligand is expected to be oriented cis to the aminocarbyne ligand and perpendicular to the plane of the aminocarbyne ligand. The aminocarbyne complexes **55** are reduced by PMe$_3$ to give complexes **56** and **57** [Eq. (46)] (83,84).

Herrmann and co-workers showed that the 17-electron complex **62** may be subjected to electrochemical oxidation as well as reduction (54,55). The redox potentials of the respective halo complexes follow the expected trend for change in the halide ligands. In accordance with these electrochemical studies, both chemical oxidation and reduction reactions of **62** were found [Eqs. (49) and (50)]. Oxidation of the rhenium complex Re(CCMe$_3$)(η^5-C$_5$Me$_5$)Cl$_2$ by dioxygen results in cleavage of the neopentylidyne ligand from the metal center as pivalic anhydride and pivaloyl chloride.

In several cases oxidation reactions of carbyne complexes were found to be accompanied by transformations of the carbyne ligands. Complex **63**

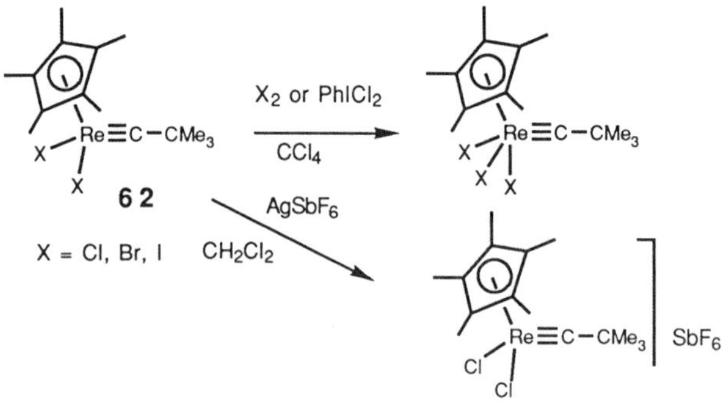

$$(49)$$

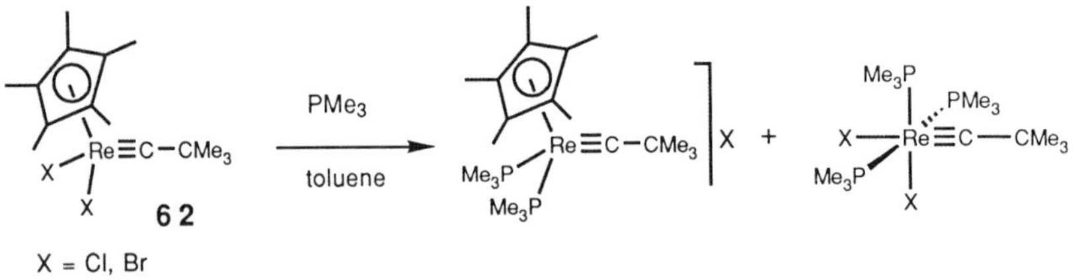

X = Cl, Br, I

$$(50)$$

X = Cl, Br

undergoes a reversible one-electron oxidation process by cyclic voltam-metry (90). However, the reaction of **63** with diazonium salt and with CF_3I [Eqs. (51) and (52)] leads to formal oxidation with concomitant loss of a

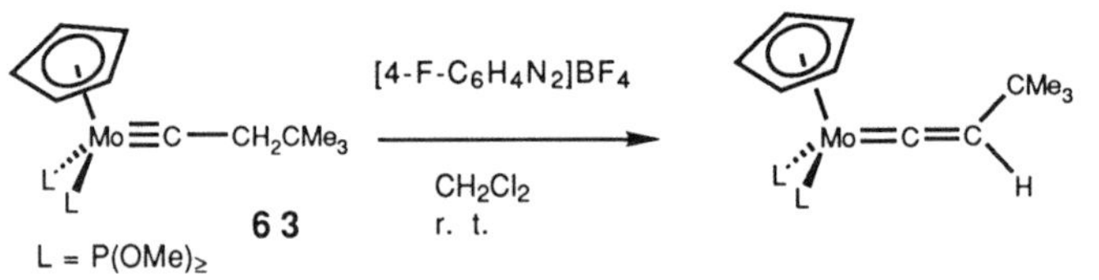

$$(51)$$

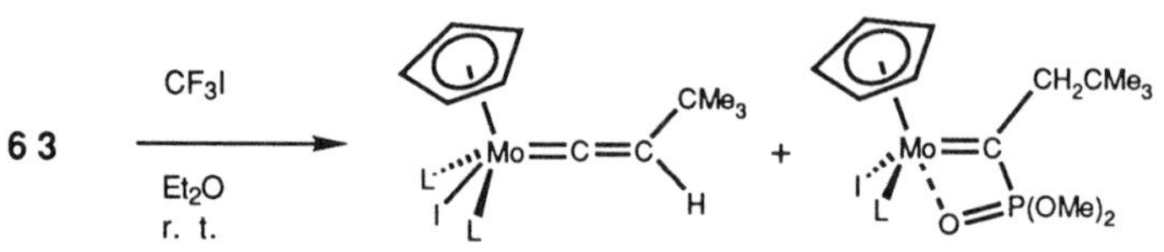

L = P(OMe)₂

$$(52)$$

proton from the β-carbon of the alkylidyne ligand (91). The second prod-uct of Eq. (52) may be considered as an example of migration of a metal-bonded ligand to the carbyne carbon. The migrating ligand [P(O)(OMe)₂] was probably formed by demethylation of a phosphite ligand.

The alkylidyne complexes **64** exhibit a reversible one-electron oxidation in cyclic voltammetry, whereas irreversible anodic processes are observed for the aminocarbyne complex **65**. In solvents such as tetrahydrofuran β-deprotonation of **64** and **65** is induced by anodic oxidation [Eqs. (53)

$$[Re(CCH_2R)Cl(dppe)]^+ \xrightarrow{\ -e^-/-H^+\ } [Re(CCHR)Cl(dppe)]^+ \quad (53)$$

64

$$[Re(CNH_2)Cl(dppe)]^+ \xrightarrow{\ -e^-/-H^+\ } [Re(CNH)Cl(dppe)]^+ \quad (54)$$

65

and (54)] (*92–94*). Cathodic reduction of **64** and **65** generates vinylidene and isocyanide complexes with concomitant evolution of hydrogen.

Cationic 2-azaallenylidene complexes **66** are subject to one-electron reductions (see also Section IV,B) [Eq. (55)] (*95,96*). The unpaired electron

$$[(CO)_5M{=}C{=}N{=}C(Mes)_2]BF_4 \xrightarrow[\ CH_2Cl_2\ ]{\ Fe(\eta^5{-}C_5H_5)_2\ } [(CO)_5M{-}C{\equiv}N{-}C(Mes)_2]^{\cdot} \quad (55)$$

M = Cr, W (Mes = mesityl)   **67**
66

in **67** is mostly localized on the $C(Mes)_2$ group of the azaallenylidene ligand, which may be best regarded as a radical derivative of an isocyanide ligand. The $CNCR_2$ ligand in **67** has very little, if any, carbyne character. When the substituents on the azaallenylidene ligand are sterically less demanding, coupling of the radical centers may occur as shown in Eq. (56).

$$[(CO)_5W{=}C{=}N{=}CR_2]AlBr_4 \xrightarrow{\ thf\ } Br(CO)_4W{\equiv}C{-}N{=}CR_2 \ +$$

R = Ph, 4-BrC₆H₄, 4-MeOC₆H₄

Mes, CR₂ = C(C₆H₄)₂O

$$(CO)_5W{-}C{\equiv}N{-}CR_2 \quad (56)$$
$$\mid \quad (R = Ph)$$
$$(CO)_5W{-}C{\equiv}N{-}CR_2$$

B. Photochemistry of Alkylidyne Complexes

The *cis*-bis(phosphine)-substituted complexes **68** were found in our laboratory to isomerize to the trans derivatives under irradiation with visible light [Eq. (57)] (*97*). The isomerization reaction was postulated to involve a stereochemically nonrigid pentacoordinate intermediate which is generated by photochemical carbyne–carbonyl coupling. (The evidence for the ligand coupling step as well as other photoinduced coupling reactions are discussed in Section IV,F.)

$$M = Mo; \ X = Br; \ R = Me$$
$$M = W; \ X = Cl; \ R = Me$$
$$M = W; \ X = Br; \ R = Ph$$

$$(57)$$

A study of the photophysical and photochemical properties of alkylidyne complexes such as $W(CPh)Cl(CO)_2(tmeda)$ was reported by Bocarsly et al. (see also Section II,B) (24). Quenching experiments with aromatic hydrocarbons indicate that a significant amount of triplet character is associated with the emissive excited state. Bimolecular oxidative and reductive quenching is observed as well. Methylviologen dichloride is reduced and N,N,N',N',-tetramethyl-p-phenylenediamine is oxidized by $W(CPh)Cl$-$(CO)_2(tmeda)$ on irradiation with 488 nm light. Photosubstitution of two CO ligands in $W(CPh)Cl(CO)(dppe)$ by dppe to give $W(CPh)Cl(dppe)_2$ was observed as well.

Photochemically induced electron transfer reactions of carbyne complexes were also reported by McElwee-White and co-workers [Eq. (58)]

$$M = Mo; \ R = Ph$$
$$M = W; \ R = cyclopropyl$$

$$(58)$$

(98,99). Irradiation of 69 in chlorinated solvents in the presence of trimethylphosphine gives rise to the oxidized products 70. The reaction was postulated to proceed via short-lived 17-electron carbyne complex intermediates. Electrochemical evidence for the existence of 17-electron species was reported. In the absence of trimethylphosphine no products could be characterized for the molybdenum phenylcarbyne complex. In the absence of trimethylphosphine the tungsten cyclopropylcarbyne complex affords $W(\eta^5\text{-}C_5H_5)Cl_3(CO)[P(OMe)_3]$ and cyclopentenone [Eq. (59)] (99).

The photochemistry of 2-azaallenylidene complexes was investigated by Seitz and Wrighton (96). These compounds undergo reduction on irradiation in methylene chloride solution. Whereas bulky mesityl groups stabilize the γ-carbon-centered radical 74, the less bulky diphenyl complex 75

$$(59)$$

gives rise to the dimer **76** [Eq. (61)]. The main product of the reaction in Eq. (60), however, is the carbonyl substitution product **73**. The initial

$$(60)$$

$$(61)$$

product of photosubstitution is the cis isomer, which isomerizes photochemically or thermally to the trans product. The relative stability of *cis*-**73** is a clear indicator that the carbyne character of the 2-azaallenylidene ligand is weak.

C. *Ligand Substitution Reactions*

The mechanism of ligand substitution reactions in the carbyne complexes *trans*-M(CR)X(CO)$_4$ (M = Cr, W; R = Me, Ph, NEt$_2$; X = Cl, Br, I, SePh) was investigated by H. Fischer and co-workers (*100*). The influence of the metal center, the trans ligand, and the carbyne substituent on the M—CO dissociation step was determined. The reactions with PPh$_3$ in 1,1,2-trichloroethane [Eq. (62)] all follow first-order kinetics, with activa-

$$(62)$$

tion enthalpies ranging from 97 to 116 kJ/mol and activation entropies ranging from 34 to 72 J/mol/K. The reaction rate is virtually independent of the carbyne substituent R, indicating that electronic variation of the carbyne ligand similarly affects ground and transition states. The rate is faster for chromium complexes than for analogous tungsten compounds as expected on the basis of generally stronger W—C(CO) versus Cr—C(CO) bonds. The reaction rate increases strongly in the series *trans*-X equals I, Br, Cl, SePh, reflecting the ability of the various X ligands to stabilize the transition state by electron pair donation (cis-labilization). The reaction rate decreases slightly with increasing steric requirements of R. The larger substituents possibly hinder the geometric relaxation required to bring the donor ligand X into a position to stabilize the transition state. The rate of thermal decomposition of metal carbyne complexes was found to be the same as the rate of CO substitution. Thus, thermolysis of tetracarbonyl-metal carbynes is initiated by loss of carbon monoxide.

Several examples of CO substitution by halide ions in aminocarbyne complexes were reported by Fischer [Eq. (63)] (*101*). The halide ligand in

$$X = Cl;\ R = c\text{-Hex, i-Pr; Kat} = PPN$$
$$X = Br;\ R = c\text{-Hex; Kat} = NEt_4$$

the anionic complexes **77** is easily substituted by triphenylphosphine. The complex [HCN(Ph)CH$_2$CH$_2$N(Ph)][W{CN(c-C$_6$H$_{11}$)$_2$}Br$_2$(CO)$_3$] was obtained from the reaction between [W{CN(c-C$_6$H$_{11}$)$_2$}Br(CO)$_4$] and the electron-rich olefin [=CN(Ph)CH$_2$CH$_2$N(Ph)]$_2$ in 42% yield (*102*). The product presumably arises from a simple substitution of CO by bromide such as in Eq. (63). [HCN(Ph)CH$_2$CH$_2$N(Ph)]Br, which may be generated by reaction of the electron-rich olefin with the solvent and by decomposition of the starting carbyne complex, could serve as the bromide source.

Bis-nitrogen donor ligand-substituted alkylidyne complexes such as **78** were found to be good alternatives as starting materials to the thermally labile tetracarbonyl complexes (*44*). Substitution of one pyridine ligand in the bispyridine-substituted complexes **78** occurs in the presence of excess chloride [Eq. (64)] (*103*). The anionic complexes **79** are stable in the presence of excess chloride but are too labile to be isolated in pure form. These complexes were used in the synthesis of stable alkylidyne alkene tungsten complexes [Eq. (180), Section IV,G]. With phosphines both pyridine ligands in **78** are substituted [Eq. (65)] (*44*). For PMe$_3$ the reaction of the tetracarbonyl complex W(CPh)Br(CO)$_4$ had previously been shown to

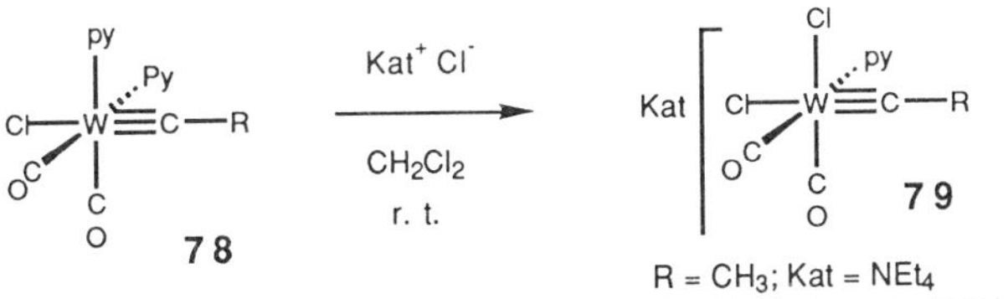

$$(64)$$

R = CH$_3$; Kat = NEt$_4$
R = Ph; Kat = NEt$_4$, PPN

$$(65)$$

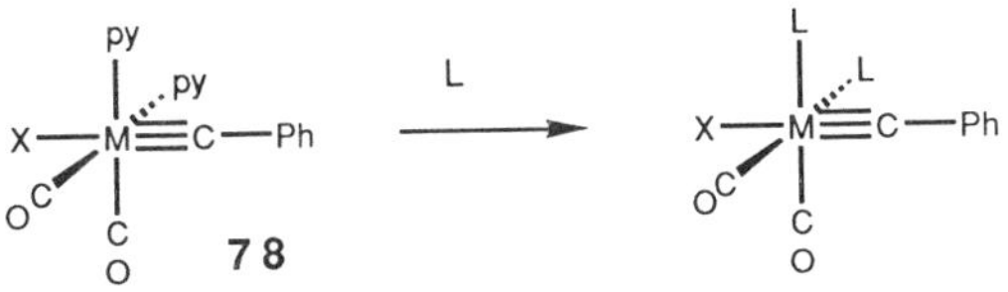

L = PMe$_3$; M = Mo; X = Br
M = W; X = Cl
L = 1/2 dppe; M = W; X = Cl

give a different result: $W\{C(PMe_3)(Ph)\}Br(CO)_2(PMe_3)_2$ (104). Apparently, the electrophilicity of the carbyne ligand in 78 is reduced by the presence of the pyridine ligands, thereby preventing attack by PMe$_3$ at the carbyne carbon.

Substitution of more than two carbonyl ligands by simple donor ligands does not occur easily. Whenever such a step was observed, special circumstances were present. Complex 80 reacts in neat trimethylphosphine to give a purple solid which very slowly redissolves, forming the tristrimethylphosphine complex 81 (82). The purple complex was characterized as the ketenyl complex $W(\eta^2\text{-PhCCO})Cl(CO)(PMe_3)_3$ (97). The reaction shown in Eq. (66) is thus not a simple ligand substitution (for other overall ligand

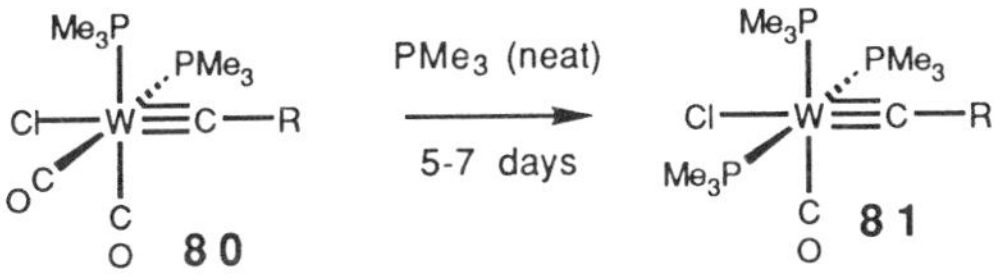

$$(66)$$

R = Ph, Tol

substitution reactions via metal ketenyl intermediates, see Section IV,F). The trans isomer, on the other hand, undergoes carbon monoxide substitution easily at slightly elevated temperatures [Eq. (67)] (97). Although no mechanistic information is available, it is likely that the substitution process of Eq. (67) is dissociative owing to the trans arrangement of the two carbonyl ligands.

In the tetracarbonyl aminocarbyne complex 82 attack of trimethylphos-

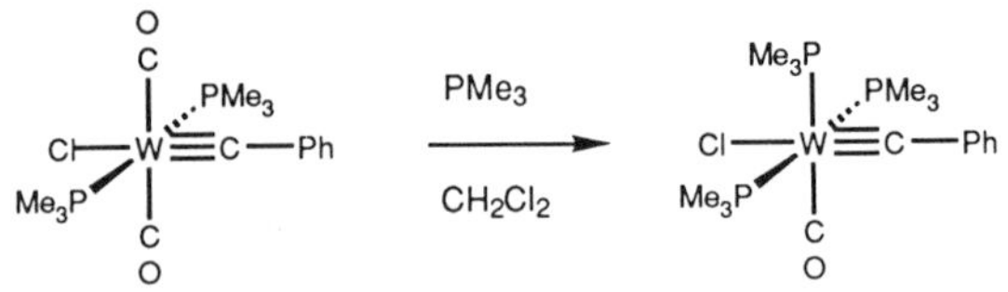

(67)

phine at the carbyne carbon is not observed (*84*). In this case electron donation by the amino group seems to reduce the electrophilicity of the carbyne carbon to prevent attack by trimethylphosphine. A stepwise substitution of two carbonyl ligands is observed [Eq. (68)]. The bis-substituted

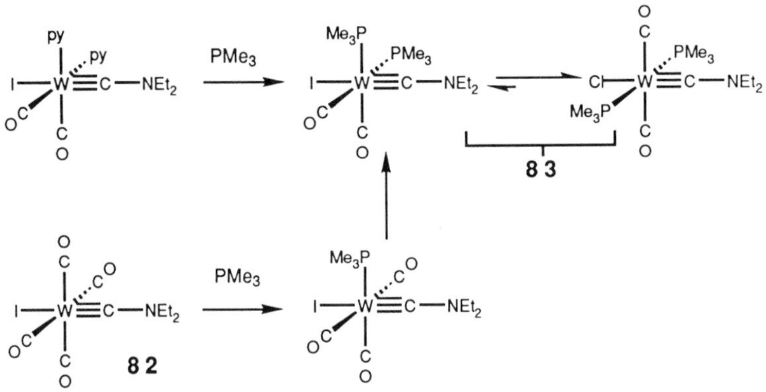

(68)

product is initially obtained as the cis product, which slowly isomerizes to the more stable trans form. At higher temperatures a mixture of the cis and trans isomers of **83** react with PMe$_3$ to form the tris-substituted complex **84** [Eq. (69)]. Presumably, substitution of the third carbonyl ligand in **83** occurs in the trans isomer.

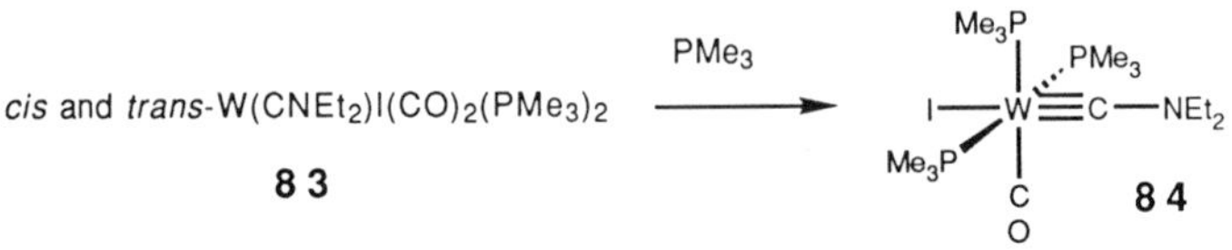

cis and *trans*-W(CNEt$_2$)I(CO)$_2$(PMe$_3$)$_2$ **83**   **84** (69)

If the substituting ligands have π acceptor properties, more than two carbonyl ligands can be replaced in simple thermal processes. We demonstrated that treatment of complexes **78** with trimethylphosphite at room temperature affords the trisphosphite-substituted complexes **85** (*105*). When M equals Mo and W the remaining carbonyl ligand may be replaced by heating in neat trimethylphosphite [Eq. (70)]. The phosphite ligands in

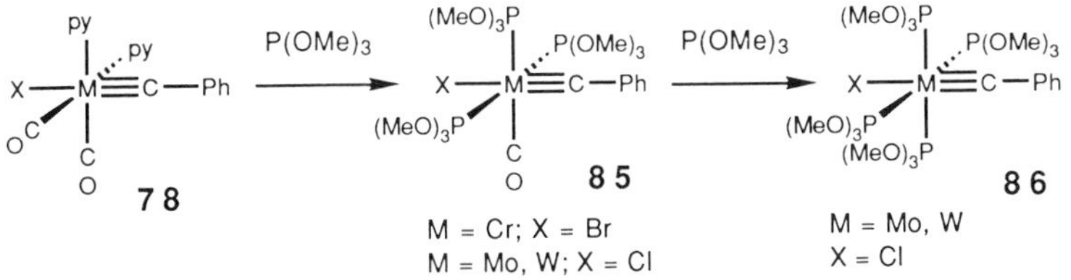

$$(70)$$

M = Cr; X = Br
M = Mo, W; X = Cl

M = Mo, W
X = Cl

the tungsten complex **86** (X = Cl) were shown to undergo substitution by dppe to afford W(CPh)Cl(dppe)[P(OMe)$_3$]$_2$ and W(CPh)Cl(dppe)$_2$.

Filippou synthesized tris- and tetrakisisocyanide derivatives. Reaction of the bispicoline phenylcarbyne complexes **87** with CNCMe$_3$ first gives the bis-substituted compounds **88**, which on further treatment with the isocyanide at elevated temperatures afford the tetrakisisocyanide complexes **89** [Eq. (71)] (*106,107*). The tris-substituted product **90** is an observable

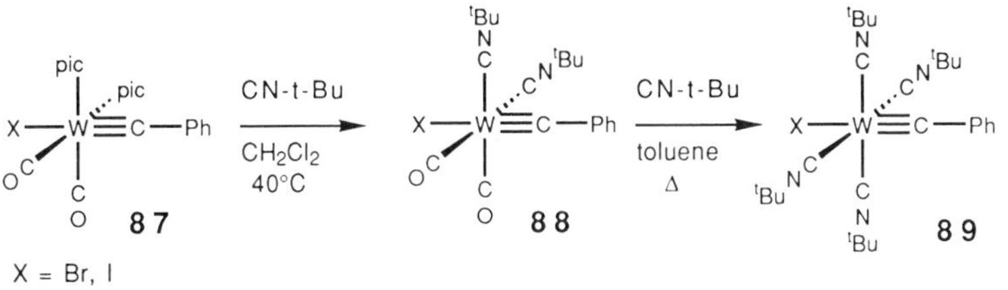

$$(71)$$

X = Br, I

intermediate in this reaction, but it forms only as a mixture with the tetrakis-substituted product **89**. It may be isolated in high yield if the reaction is conducted in the presence of CoCl$_2$ as a catalyst for CO substitution [Eq. (72)]. The aminocarbyne complex **91** behaves slightly differ-

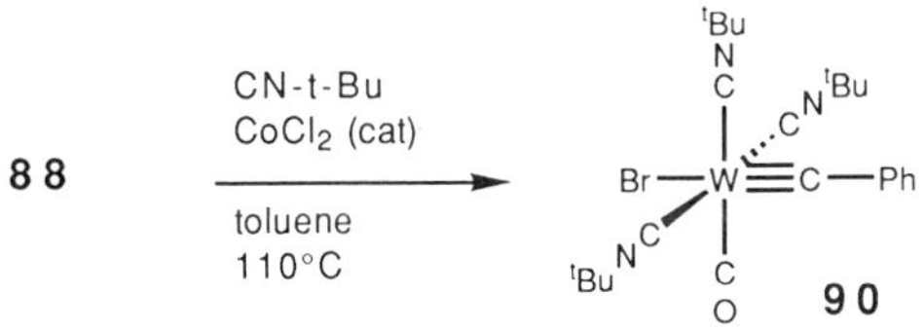

$$(72)$$

ently (*108–110*). The activation energies for the substitution of the third and fourth carbonyl ligands are sufficiently different to allow the isolation of both **92** and **93** from the direct substitution processes [Eq. (73)]. The tris-substituted complex **92**, however, is obtained as a mixture with the cationic complex [W(CNEt$_2$)(CO)(CN-*t*-Bu)$_4$]I. The two complexes may be interconverted as shown in Eq. (79) (*108,109*).

$$(73)$$

Substitution of the ligand trans to the metal–carbon triple bond, usually a halide, was observed in several situations. Substitution of halide in the tetracarbonyl systems is rare because most nucleophiles will substitute carbon monoxide. It was found that the halide is substituted by the carbonyl metallate anions $[Mo(\eta^5\text{-}C_5H_5)(CO)_3]^-$ and $[Co(CO)_4]^-$ [Eq. (74)]

$$(74)$$

R = Tol; $L_nM = (\eta^5\text{-}C_5H_5)(CO)_3Mo$

R = Me; $L_nM = (CO)_4Co$

(*111,112*). In other cases donor ligand-substituted derivatives were employed [Eqs. (75)–(79)]. Substitution of the trans halide ligands was

$$(75)$$

achieved with group 15-based ligands, for example, $NaAsPh_2$, $Li[(CO)_5CrSbPh_2]$, and PR_3 [Eqs. (75), (76), and (77), respectively] (*113–115*). The diphenylarsino-substituted complex of Eq. (75) forms $W(CNEt_2)Cl(CO)_2(bpy)$ on dissolution in CH_2Cl_2, indicating high reactivity of the arsenic lone pair. Owing to the similar σ donor/π acceptor properties of the nitrogen chelate ligands and the anionic group 15 donor ligands, cis or trans derivatives of **94** [position of anionic donor relative to the aminocarbyne ligands; in Eq. (76) only the trans isomer is shown] were obtained. For complexes **94** interconversion of the cis and trans isomers

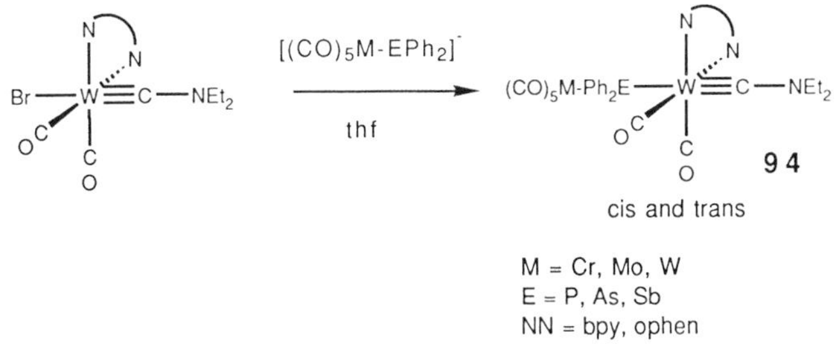

$$M = Cr, Mo, W$$
$$E = P, As, Sb$$
$$NN = bpy, ophen$$

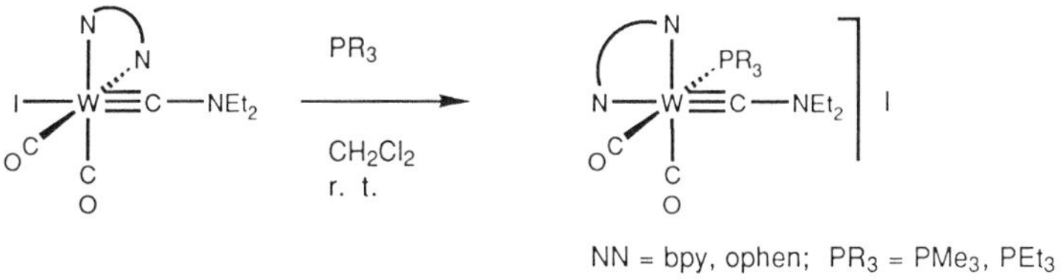

$$NN = bpy, ophen; \quad PR_3 = PMe_3, PEt_3$$

was observed by ^{1}H NMR spectroscopy. The iodide ligand in $W(CPh)I(CNCMe_3)_4$ was substituted by cyanide (*106*).

Treatment of the tetrakis-isocyanide phenylcarbyne complex **95** with $CNCMe_3$ and $TlPF_6$ affords the cationic pentakisisocyanide complex **96** [Eq. (78)] (*106*). In the absence of $TlPF_6$ only a small equilibrium concen-

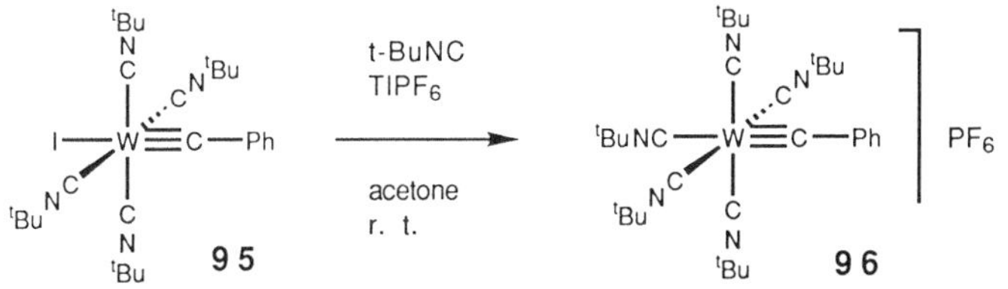

tration of the iodide salt of **96** is formed. On the other hand, the analogous diethylaminocarbyne system **97** reacts with *tert*-butylisocyanide essentially quantitatively to give **98** [Eq. (79)] (*108*). The reaction in Eq. (79) may be

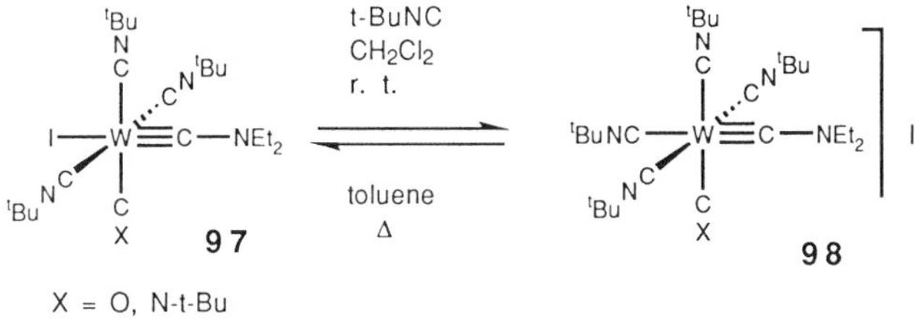

$$X = O, N\text{-}t\text{-}Bu$$

reversed by heating complex **98** in toluene. The higher lability of the isocyanide ligand trans to the phenylcarbyne ligand in **96** compared to the isocyanide trans to the aminocarbyne ligand in **98** was attributed to the stronger metal–carbon π bonds of the phenylcarbyne ligand.

The carbonyl ligand trans to the 2-azaallenylidene ligand in **99** is easily substituted by bromide [Eq. (80)] (*42*). This ligand substitution is accompa-

nied by a very interesting structural adjustment of the 2-azaallenylidene ligand. In **99** the Cr—C—N—C chain is essentially linear, whereas in **100** it is bent at the nitrogen atom. The electronic reasons for the bending were discussed by H. Fischer *et al.* and found to be analogous to those responsible for the occurrence of linear or bent forms of organic 2-azaallenium cations. Substitution of the π-acceptor carbon monoxide by the π-donor bromide causes a shift of electron density from a metal *d* orbital to a π* orbital of the adjacent C=N bond. In the bent form the 2-azaallenylidene ligand has significant iminocarbyne character.

Substitution of the trans halide ligand together with only one equatorial ligand by anionic chelate ligands is rare. Stone synthesized complex **102** by reaction of **101** with the bispyrazolylborate anion (*116*). One of the remaining two mutually trans carbon monoxide ligands is subject to further substitution [Eq. (81)].

R = Me, Tol   R = Tol; PR'$_3$ = Me, Ph

Filippou and Fischer synthesized the anionic complex **104** by reaction of **103** with disodium dicyanoethylenedithiolate [Eq. (82)] (*117*). On the basis of the ^{13}C NMR spectrum complex **104** was assigned a trigonal bipyramidal structure with the aminocarbyne ligand and one terminus of the chelate

$$\text{103} \xrightarrow[\substack{2.\ NEt_4Br \\ CH_3OH}]{\substack{1.\ Na_2[C_2S_2(CN)_2] \\ Me_2CO,\ r.\ t.}} NEt_4 \left[\text{104} \right] \qquad (82)$$

NN = bpy, ophen

ligand occupying the axial coordination sites. Although this seems to be the best fit for the experimental data, it is worthwhile to note that it is surprising to find the strongest π bonding ligand in the axial site. Complex **104** is one of very few examples of a formally unsaturated low-valent transition metal carbyne complex (*118,119*).

The substitution of the trans halide ligand and two equatorial ligands by tris-chelate ligands is a commonly employed synthetic reaction. Cyclopentadienyl derivatives were made according to Eq. (83) (*35,111,120,121*) and Eq. (84) (*69,88*). Quite a variety of trispyrazolylborate derivatives were

$$ \text{Br—M}\!\!\equiv\!\!\text{C—R} \xrightarrow{[C_5R'_5]^-} \qquad (83)$$

M = Cr; R = Tol; R' = H
M = W; R = Me, Tol; R' = Me
R = 2,6-Me$_2$C$_6$H$_3$; R' = H

$$ \text{I—W}\!\!\equiv\!\!\text{C—NEt}_2 \xrightarrow[thf]{K[C_5R_5]} \qquad (84)$$

X = O; L = py, pic; R = H, Me
X = N-t-Bu; L = CN-t-Bu; R = H

synthesized in the laboratories of Stone and Lukehart according to Eq. (85) (*121–123*). In Eq. (86) chloride and two trimethylphosphite ligands were substituted by trispyrazolylborate (*124*). Kläui and Hamers prepared the complexes **105** containing the trisoxygen chelate ligand [Co(η^5-C$_5$H$_5$){PR'$_2$(O)}$_3$]$^-$ [Eq. (87)] (*125*). Complexes **105** are noteworthy in that the trisoxygen chelate ligand is an unusually strong π-donor ligand. For

$$\text{(85)}$$

M = Cr, Mo; R = Tol; R' = H
M = W; R = Me, Ph, Tol, 2-furyl, NEt$_2$; R' = H
R = Tol; R' = pyrazolyl

$$\text{(86)}$$

$$R = \left(\text{—}\bigcirc\text{—OMe}\right)$$

$$\text{(87)}$$

M = Mo; R = Tol; R' = OCH$_3$
M = W; R = Ph, Tol; R' = OCH$_3$, O-i-Pr$_3$

105

example, the infrared absorptions of the CO ligands in **105** are at lower wave numbers (approximately 30 cm^{-1}) than the CO absorptions of the corresponding trispyrazolylborate complexes.

A family of anionic alkylidyne complexes containing dianionic six-electron donor carborane ligands was synthesized by Stone's group. Most of these new complexes contain the pentahapto carborane ligands [η^5-7,8-C$_2$B$_9$H$_9$R$_2$]$^{2-}$ (R = H, Me) [Eqs. (88) and (89)] (126–129) and some the hexahapto carborane ligand [η^6-C$_2$B$_{10}$H$_{10}$Me$_2$]$^{2-}$ [Eq. (90)] (130,131). The carborane ligands behave mostly like electron-rich cyclopentadienyl ligands, but in several instances they contribute toward the reactivity of the carbyne complexes in a unique way [Eq. (105), Section IV,D].

The one carbonyl ligand in the new oxidized forms of the aminocarbyne complexes **106–108** is easily substituted by isocyanide ligands [Eqs. (91)–(93)] (85,88,132). Substitution of iodide ligands was also observed [Eq. (91)] (71,88,132).

In the more typical high-oxidation state alkylidyne complexes substitu-

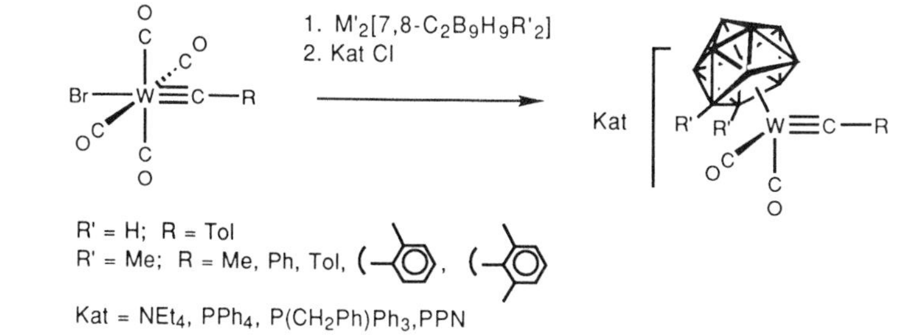
1. M'2[7,8-C2B9H9R'2]
2. Kat Cl
Kat
R' = H; R = Tol
R' = Me; R = Me, Ph, Tol,
Kat = NEt4, PPh4, P(CH2Ph)Ph3,PPN

(88)

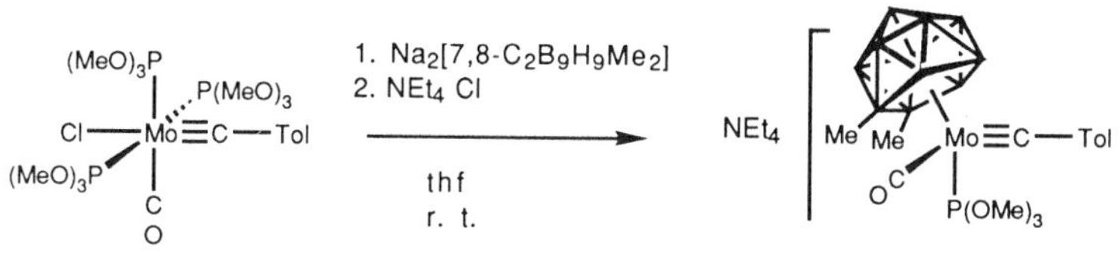
1. Na2[7,8-C2B9H9Me2]
2. NEt4 Cl
thf
r. t.
NEt4

(89)

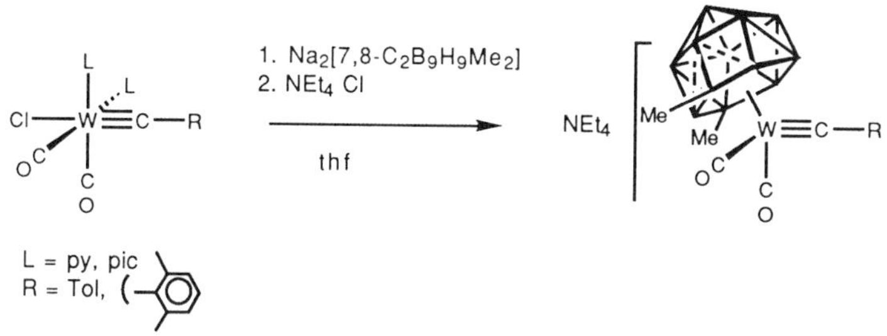
1. Na2[7,8-C2B9H9Me2]
2. NEt4 Cl
thf
NEt4
L = py, pic
R = Tol,

(90)

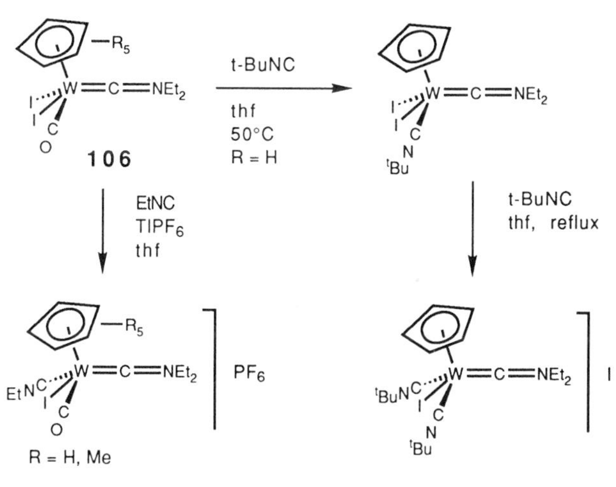
106
t-BuNC
thf
50°C
R = H
EtNC
TlPF6
thf
t-BuNC
thf, reflux
PF6
R = H, Me

(91)

W(CNEt$_2$)I$_3$(CO)(L$_2$)   t-BuNC

107 $\xrightarrow{}$ [W(CNEt$_2$)I$_2$(CN^{t}Bu)$_2$(L$_2$)]I   (92)

L$_2$ = bpy, ophen; R = t-Bu   CH$_2$Cl$_2$
L = CNR (R = Me, t-Bu   r. t.

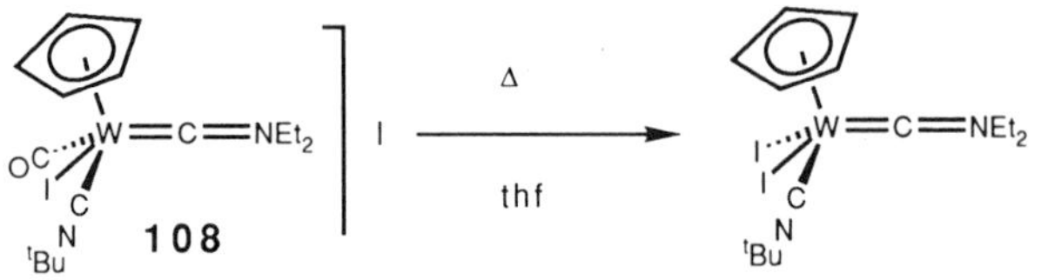

(93)

tion of halide or other anionic ligands is most commonly employed for the synthesis of new derivatives. Schrock's group prepared several new thiolate-substituted alkylidyne complexes according to the reactions shown in Eqs. (94)–(96) (*133*). The chloride ligands in the rhenium complex **109**

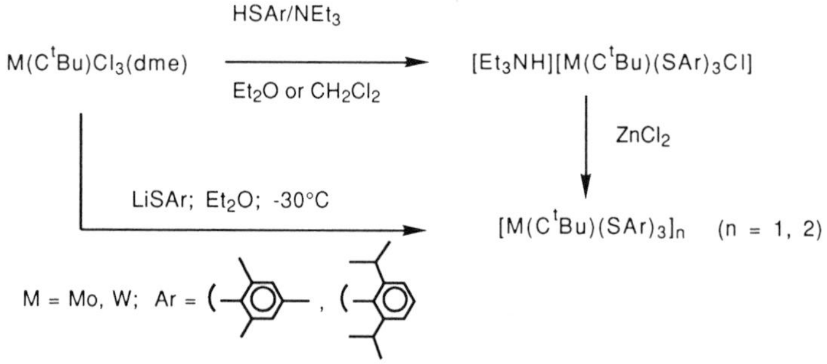

(94)

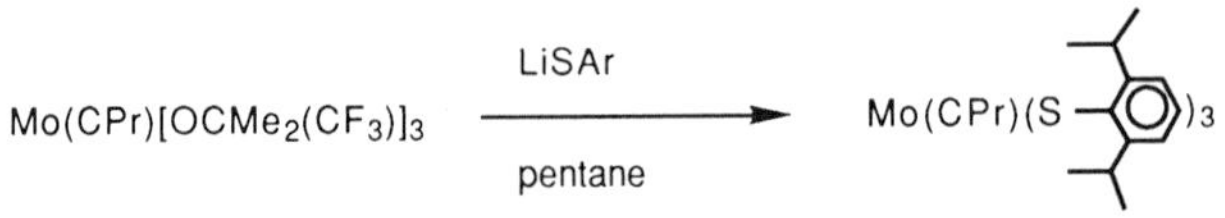

(95)

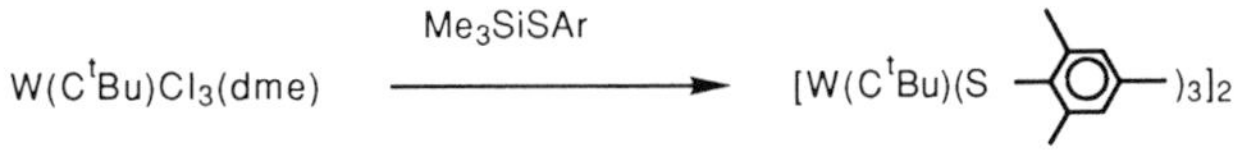

(96)

were exchanged using trimethylsilyl halide reagents [Eq. (97)] (*54,55*). The sequence of substitution reactions shown in Eq. (98) was used to synthesize the alkylidyne alkylidene rhenium complex **110** (*52*). The rhenium alkylidyne imido complexes **112** were synthesized by reaction of the amido complex **111** with lithium alkoxide reagents [Eq. (99)] (*51*). For metathesis reactions of some of these complexes, see Section IV,G.

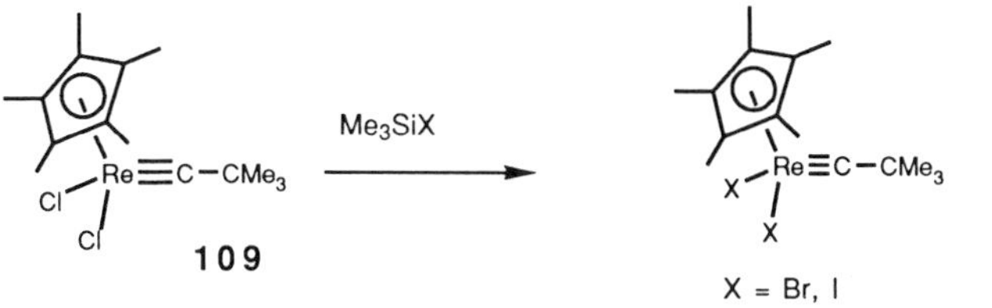

$$\text{109} \xrightarrow{\text{Me}_3\text{SiX}} \qquad \qquad \qquad (97)$$

X = Br, I

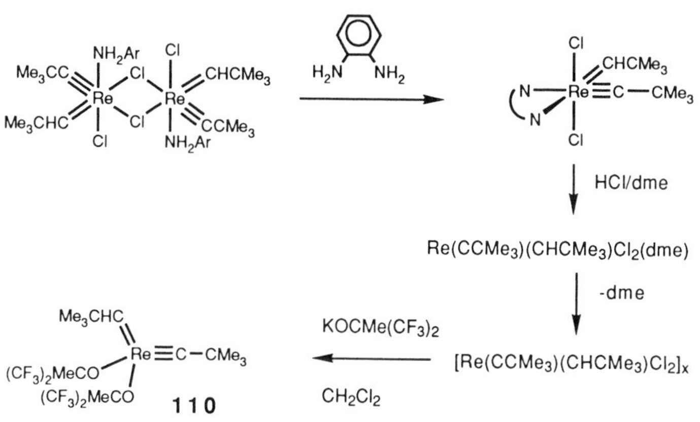

(98)

Re(CCMe₃)(CHCMe₃)Cl₂(dme)

$$[\text{Re(CCMe}_3)(\text{CHCMe}_3)\text{Cl}_2]_x$$

$$[\text{Et}_4\text{N}][\text{Re(CCMe}_3)\text{NHAr})\text{Cl}_4] \quad \xrightarrow[\text{CH}_2\text{Cl}_2]{3 \text{ LiOR}} \quad \text{112}$$

111

Ar =

R = OCMe₃
    OCMe₂CF₃
    OCMe(CF₃)₂
    O-2,6-C₆H₄-i-C₃H₇

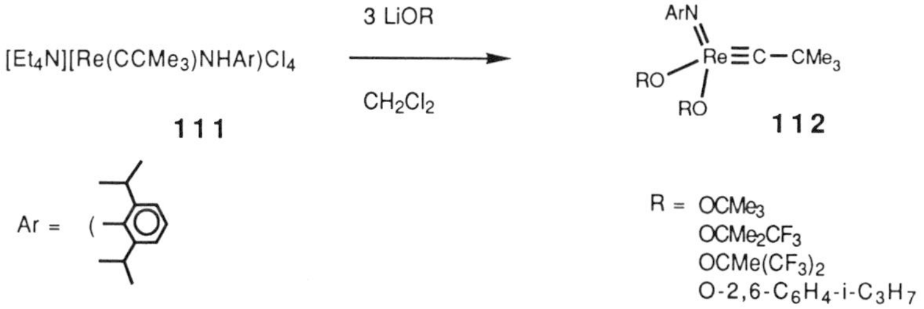

(99)

D. Reactions with Electrophiles

Protonation reactions are very interesting since their outcome can be indicative of the nucleophilic sites in alkylidyne metal complexes. Green investigated the protonation of complex **113** (*90*). Reaction with HBF₄ affords the alkylidyne hydrido complex **114** [Eq. (100)], whereas reaction with CF₃CO₂H results in protolytic cleavage of the alkylidyne ligand from the metal center and formation of Mo(η^5-C₅H₅)(CF₃CO₂)₂[P(OMe)₃]₂. The reactions both with HBF₄ and with CF₃CO₂H are believed to proceed via charge-controlled kinetic attack of the proton at the alkylidyne carbon atom to give an intermediate alkylidene ligand. Depending on the coun-

$$(100)$$

terion, different secondary transformations occur: in the presence of "non-coordinating" BF_4^- counterions the hydrogen atom migrates from carbon to the metal center to give the thermodynamically more stable alkylidyne hydride complex **114**. The more nucleophilic counterion $CF_3CO_2^-$, however, coordinates to the metal center before the hydride migration step and thus probably facilitates a second protonation of the former carbyne ligand, which may then dissociate from the metal center via additional steps, for example, rearrangement to an olefin. Reaction of **113** with diphenylphosphine also leads to cleavage of the metal–carbon triple bond, generating $Mo(\eta^5\text{-}C_5H_5)(PPh_2)[P(OMe)_3]_2$ and $Mo(\eta^5\text{-}C_5H_5)(PPh_2)[P(OMe)_3]$-$(PHPh_2)$ (90).

The trigonal bipyramidal osmium carbyne complex **115** adds HCl across the metal–carbon triple bond to give the octahedral carbene complex **116** [Eq. (101)] (56). Protonation of **115** with aqueous $HClO_4$ gives the cationic

$$(101)$$

carbene complex **117**, which contains coordinated H_2O. The aquo ligand can be replaced by chloride to afford **116**.

Protonation of complexes **118** with HCl leads to formation of the alkylidene complexes **119** [Eq. (102)] (82,97). The solid-state structure of **119**

$$(102)$$

L = PMe₃; R = Ph, Tol
L = CO; R = Ph

(R = Ph) shows that the C—H bond is activated by interaction with the metal center. Deprotonation of the benzylidene ligand was demonstrated with moderately strong bases such as pyrrolidinocyclopentene to generate the labile dichlorotungsten carbyne complex **120** [Eq. (103)]. Substitution of one chloride ligand in **120** occurs under very mild conditions [Eq. (103)].

$$\text{(103)}$$

L = CN-t-Bu, P(OMe)$_3$

Protonation of several molybdenum and tungsten alkylidyne complexes of the type M(CR)(η^5-C$_5$H$_5$)(CO)L with acids such as HCl or CF$_3$CO$_2$H was reported by Kreissl *et al.* to afford the η^2-acyl complexes **121** [Eq. (104)] (*134*). The formation of the acyl complexes **121** presumably in-

$$\text{(104)}$$

M = Mo; R = Tol; L = CO; X = Cl, CF$_3$CO$_2$
M = W; R = Me, Ph, Tol; L = CO; X = Cl, CF$_3$CO$_2$
M = W; R = c-Pr; L = CO; X = CF$_3$CO$_2$
M = W; R = Tol; L = CO; X = CCl$_3$CO$_2$
M = W; R = Tol; L = PMe$_3$; X = Cl

volves double addition of HCl across the metal–carbon triple bond accompanied by migration of the formed alkyl ligand to carbon monoxide.

The outcome of protonation of the carborane tungsten carbyne complexes **122** depends on the substituents R (H or Me) on the carborane ligand and on the nature of the acid [Eq. (105)]. For R equals H protonation with HBF$_4$ affords the alkyne-bridged dinuclear complex **123** (*135*). An analogous reaction was previously observed with the cyclopentadienyl system (*136*). For R equals Me the carborane ligand participates in a unique way. Protonation with HBF$_4$ in the presence of ligands L gives the complexes **124a** (*15*). Protonation with HX (X = Cl, I) affords the complexes **124b** (*137*). The carborane ligands in complexes **124** contain CH$_2$Tol substituents, which are possibly formed by initial protonation of the carbyne carbon followed by hydroboration. In **124b** the carborane cage has also undergone a skeletal rearrangement.

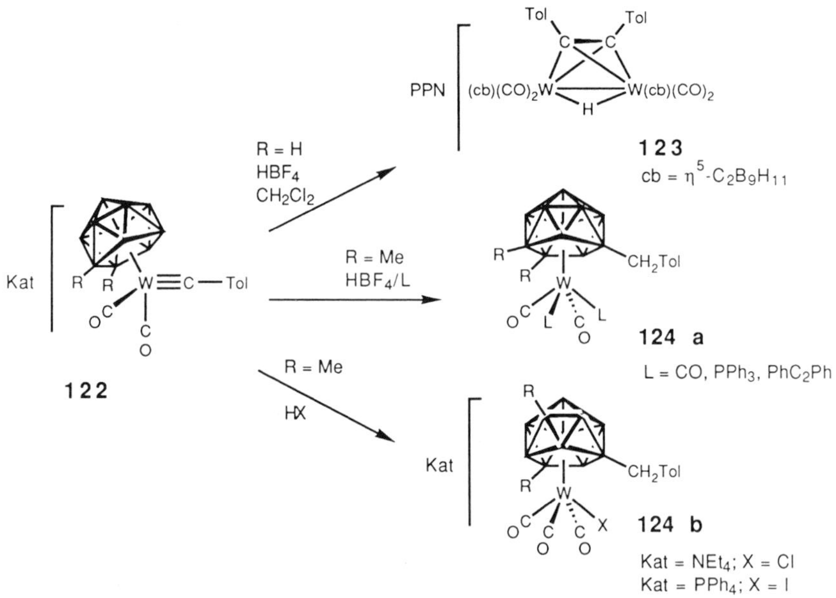

(105)

The *trans*-aminocarbyne(iodo)tungsten complexes **125** react with $CF_3SO_3CH_3$ to give the triflate derivatives **126** and methyl iodide [Eq. (106)] (*138*). This reaction is a rare example of direct attack of electrophiles at the trans halide ligand of a carbyne complex.

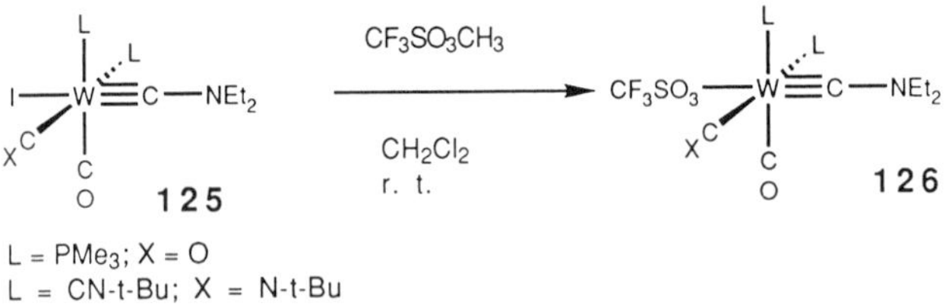

(106)

The chemistry of thiocarbyne metal complexes was pioneered by Angelici (*5,139*). Reactions of thiocarbyne complexes with electrophiles led to very interesting and unique transformations involving the metal–carbon triple bond. The sulfur substituent serves both as an electron reservoir, leading to stabilization of products, and as a leaving group, allowing for subsequent transformations. Protonation of carbyne complexes at the carbyne carbon atom generates formal 16-electron carbene complexes. Although in most cases the electron demand of the metal center is met by agostic bonding of the alkylidene C—H bonds, in protonated thiocarbyne complexes the metal center is always stabilized by electron donation from the sulfur atom [Eqs. (107) and (108)] (74,140). For this interaction to be

$$H^+ = HSO_3CF_3, \ HBF_4, \ CF_3CO_2H$$
$$\text{base} = K_2CO_3, \ Et_3N, \ KH$$

efficient the sulfur atom moves into a position to "bridge" the metal–carbon double bond. The protonation step may be reversed by the addition of base [Eq. (108)]. The carbene carbon in the protonated systems is strongly electrophilic. Addition of phosphorus-based nucleophiles gives rise to a variety of ylide complexes [Eq. (109)]. Even at the stage of the ylide products the original thiocarbyne complexes may be regenerated by reaction with base. The thiocarbene complexes react with secondary amines to afford the aminocarbyne complexes **127** [Eq. (110)] (141). With

$$PR_3 = PEt_3, \ PPh_2H, \ PPh_3, \ P(OMe)_3$$

$$(111)$$

R = Me, Et, CH$_2$CH$_2$OH, CHMe$_2$, CMe$_3$, Tol, H

primary amines formation of aminocarbyne ligands is also observed, but in these cases the aminocarbyne complexes are in equilibrium with hydrido isocyanide complexes [Eq. (111)]. For R equals H and Tol, only the aminocarbyne form is observed.

The reagent [Me$_2$SSMe]$^+$ is capable of transferring a [SMe]$^+$ group to metal–carbon triple bonds [see also Eq. (118)] (*142*). The reaction may be viewed as a direct electrophilic attack at the MC π bond. With complex **128** the dithiocarbene complex **129** is obtained [Eq. (112)]. Complex **129** reacts with nucleophiles (H$^-$, CH$_3^-$, and PMe$_3$), which add to the carbene carbon to give stable adducts [Eq. (113)]. The carbonylmetallate anions

$$(112)$$

$$(113)$$

[Mo(η^5-C$_5$H$_5$)(CO)$_3$]$^-$ and [Mn(CO)$_5$]$^-$ abstract SMe$^+$ from **129** and regenerate the thiocarbyne **128**. The addition of thiolate anions affords complexes containing η^2-(tristhiolato)methyl ligands [Eq. (114)]. Complexes

$$(114)$$

130 transform into high-oxidation state thiocarbyne complexes by the unique migration of two thiolate groups to the metal center [Eqs. (115) and (117)] (*143*). For R equals Me and Et the process may be induced thermally. Heating of the complexes with R equals Ph and Tol, however, leads to elimination of ArSSMe from the (tristhiolato)methyl ligand with regeneration of the low-valent methylthiocarbyne complex [Eq. (116)]. Irradiation with UV light induces the migration of two thiolate groups to the metal center for all R (Me, Ph, and Tol). In the photochemically induced reaction the arylthiolates are migrating to the metal center [Eq. (117)].

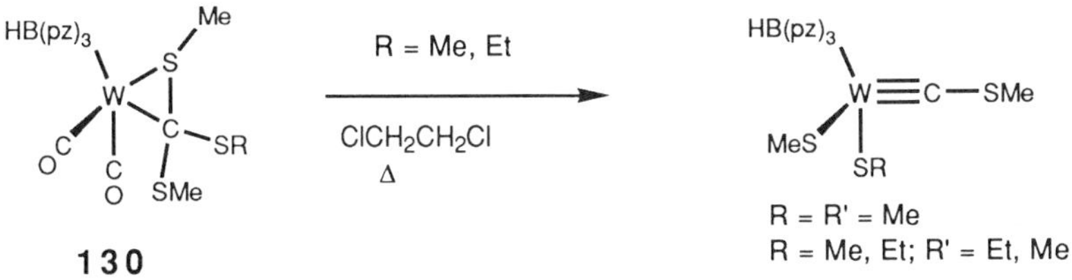

$$(115)$$

R = R' = Me
R = Me, Et; R' = Et, Me

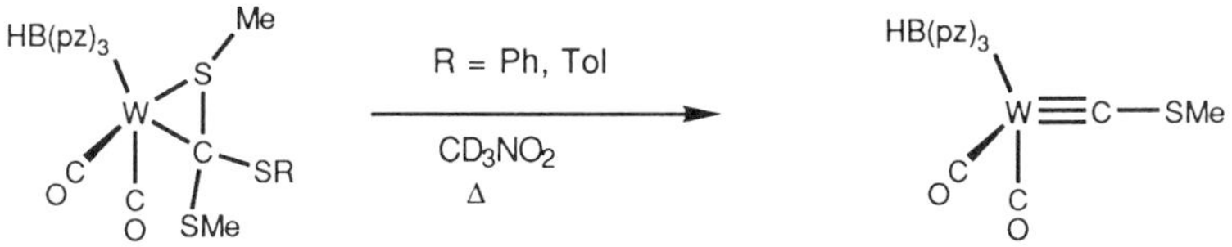

130

$$(116)$$

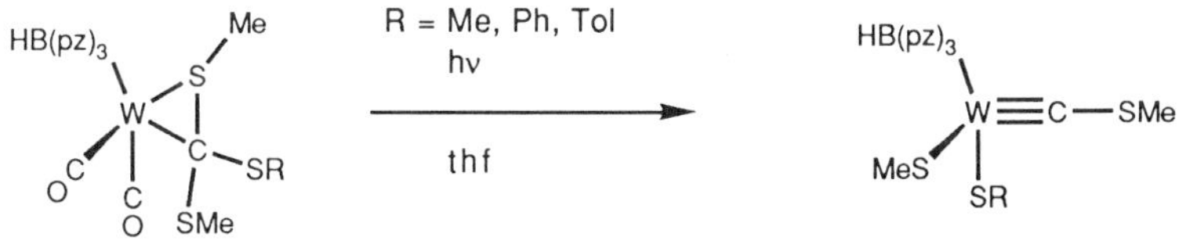

130

$$(117)$$

130

Kreissl and Keller demonstrated sequential transfer of two [SMe]⁺ groups to the metal–carbon triple bond of **131** to afford the thiocarbene complex **132** and the bismethylthio alkyl complex **133** [Eq. (118)] (*144*). In contrast to the systems obtained by Angelici from thiocarbynes, in which nucleophiles add to the carbene carbon [Eq. (109)], phosphines attack complexes **132** at the metal center, thereby inducing carbene–carbonyl coupling [Eq. (119)] (*145*). The larger steric demand of the tris-pyrazolyl-borate ligand may prevent attack at the metal center in Eq. (113). The carbene carbon in **132** may be protonated with trifluoroacetic acid [Eq. (120)]. $PClR_2$ or $SClR$ react with complex **131** to afford the ketene com-

$$(118)$$

$$(119)$$

$$(120)$$

plexes **134** [Eq. (121)] (*146,147*). The coordinated ketenes are attacked by trimethylphosphine at the ketene carbon while less nucleophilic phosphines and phosphites coordinate to the metal center with displacement of the ketene functionality from the metal center [Eq. (122)].

$$(121)$$

R = Me, Ph, Tol
EXR'_n = $PClMe_2$, $PClPh_2$, $AsIMe_2$, $SClC_6H_4NO_2$

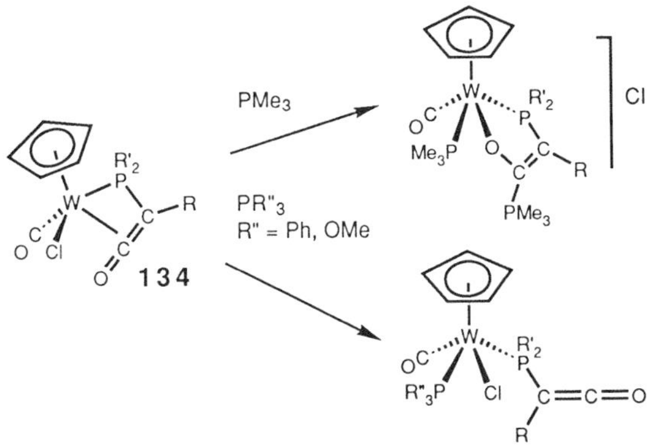

(122)

The osmium carbyne complex **115** reacts with elemental sulfur, selenium, and tellurium to afford the complexes **135** in which the element atoms "bridge" the metal–carbon triple bond [Eq. (123)] (*56*). Complex **115** also reacts with transition metal Lewis acids such as AgCl or CuI to give dinuclear compounds with bridging carbyne ligands. Reaction with elemental chlorine results in addition across the metal–carbon triple bond to generate the chlorocarbene osmium complex **136** [Eq. (124)].

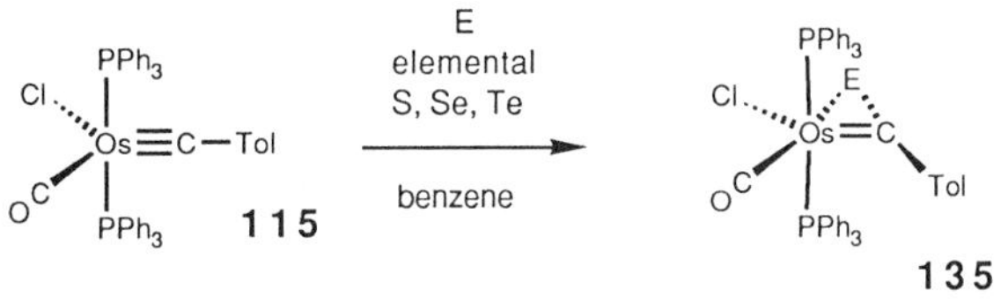

(123)

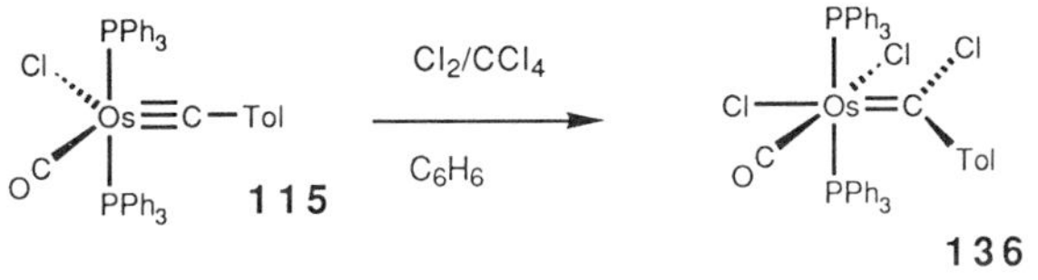

(124)

Addition of two sulfur atoms to the carbyne ligand in complexes **137** from either cyclohexene thiaepoxide or elemental sulfur results in the formation of the dithiocarboxylate complexes **138** [Eq. (125)] (*116,148*). In contrast, the thiocarbyne complex W(CSMe)(trispyrazolylborate)(CO)$_2$ does not react with elemental sulfur nor with ethylenethiaepoxide (*140*).

The incorporation of alkylidyne ligands into thioaldehyde ligands was observed in our laboratory to occur in the reaction of complexes **139** with diethylammonium diethyldithiocarbamate [Eq. (126)] (*149*). Although no

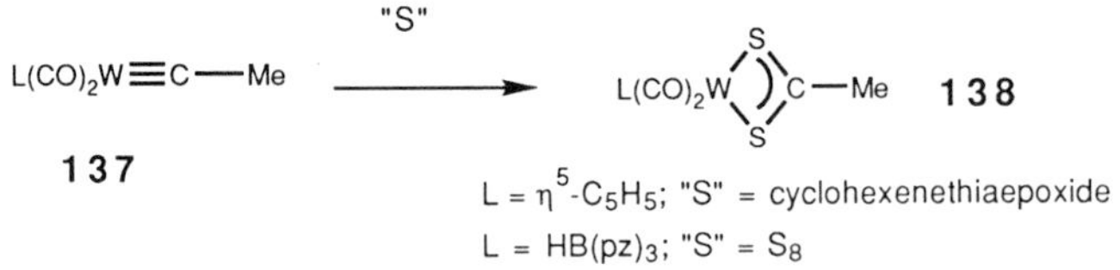
"S"
L(CO)$_2$W≡C—Me
137
L(CO)$_2$W C—Me
138
L = η^5-C$_5$H$_5$; "S" = cyclohexenethiaepoxide
L = HB(pz)$_3$; "S" = S$_8$

(125)

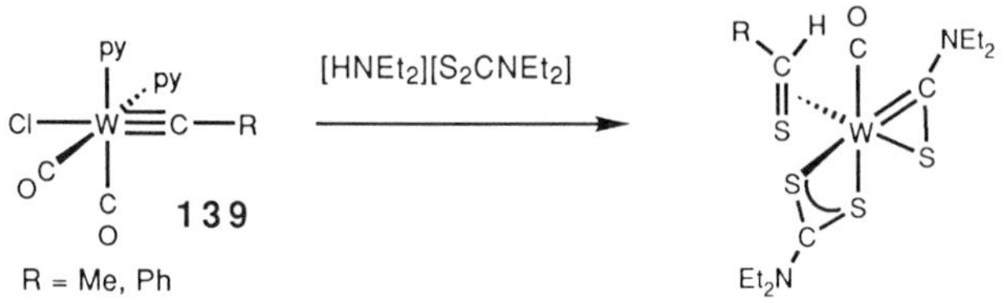
[HNEt$_2$][S$_2$CNEt$_2$]
139
R = Me, Ph
NEt$_2$
Et$_2$N

(126)

mechanistic information on this unusual transformation is available, it was postulated that after initial coordination of two dithiocarbamate ligands the alkylidyne carbon may be protonated by the ammonium ion, followed by transfer of a sulfur atom to the generated alkylidene ligand.

The phosphinidene-bridged system **140** was obtained by reaction of $W(CPh)(\eta^5\text{-}C_5H_5)(CO)_2$ with the tungsten phosphine complex shown in [Eq. (127) (*150*). CuCl-induced elimination of the substituted benzene

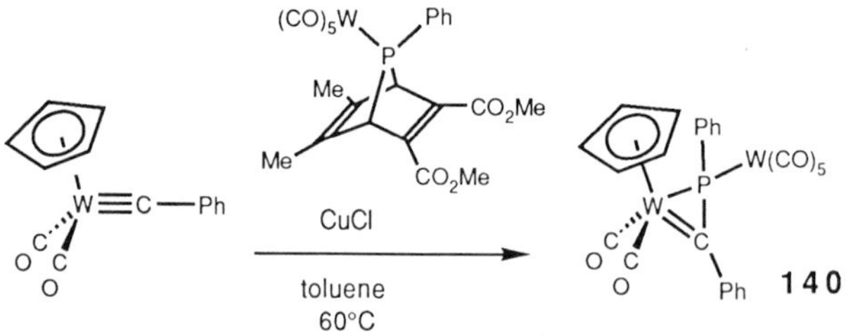
(CO)$_5$W
CO$_2$Me
CuCl
toluene
60°C
W(CO)$_5$
140

(127)

from this reagent generates the electrophilic phosphinidene complex $(CO)_5WPPh$ which adds across the metal–carbon triple bond much in the same way as the sulfur reagent in Eq. (118).

The tungsten alkylidyne complexes **141** are reactive toward borane reagents (*151*). Treatment with BH_3 affords the dinuclear complexes **142** [Eq. (128)], and reaction with borabicyclononane gives complex **143** [Eq. (129)]. Complex **143** contains both fragments of the hydroboration reagent attached to the carbyne carbon.

The trisneopentyl neopentylidyne complexes **144** of molybdenum and tungsten react with trifluoroacetic acid [Eq. (130)] and with other carboxylic acids [Eq. (131)] to afford the triscarboxylate systems **145** and **146**

$$R' = H; \quad R = Me, Ph, Tol$$
$$R' = Me; \quad R = Tol$$
(128)

(129)

$$\underset{\mathbf{144}}{M(CCMe_3)(CH_2CMe_3)_3} \xrightarrow[\text{pentane}]{\substack{3\ CF_3CO_2H \\ dme}} \underset{\mathbf{145}}{M(CCMe_3)(O_2CCF_3)_3(dme)} \quad (130)$$

$$M = Mo, W$$

$$\underset{\substack{M = Mo, W \quad \mathbf{144}}}{M(CCMe_3)(CH_2CMe_3)_3} \xrightarrow[\substack{\text{pentane} \\ -30°C}]{3\ RCO_2H} \underset{\substack{R = Me, CHMe_2, CMe_3 \quad \mathbf{146}}}{M(CCMe_3)(O_2CR)_3} \quad (131)$$

(152). The trifluoroacetate ligands in **145** act as monodentate ligands, and the complexes are stabilized by dimethoxyethane. The carboxylate ligands in complexes **146** are bidentate. Spectroscopic data of **146** are interpreted in terms of a pentagonal bipyramidal structure in which the neopentylidyne ligand occupies an axial position and one carboxylate ligand is chelating the other axial and one equatorial site while the remaining two carboxylate ligands chelate in the equatorial plane. The protolytic cleavage of the metal–neopentyl bonds probably proceeds via initial protonation at the alkylidyne carbon followed by α-elimination of neopentane. Treatment of the tris(*tert*-butoxy)molybdenum alkylidyne complex **147** with carboxylic acid gives the tris(carboxylato) derivative **148** [Eq. (132)]. However, reaction of the tris(*tert*-butoxy)tungsten alkylidyne complex **149** with neopentyl alcohol was found to result in cleavage of the metal–carbon

$$\underset{\mathbf{147}}{Mo(CCMe_3)(OCMe_3)_3} \xrightarrow[\text{pentane}]{3\ Me_2HCCO_2H} \underset{\mathbf{148}}{Mo(CCMe_3)(O_2CCHMe_2)_3} \quad (132)$$

$$W(CR)(OCMe_3)_3 \quad \xrightarrow[\substack{hexane \\ r.\,t.}]{6\ HOCH_2CMe_3} \quad W(OCH_2CMe_3)_6 + RCH_3 + 3\ Me_3COH \quad (133)$$

149

triple bond via formal triple protonation of the alkylidyne ligand [Eq. (133)] (*80*).

The transfer of a proton from an imido ligand in **150** to the alkylidyne carbon was an essential step in the synthesis of the active olefin metathesis catalysts **153** [Eq. (134)] (*153–156*). Triethylamine serves as a catalyst for

$$(134)$$

$M = Mo, W; Ar = $ (2,6-diisopropylphenyl)
$R = CMe(CF_3)_2$

the proton transfer to form **151**. The observation that the proton transfer cannot be achieved in the alkoxy derivative **152** strongly indicates that the proton transfer actually consists of sequential HCl elimination/readdition steps.

E. Reactions with Nucleophiles

The electrophilicity of the alkylidyne carbon in cationic complexes of manganese and rhenium such as **154** or **157** is well established. A study by Chen *et al.* established that the carbonyl ligands are also potential sites for nucleophilic attack (*157,158*). Reaction of **154** with the bulky carborane anion $LiC_2B_{10}H_{11}$ results in the formation of two products: the carbene complex **156**, resulting from attack at the alkylidyne carbon, and the alkylidyne acyl complex **155**, resulting from attack at a carbonyl ligand [Eq. (135)]. At room temperature complex **155** transforms into complex **156**.

Geoffroy and co-workers investigated the reactivity of complexes **157** and **159** toward nitrite and in the course of this work made the important discovery of migration of an acyl ligand to a carbonyl ligand in a 17-elec-

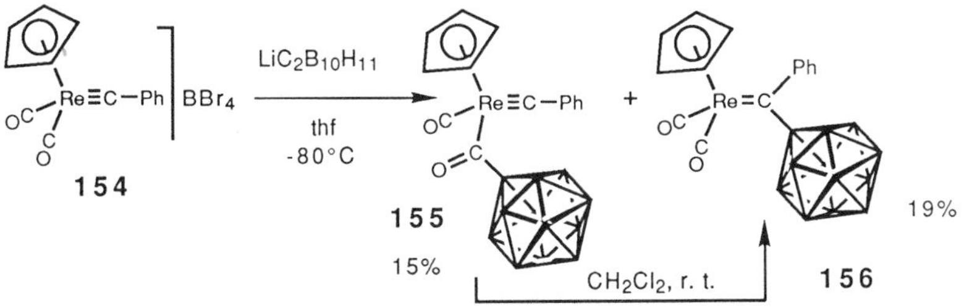

(135)

tron manganese complex (*159,160*). The isolated products are the α-keto-acyl manganese complex **158** [Eq. (136)] and the rhenium acyl complex **160** [Eq. (137)]. Low temperature spectroscopic studies revealed the initial

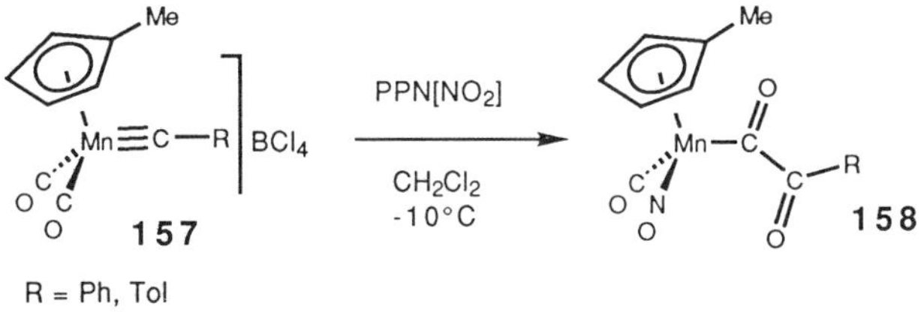

(136)

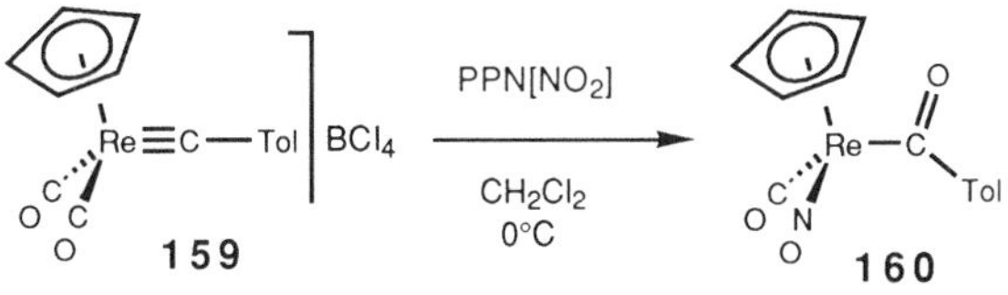

(137)

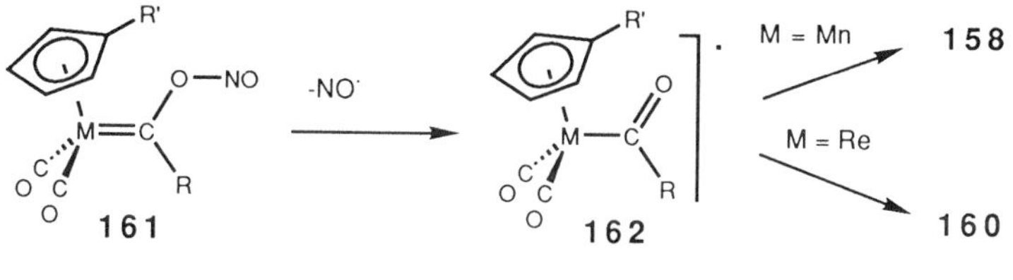

(137A)

formation of a neutral adduct which is formulated as the carbene complex **161**. Homolytic cleavage of NO was postulated to generate the 17-electron acyl complex **162** and free NO, which then recombine to afford the observed products. The intermediacy of the 17-electron acyl complexes could be verified by independent synthesis through chemical oxidation of the anionic acyl complexes $[M(\eta^5\text{-}C_5H_4R)\{C(O)R\}(CO)_2]^-$ and subsequent reaction with NO to give the same products **158** and **160** [Eq. (137a)].

The dinuclear terminal alkylidyne complexes **163** react with the hydride transfer reagent $K[BH(CHMeEt)_3]$ to give the bridged alkylidene complexes **164** (*161*). The ethylidene tungsten rhenium complex **164** (R = Me) was furthermore shown to rearrange to the $\mu-\eta^2$-vinylcomplex **165** [Eq. (138)].

$$(138)$$

The coupling of alkylidyne and acyl ligands was achieved by reaction of the tungsten complexes **166** with $Mn(CH_3)(CO)_5$ and $M(\eta^5$-$C_5H_5)(CH_3)(CO)_3$ (M = Mo, W) [Eq. (139)] (*162*). Formation of the prod-

$$(139)$$

ucts is probably initiated by migratory insertion in the methyl complexes, followed by coordination of the alkylidyne complexes to the unsaturated metal acyl species.

Reaction of the tungsten complex **167** with diphenylphosphine at elevated temperatures affords complex **168** [Eq. (140)] (*163*). The same compound could be obtained by sequential addition of $LiPPh_2$ and NH_4Br. Formation of this product is surprising considering the ease with which some phospines induce carbyne–carbonyl coupling in complex **167** (see Section IV,F). The by-product **169** could possibly be derived from an intermediate ketenyl complex. Protonation of complex **168** in acetonitrile occurs at the former carbyne carbon to give the phosphine complex **170**.

$$(140)$$

The tungsten thiocarbyne complex **171** is attacked by PR_3 (R = Me, Et) at the carbyne carbon atom (*164*). Templeton and co-workers found that with PEt_3 the phosphonium carbyne complex **172** is obtained [Eq. (141)] whereas reaction with PMe_3 gives the bisphosphonium carbene complex **173** [Eq. (142)]. One PMe_3 appears to be able to dissociate from **173**, and

$$(141)$$

$$(142)$$

on addition of MeI the phosphonium carbyne complex **174** may be isolated. Nucleophiles add to **174** at the carbyne carbon. The attack of the phosphines at the carbyne carbon of the thiocarbyne complex 171 may be compared to the reaction of the sterically less encumbered tris(pyrazolyl)borate analog W(CSMe){tris(pyrazolyl)borate}(CO)$_2$ 196. In this complex attack of the phosphine occurs at the metal center accompanied by carbyne–carbonyl coupling [Eq. (157), Section IV,F].

A related nucleophilic substitution at the carbyne carbon was demonstrated by Lalor on the chlorocarbyne complexes 175 (*165*). Reaction of 175 with stabilized carbanion reagents, for example, NaCH(CN)(CO$_2$Et), affords the anionic vinylidene complexes 176 [Eq. (143)]. Formation of 176 was postulated to occur via carbyne complexes of the type L$_n$M≡C—CHXY, which are subject to deprotonation by the reagent. Protonation of the molybdenum vinylidene complex 177 with HCl did not generate a carbyne complex, rather the vinyl complex 178 [Eq. (144)]. The unusual protonation of the vinylidene ligand at the α-carbon is probably the result of initial protonation of the metal center (compare with Scheme 1, Section III,B,1) and subsequent migration of the hydride ligand to the vinylidene ligand, which may be facilitated by formation of the chelating vinyl ligand. Other electrophiles, however, were found to add to the β-carbon of the vinylidene ligand [Eq. (145)].

M = Mo; X = Y = CN; X = Y = CO$_2$Et; X = CN, Y = CO$_2$Et
M = W; X = Y = CN

E = p-Me$_2$NC$_6$H$_4$N$_2$, Cu, HgCl

The alkylidyne ligands in the molybdenum complexes 179 were modified by deprotonation at the β-carbon with strong bases and subsequent addition of electrophiles [Eq. (146)]. Complexes 180 were desilylated by treatment with fluoride [Eq. (147)] (*48*).

HB(pz')₃

Mo≡C—CH₂R $\xrightarrow[\text{thf}]{\text{base}}$ Kat ... Mo≡C—CHRR' (146)

179 -78°C -78°C

base = BuLi or Na[N(SiMe₃)₂]

R = H; R' = Me, Et
R = Me; R' = Me

Mo≡C—CHRSiMe₃ $\xrightarrow[\text{MeCN/H}_2\text{O}]{\text{NaF}}$ Mo≡C—CH₂R (147)

180

Cationic 2-azaallenylidene complexes such as **181** are capable of accepting nucleophiles at the γ position as shown in Eq. (148) (*166*). In fact, the reverse of this conversion was applied in the synthesis of complex $[Cr(CNCCl_2)(CO)_5][AlCl_4]$ by γ-abstraction of a chloride substituent from the trichloromethylisocyanide ligand in $Cr(CNCCl_3)(CO)_5$ (*43*).

$[(CO)_5M \doteq C \equiv N \doteq C\text{-t-Bu}_2]$ [AlBr₄] $\xrightarrow[\text{-AlBr}_3]{\text{thf}}$ $(CO)_5M—C\equiv N—C(t\text{-Bu})_2Br$ (148)

181

M = Cr, W

F. Coupling Reactions of Alkylidyne Ligands

Coupling of carbyne ligands with carbonyl and isocyanide ligands has been observed in many systems. There are at least three different methods by which this bond-forming step can be induced: addition of donor ligands, addition of electrophile to the carbonyl or isocyanide ligand (accompanied by coordination of a donor ligand), and electronic excitation (accompanied by ligand addition) (*19*).

The bispyridine-substituted carbyne complexes **182** react with several anionic bidentate donor ligands to afford the anionic η^2-ketenyl complexes **183** [Eq. (149)] (*149,167*). The bis(pyrrole carboxaldiminato) complexes were further transformed into alkoxyacetylene complexes by methylation at the ketenyl oxygen atom. Addition of trimethylphosphine to the anionic carborane-substituted carbyne complex **184** at $-30°C$ gives the ketenyl complex **185** [Eq. (150)] (*135*). The hydroxy and methoxyacetylene complexes were obtained by further addition of HBF₄ and methyl triflate.

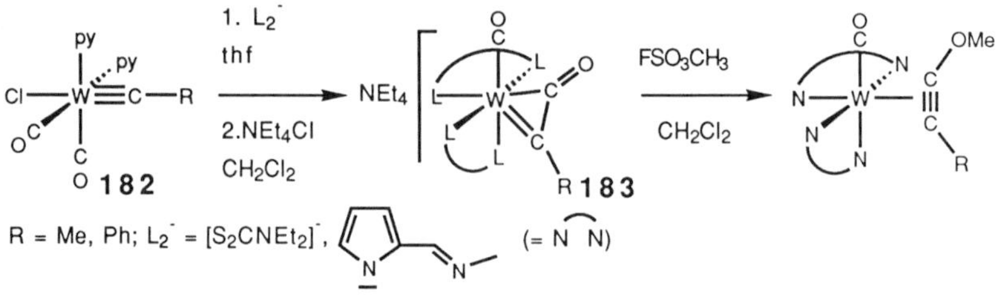

$$(149)$$

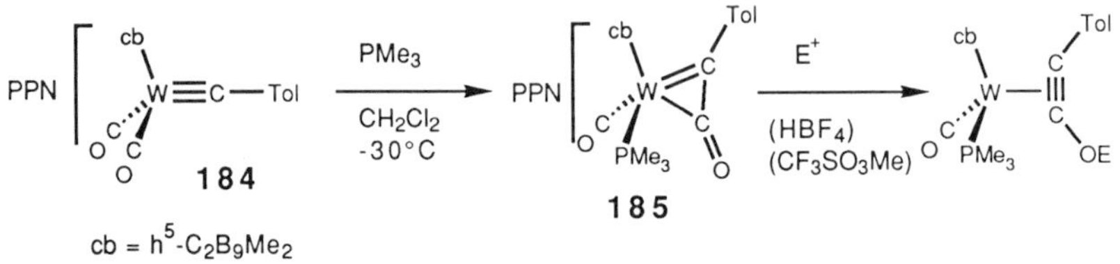

R = Me, Ph; L_2^- = $[S_2CNEt_2]^-$, (= N⌢N)

$$(150)$$

cb = h^5-$C_2B_9Me_2$

Similarly, the ketenyl complex **187** forms after addition of trimethylphosphine to the cyclopropylcarbyne complex **186** at $-15°C$ [Eq. (151)] (*168*).

Kreissl and co-workers showed that the outcome of the reaction of alkylidyne complexes such as **188** with trimethylphosphine strongly depends on the reaction conditions (*169*). Whereas at low temperatures in CH_2Cl_2 formation of ketenyl complexes is observed, at room temperature in pentane a significant amount of CO substitution product is obtained [Eq. (152)]. The trimethylphosphine-substituted complexes **189** react with carbon monoxide to form η^1-ketenyl complexes. A very similar result was obtained by Stone and co-workers in the reaction of complex **190** with PMe_3 in petroleum ether at room temperature [Eq. (153)] (*35*). A mixture of carbonyl substitution product and coupling product was isolated. Reac-

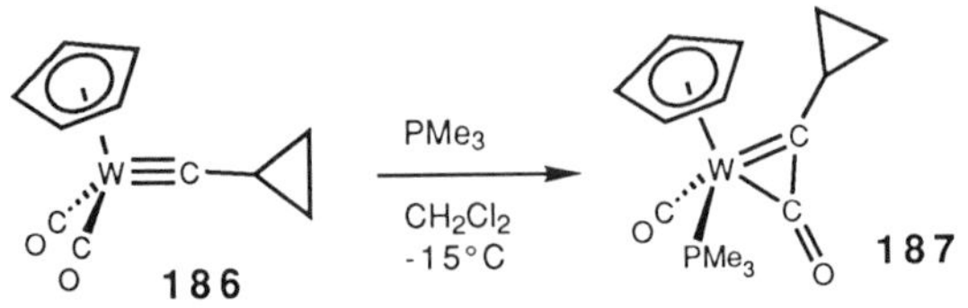

$$(151)$$

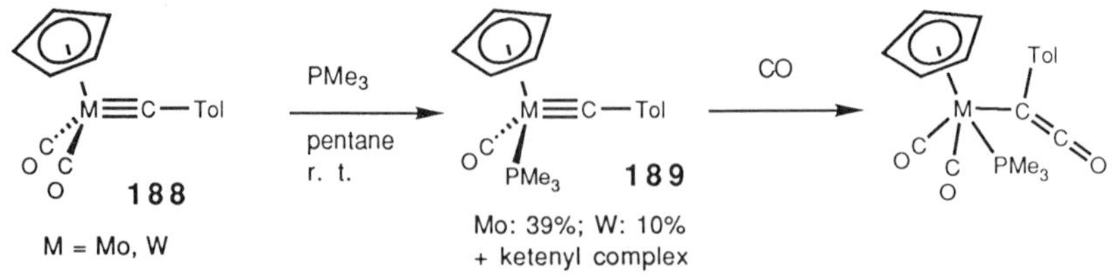

$$(152)$$

M = Mo, W

Mo: 39%; W: 10%
+ ketenyl complex

$$(153)$$

tion of the complex $W(CTol)[HB(pz)_3(CO)_2]$ with PMe_3 gives $W(TolCCO)[HB(pz)_3(CO)(PMe_3)]$, whereas no reaction occurs with PPh_3 (116). The substitution of one carbon monoxide ligand in $Mo(CTol)(\eta^5\text{-}C_5H_5)(CO)_2$ by PMe_3 was reported by Stone and co-workers (170). In this case no ketenyl complex was isolated; however, considering the behavior of closely related carbyne complexes just discussed, it is quite likely that ketenyl intermediates are involved in this reaction. The intermediacy of an observable ketenyl complex in a carbonyl substitution reaction was established for the reaction in Eq. (66) (Section IV,C).

Treatment of the molybdenum alkylidyne complex **191** with 300 atm CO affords a mixture of three products: the products of substitution of one and two phosphite ligands and complex **192** [Eq. (154)] (91). The methoxycarbonyl group in **192** may have arisen from the reaction of an intermediate ketenyl ligand with methanol (derived from hydrolysis of trimethylphosphite). Reaction of the compounds **193** with xylylisocyanide was shown to give complexes **194** [Eq. (155)] (91). The products contain the

$$(154)$$

$$(155)$$

unusual bis(imino)allyl ligand, which is the result of coupling of the alkylidyne ligand with two isocyanide ligands. NMR studies of the reaction of the molybdenum complex **193** with 1–4 equiv of 2,6-xylylisocyanide showed sequential substitution of the two phosphite ligands before formation of the final products. Formation of the bis(imino)allyl ligand may be envisaged as the result of two different C—C bond-forming steps: first carbyne–isocyanide coupling in analogy to carbonyl–carbyne coupling, then migration of an η^1-iminoketenyl ligand to coordinated isocyanide. Since an NMR study showed that dissociation of phosphite in **191** and **193** does not precede the addition of carbon monoxide or isocyanide, a novel mechanism involving the direct (cheletropic) addition of carbon monoxide and isocyanide to the metal–carbon triple bond to give η^2-ketenyl or keteniminyl complexes was proposed.

Templeton and co-workers used the alkylidyne–carbonyl coupling reaction shown in Eq. (156) for the synthesis of the asymmetric alkyne com-

$$(156)$$

plex **195** (*171*). Deprotonation of **195** generates an η^2-allenyl complex, and subsequent addition of electrophiles was found to occur with significant stereoselectivity.

Work in Angelici's group established that thiocarbyne ligands also undergo coupling reactions with carbonyl ligands (*5*). Reaction of the tungsten carbyne complex $W(CSR)(\eta^5\text{-}C_5H_5)(CO)_2$ [$R = C_6H_3\text{-}2,4\text{-}(NO_2)_2$] with PMe_3 gives the ketenyl complex $W(RSCCO)(\eta^5\text{-}C_5H_5)(CO)(PMe_3)$ (*140*). The tris(pyrazolylborate) tungsten thiocarbyne complex **196** reacts with PEt_3 to give ketenyl complex **197** [Eq. (157)], and complex **198** affords ketenyl complex **199** [Eq. (158)] (*74,140*). Methylation of the ketenyl complex in **197** generates a complex of the unique alkyne MeSC-COMe. A study of the reactivity of the cationic carbyne complex **198**

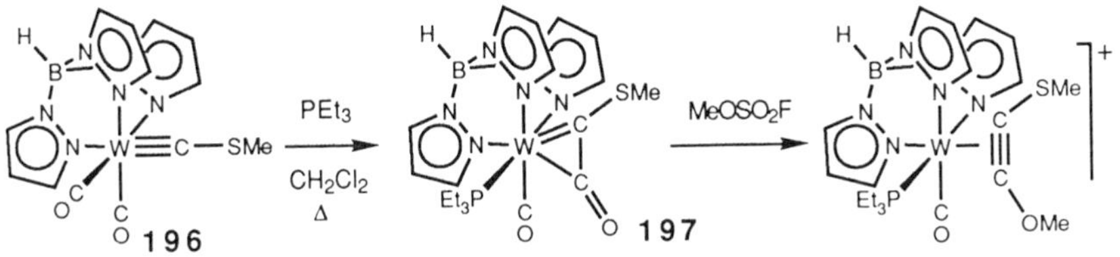

(157)

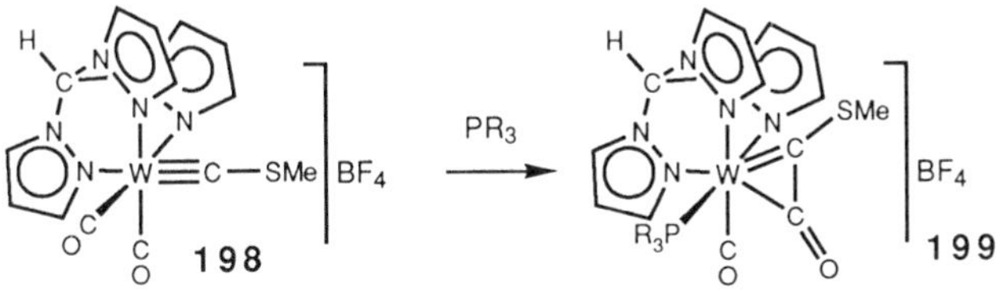

(158)

$PR_3 = PMe_3, PEt_3, PEt_2H, PMe_2Ph, P(OMe)_3, P(O-i-Pr)_3$

toward a series of nucleophiles showed that both electronic and steric factors influence the coupling reaction. The phosphines and phosphites listed in Eq. (158) were all shown to induce carbyne – carbonyl coupling. With $P(O-i-Pr)_3$ an equilibrium between the carbyne complex and the ketenyl complex was observed, which could be shifted toward the side of the ketenyl complex by increasing the phosphite concentration. Other ligands, for example, tricyclohexylphosphine or triphenylphosphite, were found to be unreactive. Tricyclohexylphosphine is basic but sterically encumbered, and triphenylphosphine is sterically nondemanding but not basic enough to induce the coupling step. Other nucleophiles such as SMe^-, CN^-, I^-, or H_2NMe did not react.

Hillhouse and co-workers discovered a new route to carbyne complexes which is based on the reverse of carbyne – carbonyl coupling (172). Reaction of complex 200 with carbon suboxide affords the ketenyl complex 201 [Eq. (159)]. Formation of the ketenyl ligand in 201 was postulated to arise from attack of phosphine on an intermediate ketenylidene complex. Warming of complex 201 results in cleavage of the ketenyl ligand and formation of the phosphonium carbyne complex 202. This mechanism was put to work in a more direct fashion by reaction of complex 200 with Ph_3PCCO [Eq. (160)].

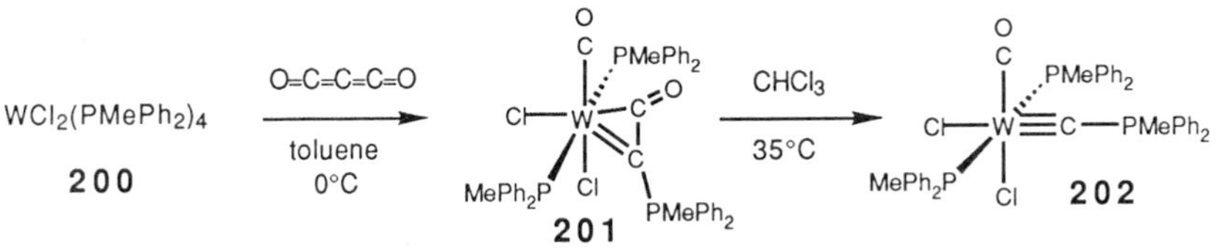

(159)

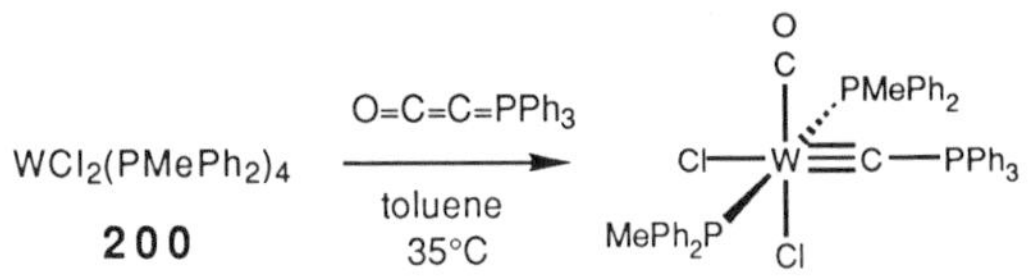

$$WCl_2(PMePh_2)_4 \quad \xrightarrow[\substack{\text{toluene} \\ 35°C}]{O=C=C=PPh_3} \quad$$

200

(160)

Work in Geoffroy's group established photoinduced carbonylation of carbyne ligands as an efficient method for the generation of ketenyl complexes (173,174). Irradiation of complex **203** in the presence of carbon monoxide gives the dinuclear complex **204** [Eq. (161)]. The bridging ligand TolCC(O)CTol was suggested to result from the interaction of the starting material with an intermediate ketenyl complex. The photogeneration of a ketenyl intermediate could be confirmed by irradiation of complex **203** in the presence of triphenylphosphine which affords complex **205** [Eq. ((162)]. In contrast, the thermal reaction of PPh$_3$ with **203** was found to produce **205** only very slowly and in low yield. Similarly, irradiation of **203** in the presence of dppe gives the η^1-ketenyl complex **206** [Eq. (163)]. Thermal reaction of **203** with dppe is slow. Chromatography on silica or Florisil leads to cleavage of the chelate ring in complex **206** and formation of a dinuclear dppe-bridged η^2-ketenyl complex. On readdition of dppe complex **206** may be regenerated.

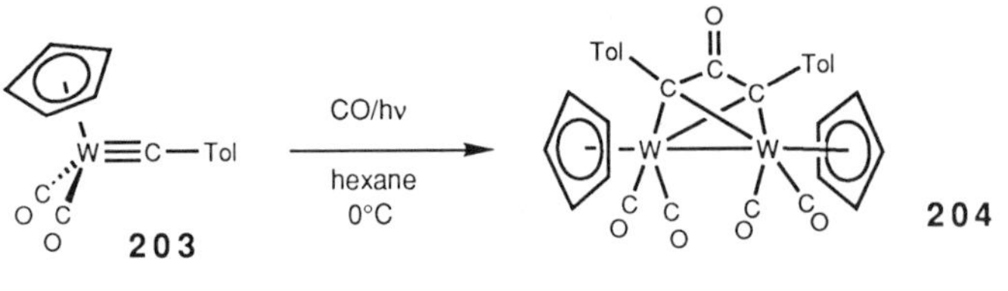

(161)

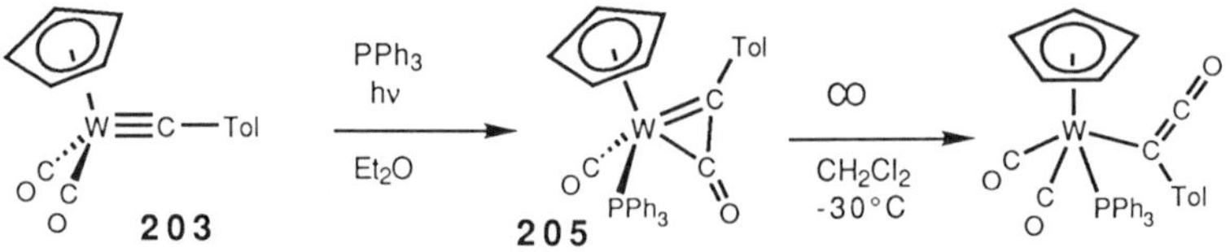

(162)

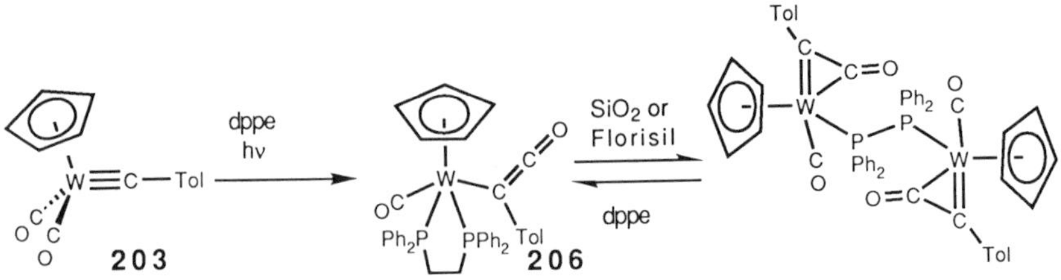

(163)

The photochemical cis–trans isomerization of bisphosphine-substituted carbyne complexes [Eq. (57), Section IV,B], which was studied in our group, was postulated to proceed via pentacoordinate metal ketenyl intermediates (*97*). Support for this postulate was found in the low-temperature characterization of ketenyl complex **208** on irradiation of complexes **207** in the presence of PMe$_3$ [Eq. (164)]. In the presence of phenylacetylene the ketenyl alkyne tungsten complex W(η^1-PhCCO)Cl(PhC$_2$H)(CO)(PMe$_3$)$_2$ forms at low temperatures.

$$(164)$$

Recent work by Stone provides additional examples of the photoreactivity of carbyne carbonylmetal complexes. Irradiation of the 2,6-xylylcarbyne molybdenum complex **209** in the presence of triphenylphosphine affords the ketenyl complexes **210,** which slowly transform into the triphenylphosphine-substituted carbyne complexes **211** [Eq. (165)] (*35*). In

$$(165)$$

contrast, the thermal reaction of complexes **209** at elevated temperatures affords directly the substitution products **211.** Templeton and co-workers investigated the reactions shown in Eqs. (166) and (167). These provide an interesting contrast between the photoreactivity of alkyl- and arylcarbyne complexes (*48*). Irradiation of the methylcarbyne complex **212** with ultraviolet light in acetonitrile solution affords the carbonyl substitution prod-

$$(166)$$

$$(167)$$

uct. Under the same conditions a ketenyl complex is obtained from the phenylcarbyne complex **213**.

Filippou demonstrated the acid-induced coupling of carbyne ligands with isocyanide ligands. Reaction of complexes **214**, which contain three or four isocyanide ligands, with triflic acid and halide ions affords the aminoalkyne complexes **215** [Eq. (168)] (*106*). The reverse of this ligand coupling reaction was utilized in what represents a novel and independent synthesis of metal–carbon triple bonds (*175,176*). The reaction scheme involves two key steps as shown in Eq. (169). First the bromo substituent on the alkyne ligand in complex **216** is replaced by a *tert*-butylamino group. In the final step the amino alkyne ligand is deprotonated with PhLi, leading to the elimination of HBr from **217** and cleavage of the aminoalkyne to generate coordinated carbyne and isocyanide. The aminocarbyne complexes **218** exhibit analogous reactivity toward acid (*109,177*). Treatment of the aminocarbyne bisisocyanide complex **218** with hydrogen iodide results in coupling of the aminocarbyne ligand with isocyanide [Eq. (170)]. In addition to the coupling product **219** the bisisocyanide complex **220** was formed as a by-product. Complex **220** was obtained in good yield from the reaction of the trisisocyanide complex **221** with HI. The coupling reaction proceeds more slowly with the dicarbonyl complex **218** than with

$$(168)$$

L = CO; X = Br
L = CN-t-Bu, X = Br,I

$$(169)$$

R = Et, t-Bu

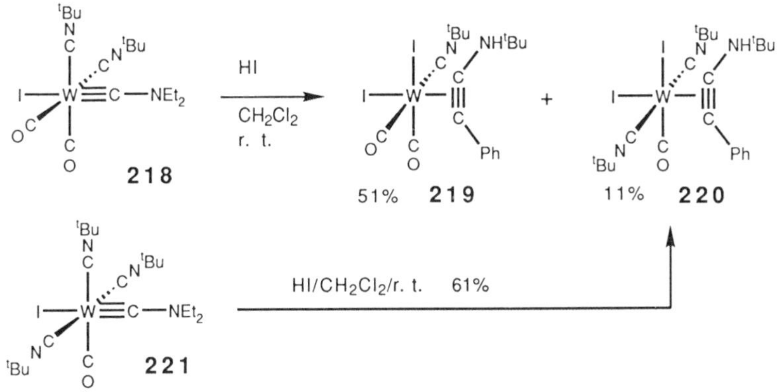

(170)

the monocarbonyl complex **221,** indicating that the coupling reaction is facilitated by increasing electron density on the metal center (*177*).

Warner and Lippard achieved the reductive coupling of two isocyanide ligands in complexes **222** by reduction with zinc in wet tetrahydrofuran [Eq. (171)] (*178*). The reaction most likely involves the coupling of an intermediate aminocarbyne ligand with isocyanide as the key carbon–carbon bond-forming step. The analogous coupling of two carbonyl ligands was achieved in the tantalum complex **223** by reduction with magnesium followed by treatment with trimethylsilyl chloride [Eq. (172)] (*179,180*). The intermediacy of a siloxycarbyne complex in the reductive coupling of two carbonyl ligands was demonstrated by independent synthesis of the siloxycarbyne complex **224** [Eq. (33), Section III,B,3] and the induction of siloxycarbyne–carbonyl coupling on addition of silylating reagents [Eq. (173)] (*73*). The coupling of methylisocyanide and carbon monoxide ligands was achieved as shown in Eq. (174) (*181*).

The formal coupling of two alkylidyne ligands was accomplished in our laboratory as shown in Eq. (175) (*182*). The reaction sequence used in

$$\underset{\mathbf{222}}{[Mo(CNR)_4(bpy)Cl]^+} \xrightarrow[\text{thf}]{Zn/H_2O} \begin{matrix} [Mo(CN^tBu)_3(^tBuNHCCNH^tBu)](pby))]^{2+} \\ + \\ [MoCl(CN^tBu)_4(^tBuNHCCNH^tBu)]^+ \end{matrix} \qquad (171)$$

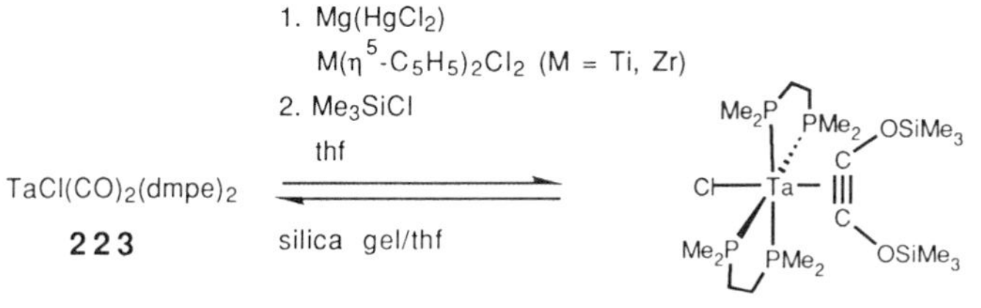

(172)

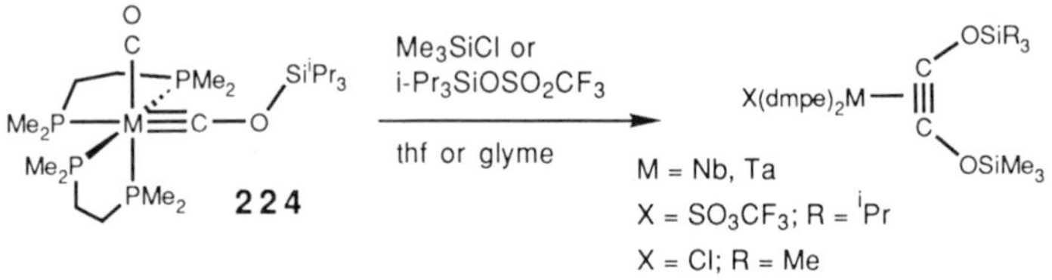

(173)

1. NaHg$_x$
2. Me$_3$SiCl

NbCl(CNMe)(CO)(dmpe)$_2$ $\xrightarrow{\hspace{3cm}}$ NbCl[(Me$_3$Si)MeNCCOSiMe$_3$](dmpe)$_2$ (174)

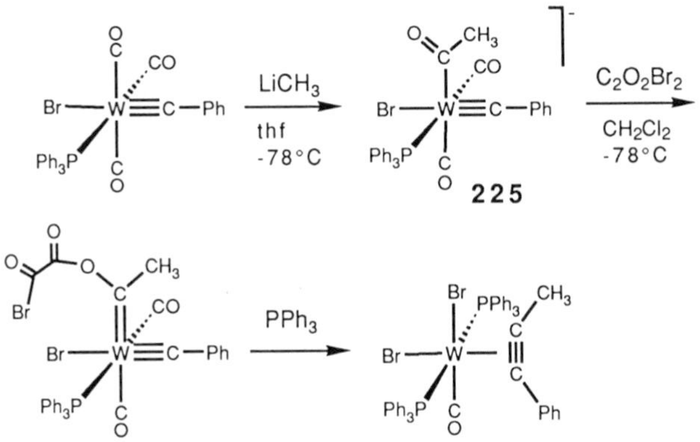

(175)

"oxide abstraction" from acyl ligands [Eq. (5), Section III,4,2] could be performed on an alkylidyne acyl tungsten complex. In analogy to the previous study (44) the reaction of the acyl complex **225** with oxalyl bromide was suggested to give a labile carbene complex intermediate. Dissociation of bromooxalate from this intermediate would generate a second alkylidyne ligand. Whether an actual bisalkylidyne intermediate is formed or whether carbon–carbon bond formation occurs simultaneously with the fragmentation of the carbene ligand is not known.

The formal coupling of two alkylidyne units had previously been achieved by Murahashi *et al.* in the reaction shown in Eq. (176) (*183*). Pyrolysis or photolysis of the bis(α-diazobenzyl) palladium complex **226** was found to result in formation of diphenylacetylene and a minor amount

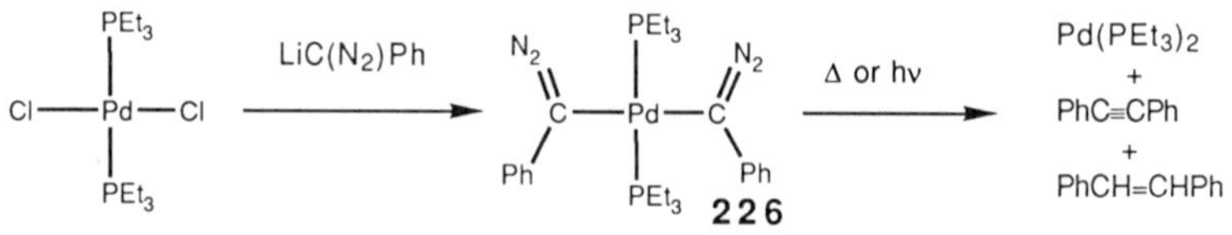

(176)

of stilbenes. A mixture of complexes **226** with R being Ph and R being Tol
did not result in the formation of unsymmetrically substituted alkyne. The
intermediacy of a bisalkylidyne palladium complex was postulated.

Chisholm and co-workers studied reactions of high-valent tungsten alkylidyne complexes with carbon monoxide (*79,184,185*). The reaction of
the ethylidyne complex **227** with CO leads to the formation of the dinuclear dimethylacetylene-bridged complex **228** [Eq. (177)], but the analogous reaction of the aminocarbyne complexes **229** affords the dinuclear
ketenyl complexes **230** and **231**. In complexes **230** and **231** two
$W(OR)_3(R'_2NCCO)(py)_n$ units ($n = 0$ and 1, respectively) are held together
via $\mu-\eta^3$-ketenyl units. There is no direct interaction between the metal
centers. The aminocarbyne–carbonyl coupling in Eq. (178) was demon-

strated to be a reversible process by incorporation of ^{13}C-labeled carbon
monoxide into the aminoketenyl ligand. Complex **231** was found to rearrange slowly at room temperature into the diaminoacetylene complex
232. The outcome of the reactions in Eqs. (177) and (178) may be compared to related coupling reactions of low-valent alkylidyne complexes. In

low-valent systems the coupling of alkylidyne (alkyl and arylcarbyne) ligands and carbonyl ligands was observed in many cases, whereas there is no single example of aminocarbyne–carbonyl coupling.

G. Reactions with Unsaturated Organic Substrates

Several alkylidyne alkene tungsten complexes were synthesized in our laboratory. The neutral complex **233** and the anionic complexes **235** were found to react with maleic anhydride and fumaronitrile to afford the products **234** and **236** [Eqs. (179) and (180)] (*103*). The tmeda-substituted

$$\text{(179)}$$

alkene = maleic anhydride
fumaronitrile

$$\text{(180)}$$

R = Me; Kat = NEt$_4$
R = Ph; Kat = NEt$_4$, PPN

derivatives W(CPh)Cl(alkene)(CO)(tmeda) could be prepared either by reaction of the tmeda complex W(CPh)Cl(CO)$_2$(tmeda) with the alkenes at elevated temperatures or by substitution of the pyridine ligands in **234**. The formation of the alkene complexes in Eqs. (179) and (180) was postulated to occur in a two-step process: initial substitution of pyridine or chloride and subsequent trans-labilization of carbon monoxide by the coordinated olefin, followed by substitution of carbon monoxide to give the final products. The molybdenum alkylidyne complex **237** was reported by Green *et al.* to react with tetracyanoethylene to give the alkylidyne alkene complex **238** [Eq. (181)] (*90*). This reaction is probably initiated by electron transfer from **237** to tetracyanoethylene. The molecular structures of **234** (alkene = maleic anhydride) and **238** reveal a perpendicular orientation of the olefins relative to the metal–carbon triple bonds. This orientation allows optimum π bonding of both alkene and carbyne ligands to the metal center.

$$L = P(OMe)_3 \tag{181}$$

Reaction of the picoline-substituted alkylidyne complex **239** with allyl bromides yields the vinylcarbene complexes **240** [Eq. (182)] (*186,187*). In

$$ \tag{182}$$

the solid-state structures of both complexes **240** (R = H, Me), the vinyl groups of the vinylcarbene ligands were found to interact weakly with the metal center. The bonding of the vinylcarbene ligand was discussed in terms of the absence of π backbonding from the metal to the vinyl group (*187*). Vinylcarbene–carbonyl ligand coupling in **240** (R = H) was found to be induced by the addition of diethyldithiocarbamate [Eq. (182)].

Several reactions of low-valent alkylidyne complexes with alkynes were reported. Treatment of the benzylidene complex **241** with base in the presence of alkynes gives the alkylidene alkyne complexes **242** [Eq. (183)]

$$ \tag{183}$$

(*188*). ^{1}H NMR shows that the alkyne ligand in **242** (R = H) rotates about the metal–alkyne bond axis with an activation barrier of only 50 kJ/mol. The structure of the diphenylacetylene complex **242** was determined by X-ray crystallography. The reaction shown in Eq. (183) was expected to generate alkylidyne alkyne complexes such as $W(CPh)Cl(PhC_2R)(CO)$-$(PMe_3)_2$. Such species are probably intermediates in the reaction; however, a π conflict between the adjacent $C\equiv C$ and $M\equiv C$ triple bond systems

(*36*) may increase the basicity of the alkylidyne carbon, thereby favoring protonation to form the alkylidene ligand.

When the tristrimethylphosphine complex **243** was allowed to react with phenylacetylene under aprotic conditions two new products could be isolated (*189*): the metallacyclopentadienone **244** and the oxocyclobutenyl tungsten complex **245** [Eq. (184)]. The formation of the two products may

$$(184)$$

be explained by the initial formation of an alkylidyne alkyne complex and rearrangement to metallacyclobutadiene and metal cyclopropenyl species. Insertion of carbon monoxide into these entities could give the observed products. Green and co-workers also postulated the formation of low-valent alkylidyne alkyne and metallacyclobutadiene intermediates to explain the formation of a cyclopropenyl molybdenum complex in the reaction of $Mo(\eta^5\text{-}C_5H_5)(PhC_2Ph)\{P(OMe)_3\}_2$ with vinylmagnesium bromide (*190*).

An actual metathesis reaction of a low-valent carbyne complex was observed by Stone and co-workers (*124*). The molybdenum neopentylidyne complex **247** could be isolated from the reaction of complex **246** with *tert*-butylphosphanitrile [Eq. (185)].

$$(185)$$

The coupling of carbyne ligands with alkynes was observed in the dinuclear terminal carbyne complexes **248** [Eq. (186)] (*191,192*). Two types of products were observed. Complexes containing a bridging C_3R_3 ligand (**I**) and the CO insertion products thereof (**II**). For example, reaction of the complex with ditolylacetylene gives only **I**, whereas reaction of the ethylidyne complex with dimethylacetylene gives only **II**. A mixture of **I** and **II** was obtained in the reactions of the alkylidyne complexes (R = Me, Tol)

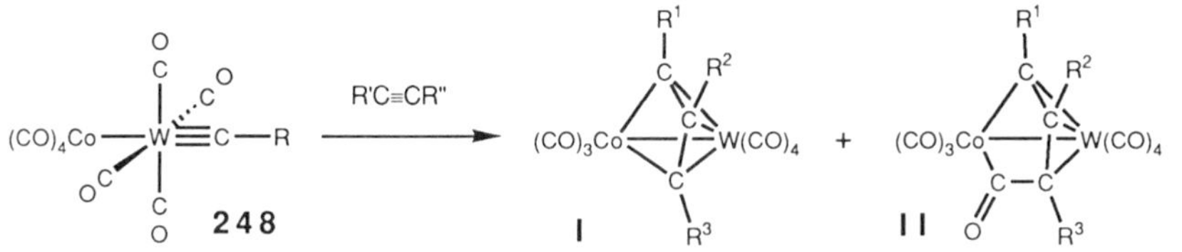

(186)

R = Tol; R' = R" = Tol: I, R^1 = R^2 = R^3 = Tol
R = Tol; R' = R" = Ph: I, R^1 = Tol, R^2 = R^3 = Ph, R^1 = R^3 = Ph, R^2 = Tol
R = Me or Tol; R' = Me, R" = SiMe$_3$:   I and II, R^1 = Me or Tol, R^2 = Me, R^3 = SiMe$_3$
R = Tol; R' = R" = Me or Et: II, R^1 = Tol, R^2 = R^3 = Me or Et
R = Tol; R' = Me, R" = Et: II, R^1 = Tol, R^2 or R^3 = Me or Et
R = Me; R' = R" = Me: II, R^1 = R^2 = R^3 = Me
R = Me; R' = Me, R" = Ph: II, R^1 = Ph, R^2 = R^3 = Me

with methyl(trimethylsilyl)acetylene. Products **I** and **II** obtained from the latter reaction were shown to be interconvertible by the application of vacuum or carbon monoxide pressure.

Katz and co-workers reported the reactions of the chromium and tungsten ethylidyne complexes **249** and **250** with 1-alkynyl-1'-vinylbiphenyl (*193,194*). For both metals the reactions lead to the incorporation of the carbyne ligand into the organic substrate, but in slightly different ways as shown by formation of the products **251–253** [Eq. (187)]. The product of

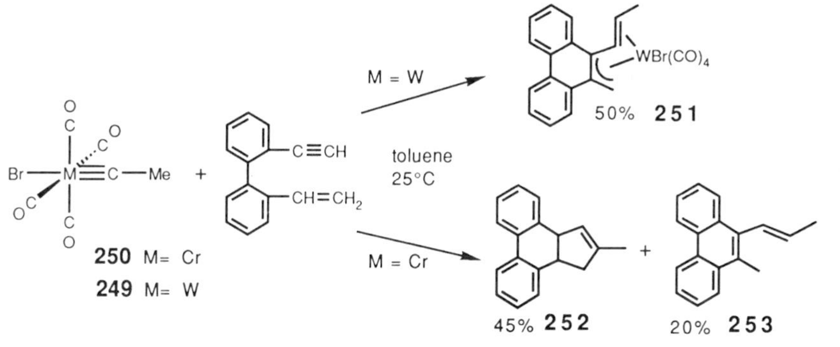

(187)

the reaction with the tungsten carbyne was characterized by X-ray crystallography.

Geoffroy and co-workers investigated the protonation of cyclopentadienyl dicarbonyltungsten alkylidyne complexes with HBF$_4$ in the presence of alkynes (*195,196*). The outcome of the reactions depends on the nature of both the alkylidyne ligand and the alkyne. Protonation of complex **254** in the presence of diphenylacetylene affords the vinylcarbene complex **255**

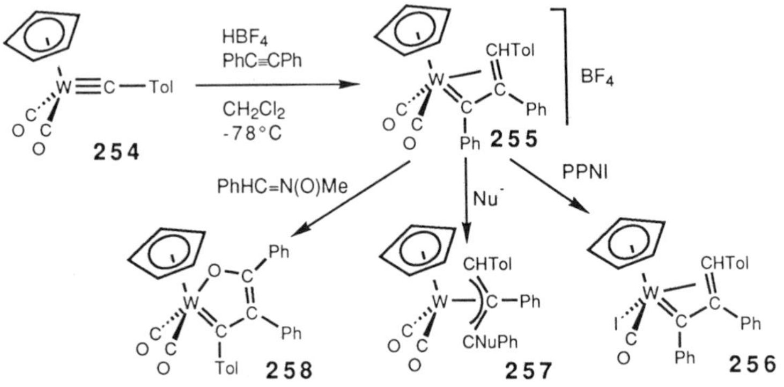

(188)

[Eq. (188)]. The formation of **255** was postulated to involve protonation of the alkylidyne carbon followed by insertion of alkyne into the generated metal–alkylidene bond. Substitution of one carbon monoxide ligand by iodide gives the neutral vinylcarbene complex **256**, whose structure was determinded by X-ray crystallography. The vinylcarbene ligand in **255** is susceptible to nucleophilic attack. With phosphines and hydride transfer reagents allyl complexes **257** were obtained. Transfer of oxygen from an imine oxide produced the metallafurane ring system **258**. With 2-butyne the naphthol complex **259** is obtained [Eq. (189)]. Together with the

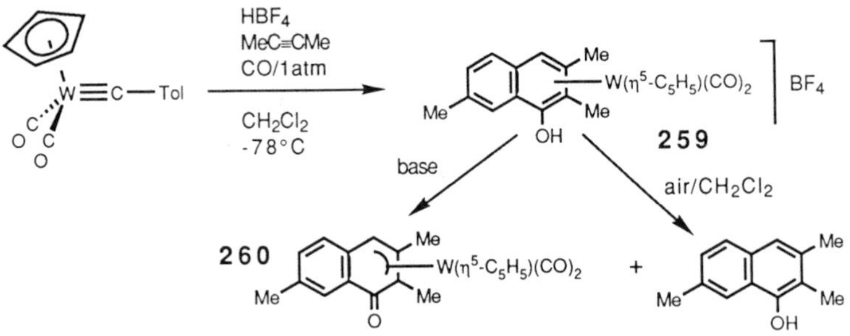

(189)

results shown in Eq. (188) formation of the naphthol ring in Eq. (189) provides evidence for the intermediacy of metal vinylcarbenes in the synthesis of naphthols from chromium carbene complexes and alkynes (*197*). Deprotonation of the naphthol ligand gives the allyl complex **260**, and air oxidation liberates the naphthol from the metal center. Protonation of the 2,6-xylylcarbyne complex **261** in the presence of 2-butyne gives complex **262** [Eq. (190)].

The cycloaddition of imines to the metal–carbon triple bond in the cationic rhenium carbyne complex **263** was reported by Geoffroy and

$$\text{(190)}$$

co-workers [Eq. (191)] (40,198). The products are the four-membered rhenacycles **264**. Irradiation of the cationic complex **264** (R = CH$_3$) results in substitution of one CO ligand by chloride, whereby the counterion BCl$_4^-$ serves as the chloride source. Hydrolysis of the cycloaddition product affords the aminocarbene complex **265** and benzaldehyde. The reaction of benzaldehyde methylimine with the manganese carbyne complexes **266** affords the products **267** [Eq. (192)]. The reactions of the manganese

$$\text{(191)}$$

$$\text{(192)}$$

and rhenium carbyne complexes with imines are probably initiated by nucleophilic attack of the imine nitrogen at the carbyne carbon atom. The electrophilic benzylidene group then adds to the metal center, in the case of rhenium, to form the 2 + 2 cycloaddition product. In the case of manganese, electrophilic attack occurs at the cyclopentadienyl ring with substitution of the substituent R, surprisingly both when R is H and when R is Me. The smaller size of manganese compared to rhenium may be one of the reasons for the difference in reactivity.

Somewhat related are the reactions of the rhenium carbyne complex **263** with nitroso compounds (40,198). Reaction of the tetrachloroborate salt

263 with monomeric *t*-BuNO affords the metallacycle **268** [Eq. (193)]. Complex **268** may be the result of a $2 + 2$ cycloaddition of the nitroso group to the metal–carbon triple bond followed by insertion of carbon monoxide into the ring. The electron-deficient complex **268** abstracts chloride from the counterion to give **269**. When the dimer (*t*-BuNO)$_2$ is allowed to react with **263** the same ring system is formed, but the metal center also scavenges a carbon monoxide ligand, which presumably stems from simultaneous decomposition of the starting material. On standing **270** transforms into **269**. When the tetraphenylborate salt **271** is allowed to react with *t*-BuNO the cycloadduct **272** can be isolated. Addition of chloride ion to **272** affords the CO insertion product **269**. The insertion step may be reversed by the addition of AgBF$_4$. Addition of trimethylphosphine to complex **272** results in replacement of the cyclopentadienyl ring by only two phosphine ligands. The ability of the five-membered as well as the four-membered rings to exist in formal 16- and 18-electron complexes (e.g., **272** and **273,** respectively) is most likely due to the ability of the ring oxygen atom to donate electrons to the metal center.

The metathesis of alkynes by metal alkylidyne complexes is now a well-established process. The reaction is useful for the catalytic metathesis of alkynes as well as for the synthesis of new metal alkylidyne complexes. The general electronic and steric factors favoring the metathesis reaction were discussed in recent reviews by Schrock (*6, 7*).

Efforts by Krouse and Schrock to prepare polyenynes by a metathetical process led to the preparation of several dinuclear complexes bridged by unsaturated bisalkylidyne ligands [Eq. (194)] (*199*). Two possible reasons for the failure to obtain the desired polymers were mentioned. The dinuclear "metal-capped" species were found to be thermodynamically more stable than mononuclear complexes with "terminal" alkylidyne ligands, and the solubility of the oligomers, if formed, appears to decrease rapidly with increasing molecular weight.

The metathesis of alkylidyne complex **274** with the ruthenium acetylide

$$(194)$$

complex **275** to generate a μ-carbido complex was accomplished by Latesky and Selegue [Eq. (195)] (*200*). The X-ray structure of complex **276**

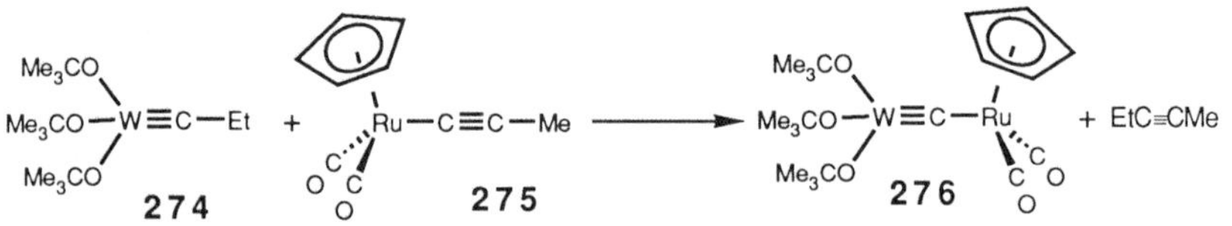

(195)

shows that the μ-carbido ligand is bonded like an alkylidyne ligand to
tungsten and like an acetylide ligand to ruthenium.

The reaction of the rhenium alkylidyne complex **277** with diisopropyl-
acetylene and with diethylacetylene [Eq. (196)] demonstrates the sensitiv-
ity of metathesis reactions toward steric factors (*51*). With diisopropylace-
tylene an alkylidyne complex is obtained whereas the reaction with
diethylacetylene gives a metallacyclobutadiene. In the metathesis reactions
the alkyne with the bulkiest groups cleaves most easily from intermediate
metallacyclobutadiene complexes. The rhenacyclobutadienes with the
smallest substituents thus become sinks and slow down the effective rate of
metathesis. The alkylidyne alkylidene rhenium complex **278** is an active
olefin metathesis catalyst (*52*). Reaction with hexene transforms the neo-
pentylidene group into a propylidene group as shown in Eq. (197).

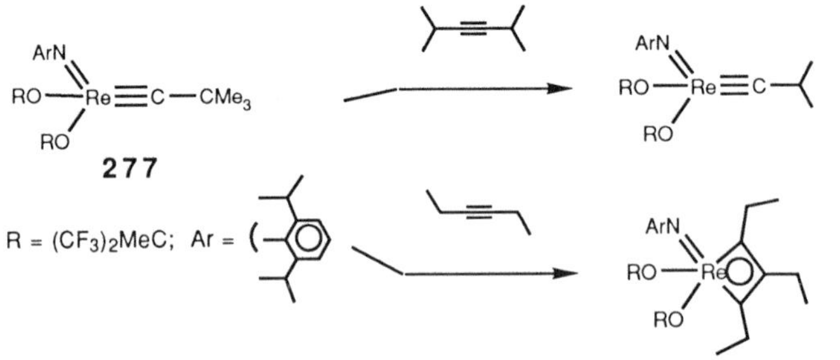

(196)

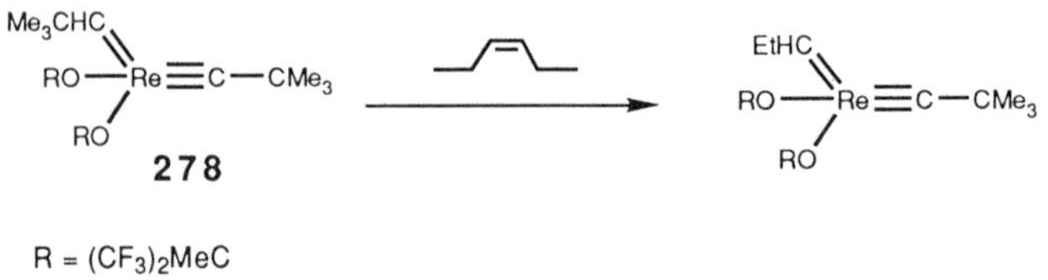

$R = (CF_3)_2MeC$; $Ar = $

(197)

$R = (CF_3)_2MeC$

Cyclooctyne reacts with hexakis(*tert*-butoxy) ditungsten (as the added
form of metathesis catalyst) to give primarily a mixture of cyclic polycy-

clooctynes [Eq. (198)] (75). On the other hand, tris(2,6-diisopropylphen-oxide)neopentylidyne tungsten adds cyclooctyne to form the metallacyclo-butadiene **279** [Eq. (199)], and tris(*tert*-butoxy)butylidyne molybdenum forms a linear ring-opening polymerization product which is capped by the molybdenum alkylidyne group [Eq. (200)] (*201*).

$$ \xrightarrow[\text{toluene}]{\substack{(Me_3CO)_3W\equiv W(OCMe_3)_3 \\ \text{(catalyst)}}} \quad CH_2\text{-}[C\equiv C\text{-}(CH_2)_6\text{-}]_nC\equiv C\text{-}(CH_2)_4\text{-}CH_2 \qquad (198) $$

$$ \text{(RO)}_3W\equiv CCMe_3 \qquad R = \left(\text{—} \bigcirc \right) \qquad \longrightarrow \qquad \text{(RO)}_3W \qquad \mathbf{279} \qquad (199) $$

$$ \text{(t-BuO)}_3Mo\equiv CPr \ + \ n \quad \bigcirc \quad \longrightarrow \quad \text{(t-BuO)}_3Mo[\equiv C\text{-}(CH_2)_6\text{-}C]_n\equiv CPr \qquad (200) $$

The triscarboxylate alkylidyne complexes **280** form cyclopropenyl metal complexes with alkynes [Eq. (201)] (*152*). The complexes **281** do not revert

$$ \substack{\text{(RCO}_2)_3M\equiv C\text{-}CMe_3 \\ \mathbf{280}} \quad \underset{X}{\overset{R'C\equiv CR'}{\rightleftharpoons}} \quad \substack{\text{(RCO}_2)_3M \\ \mathbf{281}} \text{—} CMe_3 \qquad (201) $$

M = Mo; R = CHMe$_2$, CMe$_3$; R' = Me, Et, Ph
M = W; R = CMe$_3$; R' = Me, Et

to alkylidynes and alkynes. It is believed that in these systems a planar metallacyclobutadiene is not accessible from which to lose alkyne. How-ever, when the carboxylate ligands are replaced by sterically more de-manding *tert*-butoxide ligands, alkynes are cleaved off to regenerate a metal alkylidyne [Eq. (202)]. In the presence of weak and sterically little

$$ \text{(RCO}_2)_3M\text{—}CMe_3 \quad \xrightarrow[\text{pentane}]{\text{LiOCMe}_3} \quad \substack{\text{(Me}_3CO)_3M\equiv CCMe_3 \ + \\ \text{(Me}_3CO)_3M\equiv CR'} \qquad (202) $$

M= Mo; R = CHMe$_2$; R' = Ph: Mo$\equiv$CCMe$_3$ 100%
M = W; R = CMe$_3$; R' = Me: W$\equiv$CCMe$_3$ 50%
                                      W$\equiv$CMe 50%

demanding donor ligands, such as in **282** and **283**, the alkylidyne complexes take up two alkynes, forming cyclopentadienyl ligands [Eqs. (203) and (204)] (*76,152*).

$$(CF_3CO_2)_3(dme)M\equiv C\text{-}CMe_3 \quad \xrightarrow{\;RC\equiv CR\;} \quad M(\eta^5\text{-}C_5(CMe_3)Me_4)(O_2CCF_3)_n \qquad (203)$$

282

M = Mo; R = Me; n = 3
M = W; R = Et; n = 4

$$Cl_3W \underset{Me}{\overset{Me}{\boxed{}}}\!\!-Me \quad \xrightarrow{\;MeC\equiv CMe\;} \quad [W(\eta^5\text{-}C_5Me_5)Cl_4]_2\ + \\ W(\eta^5\text{-}C_5Me_5)Cl_2(MeCCMe) \qquad (204)$$

283

Terminal acetylenes are not metathesized well. The main reason is that metallacyclobutadienes carrying a hydrogen atom at the central ring carbon atom are easily deprotonated. A few well-defined examples formed according to Eq. (205) from alkylidyne complexes and terminal acetylenes

$$(RO)_3(L)W\equiv C\text{-}CMe_3 \quad \xrightarrow{\;R'C\equiv CH\;} \quad (RO)_3W\,\overset{CMe_3}{\underset{R'}{\boxed{}}}\!\!-H$$

R = CH(CF_3)_2; L = dme; R' = CMe_3, Ph
R = 2,6-i-Pr_2C_6H_3; no L; R' = CMe_3
R = 2,6-Me_2C_6H_3; L = thf; R' CMe_3

$$\Big\downarrow NEt_3 \qquad (205)$$

$$(RO)_2(py)_nW\,\overset{CMe_3}{\underset{R'}{\boxed{}}} \quad \xleftarrow{\;py\;} \quad (RO)_2W\,\overset{CMe_3}{\underset{R'}{\boxed{}}}$$

R = CH(CF_3)_2; n = 2
R = 2,6-i-Pr_2C_6H_3, 2,6-Me_2C_6H_3; n = 1

(*202*). Triethylamine easily deprotonates the ring. The "deprotio" metallacyclobutadienes are more easily isolated after addition of pyridine ligands. Deuterium exchange on the β carbon of $W(C_3\text{-}t\text{-}Bu_2D)[OCH(CF_3)_2]_3$ with $HOCH(CF_3)_2$ was demonstrated. Proton exchange probably occurs via elimination of $DOCH(CF_3)_2$ and readdition of $HOCH(CF_3)_2$. Even the presence of dimethoxymethane causes the elimination of $HOCH(CF_3)_2$ [Eq. (206)]. In this case an equilibrium between the metallacyclobutadiene and its deprotonated counterpart could be observed. For the phenyl-sub-

$$(RO)_3W \langle \overset{CMe_3}{\underset{R'}{\bigcirc}} \rangle -H + dme \rightleftharpoons (RO)_2(dme)W \langle \overset{CMe_3}{\underset{R'}{\bigcirc}} \rangle + HOR \qquad (206)$$

R = CH(CF$_3$)$_2$ R' = CCMe$_3$: 7 : 1
                      R' = Ph: 1 : 1

stituted ring the equilibrium lies slightly more toward the side of the deprotonated product. The "deprotio" metallacyclobutadienes W(C$_3$-t-Bu$_2$)(OR$_2$) react further with $tert$-butylacetylene to give compounds containing a planar bicyclic WC$_5$ ring system [Eq. (207)]. The alkylidene-like carbon of the ring system is protonated by CF$_3$CO$_2$H and PhCO$_2$H [Eq. (207)]. The reaction of the molybdenum alkylidyne complex **284** with

$$(RO)_2W \langle \overset{CMe_3}{\underset{CMe_3}{\bigcirc}} \rangle \xrightarrow{Me_3CC\equiv CH} (RO)_2W \xrightarrow{R''CO_2H} (RO)_2(R''CO_2)W \qquad (207)$$

R = 2,6-iPr$_2$C$_6$H$_3$
    2,6-Me$_2$C$_6$H$_3$

R" = CF$_3$, Ph

terminal acetylenes directly gives the deprotonated metallacyclobutadiene [Eq. (208)] (*152*).

$$(CF_3CO_2)_3(dme)Mo\equiv CCMe_3 \xrightarrow[-CF_3CO_2H]{R'C\equiv CH} (CF_3CO_2)_2(dme)Mo \langle \overset{CMe_3}{\underset{R'}{\bigcirc}} \rangle$$

284

$$\downarrow py$$

$$(CF_3CO_2)_2(py)_2Mo \langle \overset{CMe_3}{\underset{R'}{\bigcirc}} \rangle \qquad (208)$$

Weiss and co-workers reported a series of papers on the use of alkylidyne complexes as precursors for olefin metathesis catalysts. W(CCMe$_3$)-Cl$_3$(dme) was found to initiate ring-opening metathesis polymerization of cyclopentene [Eq. (209)] and metathesis of n-hexene as well as other 1-alkenes (*203,204*). The rate of metathesis is slowed down by branching

$$\bigcirc \xrightarrow[CH_2Cl_2]{W(CCMe_3)Cl_3(dme)} \quad \text{75% trans} \qquad (209)$$

near the olefin group. The presence of small concentrations of 1-alkynes does not affect olefin metathesis, but large amounts of 1-alkyne lead to quenching of metathesis activity and to alkyne polymerization. The activity of $W(CCMe_3)Cl_3(dme)$ toward 1-alkyne polymerization is similar to that of the catalyst system $WCl_6/1$-alkyne. Deposition of the carbyne complexes $W(CPh)X(CO)_4$ (X = Cl, Br, I) on reduced Phillips catalyst (SiO_2/Cr^{II}) affords an active alkene metathesis catalyst (205). In the absence of Cr^{II} little activity is observed. On the other hand, if $W(CCMe_3)(OCMe_3)_3$ is deposited on SiO_2/Cr^{II} it is less active as an olefin metathesis catalyst than if deposited on silica alone (206). Deposition of $W(CCMe_3)(CH_2CMe_3)_3$ and $W(CCMe_3)Cl_3(dme)$ on silica also gives active catalysts.

Freudenberger and Schrock reported the metathesis of the metal–carbon triple bond in **285** with the carbon–nitrogen triple bond in acetonitrile [Eq. (210)] (207). Metathesis-like reactions of **285** also occur with organic carbonyl groups to generate oxo vinyl metal complexes [Eq. (211)]

(207). Hydrolysis of the vinyl groups from the metal center affords olefins. Weiss, Schubert, and Schrock investigated the reaction of the alkylidyne complex $W(CCMe_3)Cl_3(dme)$ with cyclohexyl isocyanate (208). Two isocyanate molecules are incorporated into the complex as shown in Eq. (212). The reaction was postulated to proceed via a cycloaddition adduct of isocyanate to the metal–carbon triple bond and cleavage of the four-membered ring into metal imido and ketenyl species. Subsequent insertion of a second isocyanate into the metal–ketenyl bond would then give the observed product.

In several reactions reported in the literature metal alkylidynes may have been involved, even though there is currently no direct evidence available.

$$(212)$$

Irradiation of the bisalkyne complex $W(S_2CNEt_2)_2(PhC_2H)$ in the presence of phenylacetylene affords the cyclopentadienyl complex $W(SCPhCHS)(S_2CNEt_2)\{\eta^5\text{-}C_5H_2Ph_2(NEt_2)\}$ (209). The cyclopentadienyl ligand is the formal coupling product of two phenylacetylene ligands and one diethylaminocarbyne entity, the latter originating from diethyldithiocarbamate. The cocondensation of iron atoms with cyclopentadiene and alkynes affords ferrocene and substituted ferrocenes (210). Formation of the substituted ferrocene ligands was postulated to involve iron alkylidyne species as reactive intermediates. Reduction of the acyl complex $Ta(\eta^2\text{-}OCCH_2CMe_3)(\eta^5\text{-}C_5Me_5)Cl_3$ with magnesium affords free pentamethyl-neopentylbenzene (211). This substituted benzene is the formal product of alkylidyne insertion into the cyclopentadienyl ring. The reaction was found to be predominantly an intramolecular process, but no evidence for a potential alkylidyne complex intermediate was found. The treatment of some ketones with metal species derived from metal hexacarbonyls or WCl_6/sec-BuLi afforded olefins (212). With benzaldehyde as the substrate some diphenylacetylene was obtained in addition to stilbene. A tungsten carbyne was postulated as an intermediate.

V

CONCLUDING REMARKS

The chemistry of metal–carbon triple bonds has developed considerably during the late 1980s. The synthetic basis was broadened, the utility of high-valent metal alkylidynes in metathesis reactions was further developed and refined, and the potential of low-valent carbyne complexes for applications in organic synthesis has become more apparent. The discovery of novel iridium alkylidyne complexes indicates that the full range of metal–carbon triple bonds is not yet known. We can therefore expect that future work in this area of organometallic chemistry will lead to new discoveries with fundamental implications and practical applications.

ACKNOWLEDGMENTS

We would like to thank Cecilia M. Bastos and Kimberly A. Belsky for helpful discussions during the preparation of the manuscript. Support for the work in our laboratory by the National Science Foundation, the Petroleum Research Fund, administered by the American Chemical Society, and by the Research Corporation is gratefully acknowledged. H.H. is grateful for support from the Degussa AG, Germany.

REFERENCES

1. E. O. Fischer, G. Kreis, C. G. Kreiter, J. Müller, G. Huttner, and H. Lorenz, *Angew. Chem.* **85**, 618 (1973); *Angew. Chem., Int. Ed. Engl.* **12**, 564 (1973).
2. E. O. Fischer, *Angew. Chem.* **86**, 651 (1974); E. O. Fischer, *Adv. Organomet. Chem.* **14**, 1 (1976).
3. E. O. Fischer and U. Schubert, *J. Organomet. Chem.* **100**, 59 (1975).
4. E. O. Fischer, U. Schubert, and H. Fischer, *Pure Appl. Chem.* **50**, 857 (1978).
5. H. P. Kim and R. J. Angelici, *Adv. Organomet. Chem.* **27**, 51 (1987).
6. H. Fischer, P. Hofmann, F. R. Kreissl, R. R. Schrock, U. Schubert, and K. Weiss, "Carbyne Complexes." VCH Publishers, Weinheim, Germany, 1988.
7. R. R. Schrock, *Acc. Chem. Res.* **19**, 342 (1986).
8. R. R. Schrock, *J. Organomet. Chem.* **300**, 249 (1986).
9. M. A. Gallop and W. R. Roper, *Adv. Organomet. Chem.* **25**, 121 (1986).
10. W. R. Roper, *J. Organomet. Chem.* **300**, 167 (1986).
11. M. Green, *J. Organomet. Chem.* **300**, 93 (1986).
12. W. E. Buhro and M. H. Chisholm, *Adv. Organomet. Chem.* **27**, 311 (1987).
13. M. H. Chisholm, B. K. Conroy, B. W. Eichhorn, K. Folting, D. M. Hoffmann, J. C. Huffman, and N. S. Marchant, *Polyhedron* **6**, 783 (1987).
14. F. G. A. Stone, *Angew. Chem.* **96**, 85 (1984); *Angew. Chem., Int. Ed. Engl.* **23**, 89 (1984). F. G. A. Stone, *NATO ASI Ser., Ser. C (Adv. Met. Carbene Chem.)*, **269**, 11 (1989).
15. F. G. A. Stone, *Adv. Organomet. Chem.* **31**, 53 (1990).
16. A. J. L. Pombeiro, *J. Organomet. Chem.* **358**, 273 (1988).
17. A. J. L. Pombeiro, *Polyhedron* **8**, 1595 (1989).
18. K. Isobe, *Kagaku Zokan (Kyoto)* **109**, 135 (1986).
19. A. Mayr, *Comments Inorg. Chem.* **10**, 227 (1990).
20. M. J. Winter, *Organomet. Chem.* **14**, 225 (1986); M. J. Winter, *Organomet. Chem.* **15**, 229 (1987); M. J. Winter, *Organomet. Chem.* **16**, 230 (1987).
21. G. Huttner, A. Frank, and E. O. Fischer, *Isr. J. Chem.* **15**, 133 (1977).
22. N. Q. Dao, A. Spasojević-de Biré, E. O. Fischer, and P. J. Becker, *C. R. Acad. Sci., Ser. 2* **307**, 341 (1988).
23. C. G. Young, E. M. Kober, and J. H. Enemark, *Polyhedron* **6**, 255 (1987).
24. A. B. Bocarsly, R. E. Cameron, A. Mayr, and G. A. McDermott, *Photochem. Photophys. Coord. Compd., Proc. Int. Symp., 7th*, 213 (1987).
25. N. Aristov and P. B. Armentrout, *J. Am. Chem. Soc.* **108**, 1806 (1986).
26. R. L. Hettich and B. S. Freiser, *J. Am. Chem. Soc.* **108**, 2537 (1986).
27. R. L. Hettich and B. S. Freiser, *J. Am. Chem. Soc.* **109**, 3543 (1987).
28. S.-C. Chang, R. H. Hauge, Z. H. Kafafi, J. L. Margrave, and W. E. Billups, *J. Chem. Soc., Chem. Commun.*, 1682 (1987).
29. S.-C. Chang, R. H. Hauge, Z. H. Kafafi, J. L. Margrave, and W. E. Billups, *J. Am. Chem. Soc.* **110**, 7975 (1988).

30. F. Garin and G. Maire, *Acc. Chem. Res.* **22**, 100 (1989).
31. S. Zyade, F. Garin, and G. Maire, *New J. Chem.* **11**, 429 (1987).
32. J. M. Poblet, A. Strich, R. Weist, and M. Bénard, *Chem. Phys. Lett.* **126**, 169 (1986).
33. C. Krüger, R. Goddard, and K. H. Claus, *Z. Naturforsch., B: Anorg. Chem., Org. Chem.* **38**, 1431 (1983).
34. A. A. Low and M. B. Hall, *Organometallics* **9**, 701 (1990).
35. S. J. Dossett, A. F. Hill, J. C. Jeffery, F. Marken, P. Sherwood, and F. G. A. Stone, *J. Chem. Soc., Dalton Trans.*, 2453 (1988).
36. D. C. Brower, J. L. Templeton, and D. M. P. Mingos, *J. Am. Chem. Soc.* **109**, 5203 (1987).
37. M. F. N. N. Carvalho, A. J. L. Pombeiro, E. G. Bakalbassis, and C. A. Tsipis, *J. Organomet. Chem.* **371**, C26 (1989).
38. K. D. Dobbs and W. J. Hehre, *J. Am. Chem. Soc.* **108**, 4663 (1986).
39. E. O. Fischer and G. Kreis, *Chem. Ber.* **109**, 1673 (1976).
40. B. M. Handwerker, K. E. Garrett, K. L. Nagle, G. L. Geoffroy, and A. L. Rheingold, *Organometallics* **9**, 1562 (1990).
41. D. Miguel, U. Steffan, and F. G. A. Stone, *Polyhedron* **7**, 443 (1988).
42. H. Fischer, F. Seitz, and J. Riede, *Chem. Ber.* **119**, 2080 (1986).
43. F. Seitz, H. Fischer, J. Riede, and J. Vogel, *Organometallics* **5**, 2187 (1986).
44. G. A. McDermott, A. M. Dorries, and A. Mayr, *Organometallics* **6**, 925 (1987).
45. I. J. Hart, A. F. Hill, and F. G. A. Stone, *J. Chem. Soc., Dalton Trans.*, 2261 (1989).
46. S. J. Davies, A. F. Hill, M. U. Pilotti, and F. G. A. Stone, *Polyhedron* **8**, 2265 (1989).
47. J. R. Fernández and F. G. A. Stone, *J. Chem. Soc., Dalton Trans.*, 3035 (1988).
48. D. C. Brower, M. Stoll, and J. L. Templeton, *Organometallics* **8**, 2786 (1989).
49. L. G. McCullough, R. R. Schrock, M. R. Churchill, A. L. Rheingold, and J. Ziller, *J. Am. Chem. Soc.* **107**, 5987 (1985).
50. R. R. Schrock, D. N. Clark, J. Sancho, J. H. Wengrovius, S. M. Rocklage, and S. F. Pedersen, *Organometallics* **1**, 1645 (1982).
51. R. R. Schrock, I. A. Weinstock, A. D. Horton, A. H. Liu, and M. H. Schofield, *J. Am. Chem. Soc.* **110**, 2686 (1988).
52. R. Toreki and R. R. Schrock, *J. Am. Chem. Soc.* **112**, 2448 (1990).
53. P. D. Savage, G. Wilkinson, M. Motevalli, and M. B. Hursthouse, *Polyhedron* **6**, 1599 (1987).
54. J. Felixberger, P. Kiprof, E. Herdweck, W. A. Herrmann, R. Jakobi, and P. Gütlich, *Angew. Chem.* **101**, 346 (1989); *Angew. Chem., Int. Ed. Engl.* **28**, 334 (1989).
55. W. A. Herrmann, J. K. Felixberger, R. Anwander, E. Herdtweck, P. Kiprof, and J. Riede, *Organometallics* **9**, 1434 (1990).
56. G. R. Clark, C. M. Cochrane, K. Marsden, W. R. Roper, and L. J. Wright, *J. Organomet. Chem.* **315**, 211 (1986).
57. W. Beck, W. Knauer, and C. Robl, *Angew. Chem.* **102**, 331 (1990); *Angew. Chem., Int. Ed. Engl.* **29**, 318 (1990).
58. A. J. L. Pombeiro, A. Hills, D. L. Hughes, and R. L. Richards, *J. Organomet. Chem.* **352**, C5 (1988).
59. M. F. N. N. Carvalho, R. A. Henderson, A. J. L. Pombeiro, and R. L. Richards, *J. Chem. Soc., Chem. Commun.*, 1796 (1989).
60. A. Hills, D. L. Hughes, N. Kashef, R. L. Richards, M. A. N. D. A. Lemos, and A. J. L. Pombeiro, *J. Organomet. Chem.* **350**, C4 (1988).
61. A. Höhn and H. Werner, *Angew. Chem.* **98**, 745 (1986); *Angew. Chem., Int. Ed. Engl.* **25**, 737 (1986).
62. A. Höhn and H. Werner, *J. Organomet. Chem.* **382**, 255 (1990).

63. E. O. Fischer and J. H. Schneider, *J. Organomet. Chem.* **295**, C29 (1985).
64. R. G. Beevor, M. Green, A. G. Orpen, and I. D. Williams, *J. Chem. Soc., Dalton Trans.*, 1319 (1987).
65. P. N. Nicklas, J. P. Selegue, and B. A. Young, *Organometallics* **7**, 2248 (1988).
66. S. R. Allen, R. G. Beevor, M. Green, A. G. Orpen, K. E. Paddick, and I. D. Williams, *J. Chem. Soc., Dalton Trans.*, 591 (1987).
67. N. M. Kostić and R. F. Fenske, *Organometallics* **1**, 974 (1982).
68. A. J. L. Pombeiro, D. L. Hughes, C. Pickett, and R. L. Richards, *J. Chem. Soc., Chem. Commun.*, 246 (1986).
69. S. Warner and S. Lippard, *Organometallics* **8**, 228 (1989).
70. D. L. Hughes, M. Y. Mohammed, and C. J. Pickett, *J. Chem. Soc., Chem. Commun.*, 1399 (1989).
71. A. C. Filippou and W. Grünleitner, *Z. Naturforsch., B: Anorg. Chem., Org. Chem.* **44**, 1572 (1989).
72. A. C. Filippou, E. O. Fischer, and W. Grünleitner, *J. Organomet. Chem.* **386**, 333 (1990).
73. R. N. Vrtis, C. P. Rao, S. Warner, and S. J. Lippard, *J. Am. Chem. Soc.* **110**, 2669 (1988).
74. R. A. Doyle and R. J. Angelici, *Organometallics* **8**, 2207 (1989).
75. S. A. Krouse and R. R. Schrock, *Macromolecules* **22**, 2569 (1989).
76. I. A. Latham, L. R. Sita, and R. R. Schrock, *Organometallics* **5**, 1508 (1986).
77. M. H. Chisholm, B. K. Conroy, K. Folting, D. M. Hoffmann, and J. C. Huffman, *Organometallics* **5**, 2457 (1986).
78. M. H. Chisholm, D. Ho, J. C. Huffman, and N. S. Marchant, *Organometallics* **8**, 1626 (1989).
79. M. H. Chisholm, B. K. Conroy, J. C. Huffman, and N. S. Marchant, *Angew. Chem.* **98**, 448 (1986); *Angew. Chem., Int. Ed. Engl.* **25**, 446 (1986).
80. M. H. Chisholm, B. K. Conroy, K. Folting, D. M. Hoffmann, and J. C. Huffman, *Organometallics* **5**, 2457 (1986).
81. A. Mayr and G. A. McDermott, *J. Am. Chem. Soc.* **108**, 548 (1986).
82. A. Mayr, M. F. Asaro, M. A. Kjelsberg, K. S. Lee, and D. Van Engen, *Organometallics* **6**, 432 (1987).
83. A. C. Filippou, E. O. Fischer, and H. G. Alt, *J. Organomet. Chem.* **344**, 215 (1988).
84. A. C. Filippou, E. O. Fischer, and J. Okuda, *J. Organomet. Chem.* **339**, 309 (1988).
85. A. C. Filippou, E. O. Fischer, and R. Paciello, *J. Organomet. Chem.* **347**, 127 (1988).
86. A. C. Filippou and E. O. Fischer, *J. Organomet. Chem.* **341**, C35 (1988).
87. A. C. Filippou and E. O. Fischer, *J. Organomet. Chem.* **349**, 367 (1988).
88. A. C. Filippou, *Polyhedron* **8**, 1285 (1989).
89. U. Schubert, D. Neugebauer, P. Hofmann, B. E. R. Schilling, H. Fischer, and A. Motsch, *Chem. Ber.* **114**, 3349 (1981).
90. M. Bottrill, M. Green, A. G. Orpen, D. R. Saunders, and I. D. Williams, *J. Chem. Soc., Dalton Trans.*, 511 (1989).
91. P. K. Baker, G. K. Barker, D. S. Gill, M. Green, A. G. Orpen, and A. J. Welch, *J. Chem. Soc., Dalton Trans.*, 1321 (1989).
92. M. A. N. D. A. Lemos and A. J. L. Pombeiro, *J. Organomet. Chem.* **356**, C79 (1988).
93. A. J. L. Pombeiro, *NATO ASI Ser., Ser. C (Adv. Met. Carbene Chem.)*, **269**, 79 (1989).
94. S. S. P. R. Almeida, M. A. N. D. A. Lemos, and A. J. L. Pombeiro, *Port. Electrochim. Acta* **7**, 91 (1989).
95. F. Seitz, H. Fischer, J. Riede, T. Schöttle, and W. Kaim, *Angew. Chem.* **98**, 753 (1986); *Angew. Chem., Int. Ed. Engl.* **25**, 744 (1986).

96. F. Seitz and M. S. Wrighton, *Inorg. Chem.* **26**, 64 (1987).
97. A. Mayr, M. A. Kjelsberg, K. S. Lee, M. F. Asaro, and T. C. Hsieh, *Organometallics* **6**, 2610 (1987).
98. C. J. Leep, K. B. Kingsbury, and L. McElwee-White, *J. Am. Chem. Soc.* **110**, 7535 (1988).
99. K. B. Kingsbury, J. D. Carter, and L. McElwee-White, *J. Chem. Soc., Chem. Commun.* 624 (1990).
100. H. Fischer and B. Bühlmeyer, *J. Organomet. Chem.* **317**, 187 (1986).
101. A. C. Filippou, E. O. Fischer, and H. G. Alt, *J. Organomet. Chem.* **310**, 357 (1986).
102. A. C. Filippou, E. O. Fischer, and H. G. Alt, *J. Organomet. Chem.* **303**, C13 (1986).
103. A. Mayr, A. M. Dorries, A. L. Rheingold, and S. J. Geib, *Organometallics* **9**, 964 (1990).
104. E. O. Fischer, A. Ruhs, and F. R. Kreissl, *Chem. Ber.* **110**, 805 (1977).
105. A. Mayr, A. M. Dorries, G. A. McDermott, and D. Van Engen, *Organometallics* **5**, 1504 (1986).
106. A. C. Filippou and E. O. Fischer, *J. Organomet. Chem.* **383**, 179 (1990).
107. A. C. Filippou and W. Grünleitner, *Z. Naturforsch., B: Anorg. Chem., Org. Chem.* **44**, 1023 (1989).
108. A. C. Filippou and E. O. Fischer, *J. Organomet. Chem.* **352**, 141 (1988).
109. A. C. Filippou and E. O. Fischer, *J. Organomet. Chem.* **365**, 317 (1989).
110. A. C. Filippou, *NATO ASI Ser., Ser. C (Adv. Met. Carbene Chem.),* **269**, 101 (1989).
111. E. Delgado, M. E. Garcia, J. C. Jeffery, P. Sherwook, and F. G. A. Stone, *J. Chem. Soc., Dalton Trans.,* 207 (1988).
112. J. A. Abad, E. Delgado, M. E. Garcia, M. J. Grosse-Ophoff, I. J. Hart, J. C. Jeffery, M. S. Simmons, and F. G. A. Stone, *J. Chem. Soc., Dalton Trans.,* 41 (1987).
113. A. C. Filippou, E. O. Fischer, K. Öfele, and H. G. Alt, *J. Organomet. Chem.* **308**, 11 (1986).
114. A. C. Filippou, E. O. Fischer, H. G. Alt, and U. Thewalt, *J. Organomet. Chem.* **326**, 59 (1987).
115. A. C. Filippou, E. O. Fischer, and H. G. Alt, *J. Organomet. Chem.* **340**, 331 (1988).
116. M. D. Bermúdez, F. P. E. Brown, and F. G. A. Stone, *J. Chem. Soc., Dalton Trans.,* 1139 (1988).
117. A. C. Filippou and E. O. Fischer, *J. Organomet. Chem.* **330**, C1 (1987).
118. M. R. Churchill and H. J. Wassermann, *Inorg. Chem.* **20**, 4119 (1981).
119. E. O. Fischer, A. C. Filippou, H. G. Alt, and U. Thewalt, *Angew. Chem.* **97**, 215 (1985); *Angew. Chem.; Int. Ed. Engl.* **24**, 203 (1985).
120. E. Delgado, J. Hein, J. C. Jeffery, A. L. Raterman, F. G. A. Stone, and L. J. Farrugia, *J. Chem. Soc., Dalton Trans.,* 1191 (1987).
121. M. D. Bermúdez, E. Delgado, G. P. Elliott, N. H. Tran-Huy, F. Mayor-Real, F. G. A. Stone, and M. J. Winter, *J. Chem. Soc., Dalton Trans.,* 1235 (1987).
122. M. Green, J. A. K. Howard, A. P. James, C. M. Nunn, and F. G. A. Stone, *J. Chem. Soc., Dalton Trans.,* 187 (1986).
123. J. H. Davis, C. M. Lukehart, and L. A. Sacksteder, *Organometallics* **6**, 50 (1987).
124. A. F. Hill, J. A. K. Howard, T. P. Spaniol, F. G. A. Stone, and J. Szameitat, *Angew. Chem.* **101**, 213 (1989); *Angew. Chem., Int. Ed. Engl.* **28**, 210 (1989).
125. W. Kläui and H. Hamers, *J. Organomet. Chem.* **345**, 287 (1988).
126. M. Green, J. A. K. Howard, A. P. James, C. M. Nunn, and F. G. A. Stone, *J. Chem. Soc., Dalton Trans.,* 61 (1987).
127. F.-E. Baumann, J. A. K. Howard, O. Johnson, and F. G. A. Stone, *J. Chem. Soc., Dalton Trans.,* 2661 (1987).
128. F.-E. Baumann, J. A. K. Howard, R. J. Musgrove, P. Sherwood, and F. G. A. Stone, *J. Chem. Soc., Dalton Trans.,* 1879 (1988).

129. D. D. Devore, C. Emmerich, J. A. K. Howard, and F. G. A. Stone, *J. Chem. Soc., Dalton Trans.,* 797 (1989).
130. D. D. Devore, S. J. B. Henderson, J. A. K. Howard, and F. G. A. Stone, *J. Organomet. Chem.* **358,** C6 (1988).
131. S. J. Crennell, D. D. Devore, S. J. B. Henderson, J. A. K. Howard, and F. G. A. Stone, *J. Chem. Soc., Dalton Trans.,* 1363 (1989).
132. A. C. Filippou, E. O. Fischer, and H. G. Alt, *Z. Naturforsch., B: Anorg. Chem., Org. Chem.* **43,** 654 (1988).
133. J. S. Murdzek, L. Blum, and R. R. Schrock, *Organometallics* **7,** 436 (1988).
134. F. R. Kreissl, W. J. Sieber, H. Keller, J. Riede, and M. Wolfgruber, *J. Organomet. Chem.* **320,** 83 (1987).
135. A. P. James and F. G. A. Stone, *J. Organomet. Chem.* **310,** 47 (1986).
136. J. A. K. Howard, J. C. Jeffery, J. C. V. Laurie, I. Moore, F. G. A. Stone, and A. Stringer, *Inorg. Chim. Acta* **100,** 23 (1985).
137. S. A. Brew, J. C. Jeffery, M. U. Pilotti, and F. G. A. Stone, *J. Am. Chem. Soc.* **112,** 6148 (1990).
138. A. C. Filippou and E. O. Fischer, *J. Organomet. Chem.* **352,** 149 (1988).
139. R. J. Angelici, *NATO ASI Ser., Ser. C (Adv. Met. Carbene Chem.),* **269,** 123 (1989).
140. H. P. Kim, S. Kim, R. A. Jacobson, and R. J. Angelici, *Organometallics* **5,** 2481 (1986).
141. H. P. Kim and R. J. Angelici, *Organometallics* **5,** 2489 (1986).
142. R. A. Doyle and R. J. Angelici, *J. Organomet. Chem.* **375,** 73 (1989).
143. R. A. Doyle and R. J. Angelici, *J. Am. Chem. Soc.* **112,** 194 (1990).
144. F. R. Kreissl and H. Keller, *Angew. Chem.* **98,** 924 (1986); *Angew. Chem., Int. Ed. Engl.* **25,** 904 (1986).
145. F. R. Kreissl, H. Keller, F. X. Müller, C. Stegmair, and N. Ullrich, *NATO ASI Ser., Ser. C (Adv. Met. Carbene Chem.),* **269,** 137 (1989).
146. M. Wolfgruber and F. R. Kreissl, *J. Organomet. Chem.* **349,** C4 (1988).
147. M. Wolfgruber, C. M. Stegmair, and F. R. Kreissl, *J. Organomet. Chem.* **376,** 45 (1989).
148. F. R. Kreissl and N. Ullrich, *J. Organomet. Chem.* **361,** C30 (1989).
149. A. Mayr, G. A. McDermott, A. M. Dorries, A. K. Holder, W. C. Fultz, and A. L. Rheingold, *J. Am. Chem. Soc.* **108,** 310 (1986).
150. N. H. T. Huy, J. Fischer, and F. Mathey, *Organometallics* **7,** 240 (1988).
151. D. Barratt, S. J. Davies, G. P. Elliott, J. A. K. Howard, D. B. Lewis, and F. G. A. Stone, *J. Organomet. Chem.* **325,** 185 (1987).
152. R. R. Schrock, J. S. Murdzek, J. H. Freudenberger, M. R. Churchill, and J. W. Ziller, *Organometallics* **5,** 25 (1986).
153. C. J. Schaverien, J. C. Dewan, and R. R. Schrock, *J. Am. Chem. Soc.* **108,** 2771 (1986).
154. J. S. Murdzek and R. R. Schrock, *Organometallics* **6,** 1373 (1987).
155. R. R. Schrock, R. T. DePue, J. Feldman, C. J. Schaverien, J. C. Dewan, and A. H. Liu, *J. Am. Chem. Soc.* **110,** 1423 (1988).
156. R. R. Schrock, J. S. Murdzek, G. C. Bazan, J. Robbins, M. DiMare, and M. O'Regan, *J. Am. Chem. Soc.* **112,** 3875 (1990).
157. J. Chem, G. Lei, X. Xu, and Y. Tang, *Huaxue Xuebao (Acta Chim. Sin.)* **45,** 754 (1987).
158. J. Chen, G. Lei, W. Xu, Z. Zhang, X. Xu, and Y. Tang, *Sci. Sin., Ser. B* **3,** 24 (1987).
159. J. B. Sheridan, G. L. Geoffroy, and A. L. Rheingold, *J. Am. Chem. Soc.* **109,** 1584 (1987).
160. J. B. Sheridan, J. R. Johnson, B. M. Handwerker, G. L. Geoffroy, and A. L. Rheingold, *Organometallics* **7,** 2404 (1988).
161. J. C. Jeffery, A. G. Orpen, F. G. A. Stone, and M. J. Went, *J. Chem. Soc., Dalton Trans.,* 173 (1986).

162. I. J. Hart, J. C. Jeffery, R. M. Lowry, and F. G. A. Stone, *Angew. Chem.* **100,** 1769 (1988); *Angew. Chem., Int. Ed. Engl.* **27,** 1703 (1988).

163. G. A. Carriedo, V. Riera, M. L. Rodriguez, and J. C. Jeffery, *J. Organomet. Chem.* **314,** 139 (1986).

164. A. E. Bruce, A. S. Gamble, T. L. Tonker, and J. L. Templeton, *Organometallics* **6,** 1350 (1987).

165. S. Chaona, F. J. Lalor, G. Ferguson, and M. M. Hunt, *J. Chem. Soc., Chem. Commun.,* 1606 (1988).

166. H. Fischer, F. Seitz, and G. Müller, *Chem. Ber.* **120,** 811 (1987).

167. A. Mayr, McDermott, A. M. Dorries, and D. Van Engen, *Organometallics* **6,** 1503 (1987).

168. W. J. Sieber, M. Wolfgruber, N. H. Tran-Huy, H. R. Schmidt, H. Heiss, P. Hofmann, and F. R. Kreissl, *J. Organomet. Chem.* **340,** 341 (1988).

169. F. R. Kreissl, W. Uedelhofen, and D. Neugebauer, *J. Organomet. Chem.* **344,** C27 (1988).

170. P. G. Byrne, M. E. Garcia, N. H. Tran-Huy, J. C. Jefferey, and F. G. A. Stone, *J. Chem. Soc., Dalton Trans.,* 1243 (1987).

171. A. S. Gamble, K. R. Birdwhistell, and J. L. Templeton, *J. Am. Chem. Soc.* **112,** 1818 (1990).

172. A. K. List, G. L. Hillhouse, and A. L. Rheingold, *Organometallics* **8,** 2010 (1989).

173. J. B. Sheridan, G. L. Geoffroy, and A. L. Rheingold, *Organometallics* **5,** 1514 (1986).

174. J. B. Sheridan, D. B. Pourreau, G. L. Geoffroy, and A. L. Rheingold, *Organometallics* **7,** 289 (1988).

175. A. C. Filippou, C. Völkl, W. Grünleitner, and P. Kiprof, *Angew. Chem.* **102,** 224 (1990); *Angew. Chem., Int. Ed. Engl.* **29,** 207 (1990).

176. A. C. Filippou, C. Völkl, W. Grünleitner, and P. Kiprof, *Z. Naturforsch., B: Anorg. Chem., Org. Chem.* **45,** 351 (1990).

177. A. C. Filippou, *Polyhedron* **9,** 727 (1990).

178. S. Warner and S. J. Lippard, *Organometallics* **5,** 1716 (1986).

179. P. A. Bianconi, I. D. Williams, M. P. Engeler, and S. J. Lippard, *J. Am. Chem. Soc.* **108,** 311 (1986).

180. P. A. Bianconi, R. N. Vrtis, C. P. Rao, I. D. Williams, M. P. Engeler, and S. J. Lippard, *Organometallics* **6,** 1968 (1987).

181. E. M. Carnahan and S. J. Lippard, *J. Am. Chem. Soc.* **112,** 3230 (1990).

182. G. A. McDermott and A. Mayr, *J. Am. Chem. Soc.* **109,** 580 (1987).

183. S.-I. Murahashi, Y. Kitani, T. Uno, T. Hosokawa, K. Miki, T. Yonezawa, and N. Kasai, *Organometallics* **5,** 356 (1986).

184. M. H. Chisholm, J. C. Huffman, and N. S. Marchant, *J. Chem. Soc., Chem. Commun.,* 717 (1986).

185. M. H. Chisholm, D. Ho, J. C. Huffman, and N. S. Marchant, *Organometallics* **8,** 1626 (1989).

186. A. Mayr, M. F. Asaro, and T. J. Glines, *J. Am. Chem. Soc.* **109,** 2215 (1987).

187. A. Mayr, M. F. Asaro, and D. Van Engen, *NATO ASI Ser., Ser. C (Adv. Met. Carbene Chem.),* **269,** 167 1989.

188. A. Mayr, K. S. Lee, M. A. Kjelsberg, and D. Van Engen, *J. Am. Chem. Soc.* **108,** 6079 (1986).

189. A. Mayr, K. S. Lee, and B. Kahr, *Angew. Chem.* **100,** 1798 (1988); *Angew. Chem., Int. Ed. Engl.* **27,** 1730 (1988).

190. F. J. Feher, M. Green, and A. G. Orpen, *J. Chem. Soc., Chem. Commun.,* 291 (1986).

191. I. J. Hart, J. C. Jeffery, M. J. Grosse-Ophoff, and F. G. A. Stone, *J. Chem. Soc., Dalton Trans.,* 1867 (1988).

192. I. J. Hart and F. G. A. Stone, *J. Chem. Soc., Dalton Trans.*, 1899 (1988).
193. T. J. Katz, *NATO ASI Ser., Ser. C (Adv. Met. Carbene Chem.)*, **269**, 293 (1989).
194. T. M. Sivavec, T. J. Katz, M. Y. Chiang, and G. X.-Q. Yang, *Organometallics* **8**, 1620 (1989).
195. K. E. Garrett, J. B. Sheridan, D. B. Pourreau, W. C. Feng, G. L. Geoffroy, D. L. Staley, and A. L. Rheingold, *J. Am. Chem. Soc.* **111**, 8383 (1989).
196. G. L. Geoffroy, J. B. Sheridan, K. E. Garrett, and D. B. Pourreau, *NATO ASI Ser., Ser. C (Adv. Met. Carbene Chem.)*, **269**, 189 (1989).
197. K. H. Dötz, *Angew. Chem.* **96**, 573 (1984); *Angew. Chem., Int. Ed. Engl.* **23**, 587 (1984).
198. B. M. Handwerker, K. E. Garrett, and G. L. Geoffroy, *J. Am. Chem. Soc.* **111**, 369 (1989).
199. S. A. Krouse and R. R. Schrock, *J. Organomet. Chem.* **355**, 257 (1988).
200. S. L. Latesky and J. P. Selegue, *J. Am. Chem. Soc.* **109**, 4731 (1987).
201. S. A. Krouse, R. R. Schrock, and R. E. Cohen, *Macromolecules* **20**, 903 (1987).
202. J. H. Freudenberger and R. R. Schrock, *Organometallics* **5**, 1411 (1986).
203. K. Weiss, *Angew. Chem.* **98**, 350 (1986); *Angew. Chem., Int. Ed. Engl.* **25**, 359 (1986).
204. K. Weiss, R. Goller, and G. Loessel, *J. Mol. Catal.* **46**, 267 (1988).
205. K. Weiss and M. Denzer, *J. Organomet. Chem.* **355**, 273 (1988).
206. K. Weiss and G. Lössel, *Angew. Chem.* **101**, 75 (1989); *Angew. Chem., Int. Ed. Engl.* **28**, 62 (1989).
207. J. H. Freudenberger and R. R. Schrock, *Organometallics* **5**, 398 (1986).
208. K. Weiss, U. Schubert, and R. R. Schrock, *Organometallics* **5**, 397 (1986).
209. J. R. Morrow, J. L. Templeton, J. A. Bandy, C. Bannister, C. K. Prout, *Inorg. Chem.* **25**, 1923 (1986).
210. R. Cantrell and P. B. Shevlin, *J. Am. Chem. Soc.* **111**, 2348 (1989).
211. T. Y. Meyer and L. Messerle, *J. Am. Chem. Soc.* **112**, 4564 (1990).
212. I. A. Kopéva, I. A. Oreshkin, and B. A. Dolgoplosk, *Bull. Acad. Sci. U.S.S.R., Div. Chem. Sci. (Engl. Transl.)* **37**, 1712 (1988).
213. O. Johnson, J. A. K. Howard, M. Kapan, and G. M. Reisner, *J. Chem. Soc., Dalton Trans.*, 2903 (1988).

ADVANCES IN ORGANOMETALLIC CHEMISTRY, VOL. 32

Chemistry of Cationic Dicyclopentadienyl Group 4 Metal– Alkyl Complexes

RICHARD F. JORDAN

Department of Chemistry
University of Iowa
Iowa City, Iowa 52242

I

INTRODUCTION AND SCOPE OF REVIEW

Neutral, d^0 group 4 metal alkyl complexes of general form $Cp_2M(R)_2$ (R = H, hydrocarbyl) or $Cp_2M(R)(X)$ (X = anionic, two-electron donor) comprise the most extensively studied class of group 4 organometallics (1). In the late 1970s interest in these bent metallocene systems was fueled by both the rich CO and N_2 reduction chemistry exhibited by hydrides such as $(C_5Me_5)_2ZrH_2$ (2) as well as the synthetic organic applications developed for the hydrozirconation reagent $Cp_2Zr(H)Cl$ (3). The insertion, hydrogenolysis, and transmetallation reactions of $Cp_2Zr(R)X$ complexes have been extensively studied and utilized in synthesis (3–6). More recently attention has focused on the C—C coupling chemistry of $Cp_2Zr(benzyne)$, $Cp_2Zr(diene)$, and related species (7). Neutral Ti alkyls $Cp_2Ti(R)(X)$ are generally less stable than the Zr analogs and less well-studied. However, titanacyclobutanes undergo thermolysis under mild conditions to alkylidenes and have been exploited in organic synthesis and ring-opening olefin metathesis polymerization (8,9).

Since the mid-1980s the chemistry of related cationic 16-electron $Cp_2M(R)(L)^+$ complexes (1) and base-free 14-electron $Cp_2M(R)^+$ complexes (2) (M = Ti, Zr) has been developed (10). These complexes are considerably more reactive than their neutral counterparts as a result of the increased Lewis acidity of the cationic metal center, as well as the presence of the labile ligand L in 1 and the increased unsaturation of 2. These features promote coordination and activation of olefins, acetylenes, H_2, C—H bonds, and other substrates, and they open reaction pathways which are unavailable to neutral 16-electron analogs. Cationic complexes 1 and 2 are thus classified as highly electrophilic metal alkyls.

A 16-electron, d^0 $Cp_2M(R)X$ species contains one low-lying metal-centered LUMO localized in the "equatorial" plane between the Cp ligands which is utilized for substrate coordination (Fig. 1) (11). Zr—X π-bonding (X = OR, halide, etc.) utilizes this orbital and generally results in decreased reactivity (5). The electronic structure of a $Cp_2Zr(R)(L)^+$ complex is similar to that of neutral analogs, but the LUMO is stabilized by the metal

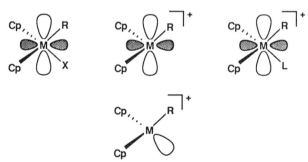

FIG. 1. Low-lying empty orbitals of d^0 $Cp_2M(R)(X)$, $Cp_2M(R)^+$, and $Cp_2M(R)(L)^+$ complexes.

charge, and the metal is thus more Lewis acidic. The base-free species **2** has two low-lying metal-centered empty orbitals and is isolobal with AlR_2^+ and d^0 $Cp_2M(R)$ (M = group 3, lanthanide).

This review focuses on the chemistry of well-characterized $Cp_2M(R)(L)^+$ and $Cp_2M(R)^+$ (M = Ti, Zr) alkyl complexes. Cationic $Cp_2Hf(R)(L)^+$ systems are mentioned only briefly. The main driving force behind studies of these systems has been their postulated role as active species in Cp_2MX_2-based olefin polymerization catalyst systems, which is discussed in Sections II and VI,A. This has prompted the development of efficient synthetic routes to $Cp_2M(R)(L)^+$ complexes (Section III) and extensive studies of their structures and bonding (Section IV) as well as reactivity (Section V). There is a significant effort directed to utilizing $Cp_2M(R)^+$ species as olefin polymerization catalysts and catalyst models (Section VI). More recently attention has focused on applications of cationic complexes **1** as reagents or catalysts for new types of C—H activation and C—C coupling reactions (Section VI).

This review does not cover several closely related areas including Richardson's gas-phase studies of $Cp_2Zr(R)^+$ species (*12*), Thewalt's extensive studies of cationic Cp_2M complexes containing inorganic ligands (*13*), "early–late" bimetallics which exhibit some ionic behavior (*14*), or other types of electrophilic metal alkyl species (*15*). Abbreviations used in the text have the following meanings: Cp, C_5H_5 or Cp ligands in general; Cp′, C_5H_4Me; Cp*, C_5Me_5; Ind, indenyl; EBTHI, ethylenebis(tetrahydroindenyl); THF, tetrahydrofuran.

II

HISTORICAL INTEREST IN THE ROLE OF CATIONIC SPECIES IN ZIEGLER–NATTA OLEFIN POLYMERIZATION

For many years it has been suspected that cationic d^0 metal alkyl species $Cp_2M(R)^+$ are involved in metallocene-based Ziegler–Natta olefin polymerization catalyst systems of the general type Cp_2MX_2/AlR_nX_{3-n}. In the early 1960s Shilov proposed on the basis of conductivity and electrodialysis experiments that the active species in the soluble Cp_2TiCl_2/AlR_nCl_{3-n} catalyst systems is cationic (*16*). At about the same time Breslow, Long, and Newburg concluded on the basis of UV–VIS spectroscopic and chemical studies that the active species in these systems is the chloride-bridged adduct $Cp_2Ti(R)$—Cl—$AlR_{n-1}Cl_{4-n}$, or a species derived therefrom, most likely the ion pair $[Cp_2Ti(R)][AlR_{n-1}Cl_{5-n}]$ (*17*). This early work suggested that the role of the Al cocatalyst in these systems is to alkylate the transition metal and to activate the resulting $Cp_2M(R)(Cl)$ species for olefin coordination and insertion (18) by Lewis acid complexation or halide abstraction. However, it was later realized that these early studies were complicated by adventitious water, which can promote polymerization via formation of alumoxanes, $[Al(CH_3)(\mu\text{-}O)]_n$, which are extremely effective cocatalysts in Cp_2MX_2-based systems (19,20). A variety of other mechanisms were also proposed (21).

Several key advances in the early 1980s generated renewed interest in the proposal that $Cp_2M(R)^+$ ions are active species in the soluble catalyst systems. Watson, Bercaw, Marks, and co-workers demonstrated that neutral group 3 and lanthanide metal systems of general type $Cp^*_2M(R)$ and $[Cp^*_2M(H)]_2$, which in monomeric form are isoelectronic (neglecting f electrons) with $Cp_2M(R)^+$ (M = Ti, Zr), are active ethylene polymerization catalysts [Eq. (1)] (22). In particular, Watson established that

$$H_2C{=}CH_2 \xrightarrow[\text{M = group 3, lanthanide}]{Cp^*_2MR} \ \ldots\ \wedge\wedge\wedge\ \ldots \qquad (1)$$

$(C_5Me_5)_2Lu(R)$ complexes undergo all of the key reactions involved in olefin polymerization including olefin insertion (chain propagation), β-H and β-alkyl elimination, and Lu—R bond hydrogenolysis (chain transfer/termination) (23).

In 1985, Eisch et al. reported the isolation of the ionic complex $[Cp_2Ti\{C(SiMe_3){=}C(Me)(Ph)\}][AlCl_4]$ from the reaction of $(Me_3Si)\text{-}C{\equiv}C(Ph)$ with $Cp_2TiCl_2/AlMeCl_2$ (24). It was proposed that the cation $(C_5H_5)_2Ti(CH_3)^+$ is generated initially by alkylation and Cl^- abstraction and inserts the acetylene as indicated in Eq. (2), and that similar reaction sequences lead to active cationic species in olefin polymerization reactions [Eq. (3)]. During this period, $Cp_2M(R)^+$ species were considered as possi-

$$\underset{Cl}{\overset{Cl}{Cp_2Ti}} + (AlMeCl_2)_2 \ \rightleftharpoons\ \underset{Cl}{\overset{Cl}{Cp_2Ti}}AlMeCl_2 \qquad (2)$$

$$\underset{Me}{\overset{Me_3Si}{Cp_2Ti^+}} \ AlCl_4^- \ \longleftarrow \ Cp_2Ti\text{-}CH_3{}^+ \ AlCl_4^-$$

$$Cp_2MX_2 + \text{Al cocatalyst} \longrightarrow Cp_2\overset{+}{M}R \ \rightleftharpoons\ \ldots\ \wedge\wedge\wedge\ \ldots \qquad (3)$$

ble active species in Cp_2MX_2/alumoxane and supported Cp_2MR_2/ alumina catalysts $(25-27)$.

Further indirect support for the cation proposal was provided by extensive studies of early-metal metallocene alkyl complexes. In general it was observed that four-coordinate d^0 metal alkyls $Cp_2M(R)(X)$ (M = Ti, Zr, X = alkyl, aryl, halide, etc.) (1), three-coordinate d^1 complexes $Cp_2Ti(R)$ (28), and neutral and cationic d^2 olefin alkyl complexes $Cp_2M(R)(olefin)^{n+}$ (M = group 5, 6) do not undergo olefin insertion (29). Together, these observations strongly implied that a three-coordinate, 14-electron, d^0 $Cp_2M(R)^{n+}$ species ($n = 0$, group 3, lanthanide; $n = 1$, group 4) or a $Cp_2M(R)(L)^{n+}$ species containing a labile ligand L is required for olefin insertion and polymerization in metallocene-based catalysts, and they prompted efforts to synthesize and isolate cationic group 4 metal alkyls (10).

III

SYNTHESIS OF $Cp_2M(R)(L)^+$ AND $Cp_2M(R)^+$ COMPLEXES

There are two examples of well-characterized $Cp_2M(R)^+$ complexes which have been isolated from Ziegler–Natta type systems. In 1976, Kaminsky *et al.* reported the isolation of $[(C_5H_5)_2Zr\{CH_2C(H)\text{-}(AlEt_2)_2\}][C_5H_5]$ from the reaction of $(C_5H_5)_4Zr$ and $AlEt_3$ [Eq. (4)] (30).

$$Cp_4Zr \quad + \quad AlEt_3 \quad \longrightarrow \quad Cp_2\overset{+}{Zr} \overset{AlEt_2}{\underset{AlEt_2}{\diagdown\diagup}} Cp^- \qquad (4)$$

An X-ray analysis of this remarkable compound, which is discussed in detail later, established the ionic structure in which the $C_5H_5^-$ counterion is tightly ion-paired with the $AlEt_2$ substituents on the β carbon of the $CH_2CH(AlEt_2)_2$ ligand. Also, as mentioned above, Eisch *et al.* isolated $[(C_5H_5)_2Ti\{C(SiMe_3)=C(Me)(Ph)\}][AlCl_4]$ from the $(C_5H_5)_2TiCl_2/[Al(CH_3)Cl_2]_2$ system in the presence of the silylacetylene $Me_3SiC\equiv CPh$ (24). There are now two general, Al-free routes to $Cp_2M(R)^+$ and $Cp_2M(R)(L)^+$ complexes which involve either oxidative or protolytic cleavage of Zr—R bonds in $Cp_2M(R)_2$ complexes. These reactions are discussed below. Additionally, ligand exchange, hydrogenolysis, insertion, and C—H activation reactions of these cationic complexes lead to a wide variety of other cationic hydrides, alkyls, alkenyls, acyls, etc., as discussed in subsequent sections.

A. Oxidative Cleavage of M—R Bonds of $Cp_2M(R)_2$ Complexes

Cationic $Cp_2Zr(R)(L)^+$ complexes are formed by reaction of neutral $Cp_2Zr(R)_2$ complexes with one-electron oxidants such as $(C_5H_4R)_2Fe^+$ or Ag^+ in the presence of the ligand L (31–33). This reaction is quite general for a variety of Cp_2Zr systems including $(C_5H_5)_2Zr$, $(C_5H_4Me)_2Zr$, $(C_5H_5)(C_5Me_5)Zr$, Me_2SiCp_2Zr (34), and chiral systems such as (EBTHI)Zr (35), as illustrated by representative examples in Eqs. (5)–(7).

$$Cp_2Zr\begin{matrix} CH_3 \\ \\ CH_3 \end{matrix} \; + \; Ag^+BPh_4^- \; \xrightarrow{CH_3CN} \; [Cp_2Zr(CH_3)(CH_3CN)_2][BPh_4] \qquad (5)$$

$$+ \; 1/2\ CH_3CH_3 \; + \; Ag^0$$

$$Cp'_2Zr\begin{matrix} \\ \\ \end{matrix} \; + \; Cp'_2Fe^+BPh_4^- \; \xrightarrow{THF} \; Cp'_2Zr\overset{+}{\big\langle}\;\;\;BPh_4^- \qquad (6)$$

$$+ \; 1/2\ PhCH_2CH_2Ph$$

$$+ \; Cp'_2Fe$$

$$Cp^*CpZr\begin{matrix} CH_3 \\ \\ CH_3 \end{matrix} \; \xrightarrow[THF]{Cp'_2Fe^+} \; Cp^*Cp\overset{+}{Zr}\begin{matrix} CH_3 \\ \\ \end{matrix} \qquad (7)$$

A simple but key point in the development of this chemistry was the realization that noncoordinating, nonreactive counterions such as BPh_4^- are required for the isolation of stable salts. Anions such as $CF_3SO_3^-$ and BF_4^- coordinate strongly or react with $Cp_2M(R)^+$ and $Cp_2M(R)(THF)^+$ (36,37). The counterion is BPh_4^- in all cases discussed in this article unless indicated otherwise. A useful reagent for these reactions is $[(C_5H_4Me)_2Fe][BPh_4]$, which is thermally stable and reacts with $Cp_2Zr(R)_2$ complexes in THF as well as in CH_2Cl_2 and toluene. The unsubstituted ferrocenium salt $[(C_5H_5)_2Fe][BPh_4]$ is less convenient because it is thermally unstable. $Ag[BPh_4]$ is effective only in CH_3CN and is thus less useful because the $(C_5H_5)_2Zr(R)(CH_3CN)_n^+$ complexes that are initially formed can undergo rapid CH_3CN insertion (R = H, Ph) or may be resistant to subsequent ligand exchange (R = CH_2Ph).

The general trend for ease of Zr—R bond cleavage by Cp_2Fe^+ reagents is Zr—CH_2Ph > Zr—CH_3 > Zr—Ph (38). Thus, $Cp_2Zr(CH_2Ph)_2$ reacts with $[(C_5H_4Me)_2Fe][BPh_4]$ in THF rapidly (minutes) at low temperature, $(C_5H_5)_2Zr(CH_3)_2$ reacts slowly (hours) at room temperature, and $(C_5H_5)_2Zr(Ph)_2$ does not react. However, the cationic phenyl complex $Cp_2Zr(Ph)(THF)^+$ is formed by selective cleavage of the Zr—CH_2Ph bond of the mixed-alkyl complex $Cp_2Zr(CH_2Ph)(Ph)$ [Eq. (8)]. One ex-

$$
\text{Cp}_2\text{Zr} \xrightarrow[\substack{-\text{Cp}'_2\text{Fe} \\ -1/2\ \text{PhCH}_2\text{CH}_2\text{Ph}}]{\text{Cp}'_2\text{Fe}^+,\ \text{THF}} \text{Cp}_2\text{Zr}^+ \xleftarrow[\text{THF}]{\overset{\text{Cp}'_2\text{Fe}}{\underset{\times}{}}} \text{Cp}_2\text{Zr} \qquad (8)
$$

ception to this trend is the reaction of $(C_5H_5)_2Zr(CH_3)(Ph)$ with $(C_5H_5)_2Fe^+$ which results in predominant Zr—Ph bond cleavage.

One-electron oxidation of $(C_5H_5)_2Ti(R)_2$ complexes does not provide general access to $(C_5H_5)_2Ti(R)(L)^+$ complexes. $(C_5H_5)_2TiMe_2$ does not react efficiently with $Ag[BPh_4]$ (39). The benzyl complex $(C_5H_5)_2Ti(CH_2Ph)_2$ reacts with Ag^+ in CH_3CN to yield a Ti(III) product and bibenzyl by an oxidatively induced *net* reductive elimination process [Eq. (9)] (40). Similar reactions of titanacyclobutanes with Ag^+, Cp_2Fe^+,

$$
\text{Cp}_2\text{Ti} \xrightarrow[\substack{2)\ \text{THF}}]{1)\ \text{Ag}^+,\ \text{CH}_3\text{CN}} \text{Cp}_2\text{Ti}^+ \underset{O}{\overset{\text{NCCH}_3}{}} \ +\ \text{PhCH}_2\text{CH}_2\text{Ph}\ +\ \text{Ag}^0 \qquad (9)
$$

DDQ, TCNQ, etc., yield cyclopropanes; stereochemical studies show that the reductive elimination is either concerted with the oxidation step or that C—C coupling following the first Ti—C bond cleavage is faster than C—C bond rotations (41). It is currently believed that these reactions proceed via outer-sphere one-electron oxidation from a metal–ligand bond. In the case of Zr, this is followed by loss of an R radical and trapping

$$Cp_2ZrR_2 \;+\; ox^+ \;\longrightarrow\; [Cp_2ZrR_2]^+ \;\longrightarrow\; [Cp_2ZrR^+] \;+\; R\cdot$$

$$+ \; red$$

$$\downarrow L$$

$$Cp_2\overset{+}{Zr}\overset{\displaystyle R}{\underset{\displaystyle L}{<}}$$

(10)

of the resulting $Cp_2Zr(R)^+$ cation by L [Eq. (10)]. In the case of Ti, oxidation induces *net* R—R reductive elimination. The HOMO of $(C_5H_5)_2M(R)_2$ has M—R bonding character (*11*).

Burk *et al.* measured the rates of electron transfer from $(C_5H_5)Zr(CH_2Ph)_2$ to several substituted ferroceniums and found a good correlation with driving force (*41*). More detailed studies of the titanacyclobutane $(C_5H_5)_2Ti\{(CH_2)_2CH(^tBu)\}$ showed that Marcus plots of ln k (electron transfer) versus E (driving force) are linear with the same slope for both homogeneous (to substituted ferroceniums) and heterogeneous (rotating Pt electrode) electron transfers and strongly support an outer-sphere electron transfer process. More qualitative observations made during the course of Zr cation syntheses, including (i) the facile cleavage of weak Zr—CH_2Ph bonds, (ii) the lack of reaction of $(C_5H_5)_2Zr(Ph)_2$, which contains strong Zr—Ph bonds and exhibits a relatively high (irreversible) $E_{\frac{1}{2}}$ values (oxidation) (*41*) and (iii) the observation of alkyl ferrocenes (derived from coupling of R $\cdot$ with Cp_2Fe^+) (*42*) as minor products in some cases, are all consistent with this mechanism. Mechanisms involving intermediate Cp_2Zr^{2+} species were discounted on the basis of studies of isolable $Cp_2Zr(L)_3^{2+}$ complexes (*43*). Mechanisms involving intermediate Ag—R species are also possible for the Ag^+ reactions.

Initial indications are that this oxidative route to cationic early metal alkyl complexes will be fairly general. Cationic $Cp_2Hf(CH_3)(L)^+$ complexes, monocyclopentadienyl systems $(C_5Me_5)Zr(R)_2(L)_2^+$, and trialkyl complexes $Zr(CH_2Ph)_3(THF)_3^+$ have recently been prepared by this method (*44,45*).

B. *Protonolysis Routes to* $Cp_2M(R)^+$ *Complexes*

In 1986 Bochmann and co-workers reported the synthesis of $[(C_5H_5)_2Ti(CH_3)(NH_3)]X$ ($X = PF_6^-$, ClO_4^-) by reaction of $(C_5H_5)_2$-$Ti(CH_3)_2$ with $[NH_4]X$ in THF at ambient temperature [Eq. (11)] (*46*). The use of bulkier ammonium reagents $HN(^nBu)_3^+$ and $HNMe_2Ph^+$, which are available as the BPh_4^- salts, in this reaction provides a general synthesis of 16-electron $Cp_2M(R)(L)^+$ and base-free 14-electron

$$Cp_2Ti\begin{smallmatrix}CH_3\\ \\CH_3\end{smallmatrix} + NH_4^+PF_6^- \xrightarrow{-CH_4} Cp_2\overset{+}{Ti}\begin{smallmatrix}CH_3\\ \\NH_3\end{smallmatrix} \quad PF_6^- \qquad (11)$$

$Cp_2M(R)^+$ complexes. Hlatky and Turner reported that the reaction of $Cp^*_2Zr(CH_3)_2$ with $[HN(^nBu)_3][BPh_4]$ in toluene yields the zwitterionic complex $Cp^*_2Zr^+ C_6H_4BPh_3^-$ via initial $Zr-CH_3$ bond protonolysis followed by subsequent BPh_4^- aryl $C-H$ bond activation [Eq. (12)] (47). An

$$Cp^*_2Zr\begin{smallmatrix}CH_3\\ \\CH_3\end{smallmatrix} + HNBu_3^+BPh_4^- \xrightarrow{-CH_4} [Cp^*_2\overset{+}{Zr}CH_3 \ \overset{-}{B}Ph_4]$$

$$\downarrow \qquad (12)$$

$$Cp^*_2\overset{+}{Zr}\text{—}\overset{\displaystyle H}{\diagup}\overset{\displaystyle \overset{-}{B}Ph_3}{\diagdown}$$

analogous complex containing a metallated $B(C_6H_4\text{-}p\text{-Et})_4^-$ anion was also prepared and characterized by X-ray diffraction.

In a conceptually similar reaction, the acidic carborane $C_2B_9H_{13}$ was utilized to protonate a $Zr-CH_3$ bond of $(C_5Me_4R)_2Zr(CH_3)_2$ (R = Me, Et), yielding the complex $[(C_5Me_4R)_2Zr(CH_3)][C_2B_9H_{12}]$ in which $C_2B_9H_{12}^-$ acts as a weakly coordinating ligand via a $B-H-Zr$ bridge [Eq. (13)]. Similar reactions have been used to prepare main group dicar-

$$Cp^*_2Zr\begin{smallmatrix}CH_3\\ \\CH_3\end{smallmatrix} + C_2B_9H_{13} \longrightarrow Cp^*_2Zr\begin{smallmatrix}CH_3\\ \\H\text{-}B_9C_2H_{11}\end{smallmatrix} \qquad (13)$$

bollide compounds (48). Turner et al. have used similar protonation reactions for the in situ generation of cationic base-free species that are active olefin polymerization catalysts (49). In this work, other noncoordinating anions, including (dicarbollide)$_2$M$^-$ (M = Fe, Co), $B(C_5F_5)_4^-$, and $CB_{11}H_{12}^-$, were also used. Marks and co-workers have prepared related actinide complexes $Cp^*_2Th(Me)(L)_n^+$ (L = NR$_3$, $n = 1$; L = THF, $n = 2$) as well as $Cp^*_2Th(Me)^+$ BPh$_4^-$ (which is strongly ion-paired) by this route (50).

Low-temperature NMR studies by Bochmann of reactions of $Cp_2Ti(CH_3)_2$ and $Cp_2Zr(CH_3)_2$ with $[HNMe_2Ph][BPh_4]$ in CD_2Cl_2 solvent reveal rapid $M-CH_3$ bond protolysis and formation of $Cp_2M(CH_3)^+$ species that are stabilized by strong ion pairing with the BPh_4^- anion and/or by adduct formation with solvent (51). These species catalyze ethylene and propylene polymerization. The use of $[HN(^nBu)_3][BPh_4]$ results in species which are significantly less active catalysts; it is suggested that in these cases NBu_3 complexes are formed. Addition of Et_2O or CH_3CN to the $(C_5H_5)_2Ti(CH_3)^+$ species yields the adducts $(C_5H_5)_2Ti(CH_3)(L)^+$. The reaction of $(C_5H_5)_2Ti(CH_3)_2$ with $[HNMe_2Ph][BPh_4]$ in CH_2Cl_2 in the presence of ethers yields $(C_5H_5)_2$-$Ti(CH_3)(L)^+$ complexes directly (52). In a similar manner Teuben and co-workers have prepared $Cp*_2M(CH_3)(THT)^+$ (M = Ti, Zr, Hf, THT = tetrahydrothiophene) complexes by reaction of $Cp*_2M(CH_3)_2$ with $[HNEt_3][BPh_4]$ in THT (53). Petersen has found that protonation of the metallacycle $Cp*_2Zr\{(CH_2)_2SiMe_2\}$ with $[HNEt_3][BPh_4]$ yields $(C_5Me_5)_2Zr(CH_2SiMe_3)(THF)^+$ [Eq. (14)] (54).

$$Cp*_2Zr \overset{\triangle}{\underset{\triangledown}{}} SiMe_2 \xrightarrow[\text{THF}]{\overset{+}{HNEt_3}} Cp*_2\overset{+}{Zr} \overset{\diagup SiMe_3}{\underset{O}{\diagdown}} \qquad (14)$$

The reaction of $(C_5H_5)_2Zr(R)_2$ (R = Ph, CH_3, CH_2Ph) with HNR_3^+ reagents in THF provides an alternate route to the THF complexes $Cp_2Zr(R)(THF)^+$ [Eq. (15)] $(38,40)$. With the unhindered reagent

$$Cp_2M \overset{\diagup R}{\underset{\diagdown R}{}} \xrightarrow[\text{THF}]{\overset{+}{HNR'_3}} Cp_2\overset{+}{M} \overset{\diagup R}{\underset{O}{\diagdown}} \qquad (15)$$

M=Zr, R=CH$_3$, R'$_3$ =Bu$_3$
M=Zr, R=Ph, R'$_3$ =Me$_2$Ph
M=Hf, R=CH$_3$, R'$_3$ =Bu$_3$

$[HNMe_3][BPh_4]$, the general order of reactivity is $Zr-Ph > Zr-CH_3 > Zr-CH_2Ph$, which parallels the order generally observed in acidolysis reactions of main group organometallics (55). In this case, subsequent THF ring-opening reactions involving the nucleophilic NMe_3 occur, leading to $(C_5H_5)_2Zr(R)(OCH_2CH_2CH_2CH_2NMe_3^+)$ complexes $(vide\ infra);$ however, this secondary reaction does not occur with $N(^nBu)_3$ or NMe_2Ph. The reactions in Eq. (15) are the most efficient routes to $Cp_2Zr(CH_3)(THF)^+$ and $Cp_2Zr(Ph)(THF)^+$, and to cationic Hf complexes

(56). However, the oxidative cleavage route is preferred for the synthesis of cationic benzyl complexes [Eq. (6)].

The protolysis route also promises to provide general access to cationic early-metal alkyl complexes in non-Cp_2M environments. Bochmann has recently reported the synthesis of $[Zr(CH_2Ph)_3][BPh_4]$, which has a zwitterionic structure with η^6-bonding of one counterion phenyl ring to Zr, as well as $Zr(CH_2Ph)_3(L)_3^+$ and $(C_5H_5)Ti(CH_3)_2(L)_2^+$ complexes by this method (57).

C. Other Routes

Several other routes to cationic Cp_2Ti alkyls have been reported. Bochmann *et al.* have found that displacement of Cl^- from $(C_5H_5)_2Ti(CH_3)Cl$ and $(Ind)_2Ti(CH_3)Cl$ by strong neutral donor ligands provides a general route to cationic $Cp_2Ti(CH_3)(L)^+$ complexes (39,46). The reaction of $(C_5H_5)_2Ti(CH_3)Cl$ with $Na[BPh_4]$ in neat CH_3CN yields the CH_3-CN complex $(C_5H_5)_2Ti(CH_3)(CH_3CN)^+$. Similarly, the reaction of $(C_5H_5)_2Ti(CH_3)Cl$ with 1 equiv each of $Na[BPh_4]$ and ligand (L = RCN, PR_3, or pyridine) in THF yields cationic products [Eq. (16)]. A variety of

$$Cp_2Ti\begin{matrix}CH_3\\\\Cl\end{matrix} \quad + \quad Na^+BPh_4^- \quad \xrightarrow[-NaCl]{L} \quad Cp_2\overset{+}{Ti}\begin{matrix}CH_3\\\\L\end{matrix} \quad BPh_4^- \qquad (16)$$

$(Ind)_2Ti(CH_3)(L)^+$ complexes have also been prepared by this general route.

More crowded ligands such as PPh_3 and dmpe do not displace Cl^- from $(C_5H_5)_2Ti(CH_3)Cl$ under these conditions. Surprisingly, the THF complex $(C_5H_5)_2Ti(CH_3)(THF)^+$ is also not formed by similar reactions with THF, nor by reaction of $(C_5H_5)_2Ti(CH_3)Cl$ with $Ag[BPh_4]$ or $Tl[BPh_4]$ in THF. Evidently, THF is insufficiently basic to displace Cl^-, and the insolubility of $Ag[BPh_4]$ inhibits Cl^- abstraction. The general approach of halide displacement has not been useful for the synthesis of $Cp_2Zr(R)(L)^+$ complexes.

One-electron oxidation of the stable Ti(III) complex $(C_5Me_5)_2Ti(CH_3)$ by $Ag[BPh_4]$ in THF yields $(C_5Me_5)_2Ti(CH_3)(THF)^+$ [Eq. (17)] (58).

$$Cp^*_2Ti-CH_3 \quad + \quad Ag^+BPh_4^- \quad \xrightarrow[-Ag^0]{THF} \quad Cp^*_2\overset{+}{Ti}\begin{matrix}CH_3\\\\O\end{matrix} \quad BPh_4^- \qquad (17)$$

Cuenca and Royo have prepared $[Cp_2Zr(Cl)(L)][x]$ ($[x] = BF_4^-$, ClO_4^-) salts via a similar oxidation of $[Cp_2ZrCl]_2$ by Ag^+ (59). Methylation of the thioformaldehyde complex $(C_5H_5)_2Ti(\eta^2\text{-}CH_2S)(PMe_3)$ followed by anion exchange yields $(C_5H_5)_2Ti(\eta^2\text{-}CH_2SMe)(PMe_3)^+$ complexes (60). Tilley *et al.* have reported that $(C_5H_5)_2Zr(CH_3)(CD_3CN)_2^+$ is formed by reaction of $(C_5H_5)_2Zr(CH_3)_2$ with $[CPh_3][BPh_4]$ in CD_3CN (61). This reaction is similar to Schrock's synthesis of $(C_5H_5)_2Ta(CH_3)_2^+$ via CH_3^- abstraction from $(C_5H_5)_2Ta(CH_3)_3$ using trityl cation (62). This route should also prove to be generally useful for the synthesis of cationic early-metal complexes.

D. Synthesis of Higher Alkyl, Alkenyl, and Allyl Complexes from Cationic Hydrides

The reactions of $(C_5H_5)_2Zr(CH_3)(THF)^+$ and $Cp'_2Zr(R)(THF)^+$ ($R = CH_3$, CH_2Ph) with H_2 in THF yield the cationic hydrides

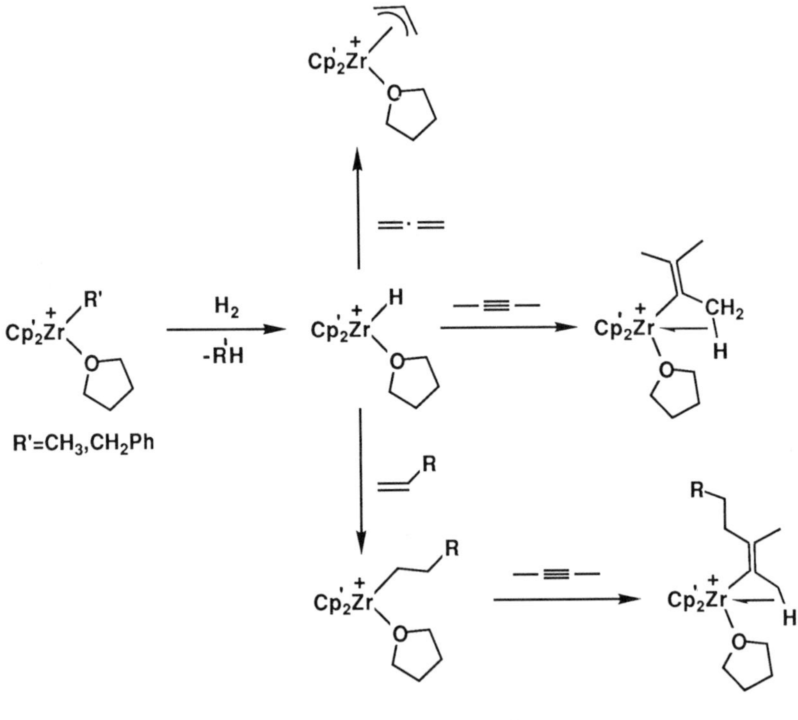

R = H, alkyl, aryl, SiMe₃, etc...

SCHEME 1

$(C_5H_5)_2Zr(H)(THF)^+$ and $Cp'_2Zr(H)(THF)^+$ (63,64). The latter is particularly useful as it is soluble in THF and highly reactive with olefins, as illustrated in Scheme 1. Reaction of this hydride with α-olefins and styrenes proceeds by normal 1,2-insertion and yields the corresponding alkyls $Cp'_2Zr(CH_2CH_2R)(THF)^+$, which are stable. These complexes are not available via the routes discussed above because the required neutral dialkyl precursors are not stable. Reaction of $Cp'_2Zr(H)(THF)^+$ with excess isobutylene or cyclohexene at low temperature yields the corresponding alkyls, but these undergo β-H elimination below room temperature (65). Reaction of $Cp'_2Zr(H)(THF)^+$ with allene yields the cationic allyl complex $Cp'_2Zr(C_3H_5)(THF)^+$, which has a dynamic η^3 structure. A variety of alkenyl complexes have been prepared by reaction of these hydrides or alkyls with acetylenes.

IV

STRUCTURES

A. $Cp_2M(R)(THF)^+$ Complexes

Cationic $Cp_2M(R)(L)^+$ complexes exhibit a variety of interesting structural features which result from the high Lewis acidity of the cationic d^0 metal centers. The first simple alkyl complex of this type to be structurally characterized was $[(C_5H_5)_2Zr(CH_3)(THF)][BPh_4]$ (3, Fig. 2) (32). The

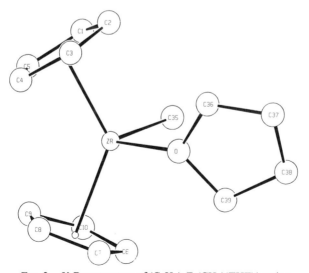

FIG. 2.   X-Ray structure of $(C_5H_5)_2Zr(CH_3)(THF)^+$ cation.

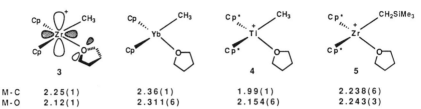

| M-C | 2.25(1) | 2.36(1) | 1.99(1) | 2.238(6) |
| M-O | 2.12(1) | 2.311(6) | 2.154(6) | 2.243(3) |

FIG. 3. Comparison of structures and M—C and M—O bond lengths of $Cp_2M-(CH_3)(THF)^{n+}$ ($n = 0,1$) complexes. M—O π bonding is possible for 3, but not for $(C_5H_5)Yb(CH_3)(THF)$, 4 or 5.

cation adopts a normal four-coordinate bent-metallocene structure (centroid–Zr–centroid 129.6°), and there are no significant cation–anion interactions. The THF is oriented nearly perpendicular to the plane between the Cp ligands (C-36—O—C-39/O—Zr—CH_3 dihedral angle 77.7°) in an orientation which allows O—Zr π bonding via interaction of the filled O p (b_1) orbital with the Zr LUMO (Fig. 3) (11). The Zr—O bond distance is quite short [2.12(1) Å versus Zr—C 2.25(1) Å], indicating that such π bonding is effective and that the THF is best considered a four-electron (σ,π) donor in this case. For comparison, the THF ligand in the lanthanide complex $(C_5H_5)_2Yb(CH_3)(THF)$, which is isoelectronic with 3 (neglecting f electrons), lies nearly in the plane between the Cp ligands (C—O—C/C—Yb—O dihedral angle 16°, Fig. 3) (66). As steric effects dominate directional bonding/overlap effects in lanthanide structures, this result implies that the in-plane THF orientation is sterically preferred for a d^0 $(C_5H_5)_2M(CH_3)(THF)^{n+}$ structure and that the perpendicular THF orientation in 3 indeed results from an electronic effect. Bochmann and Hursthouse have reported the structure of the related Ti complex $Cp^*_2Ti(CH_3)(THF)^+$ (4), again as the BPh_4^- salt (58). In this case the THF lies parallel to the plane between the Cp^* ligands in an orientation which precludes Ti—O π bonding, and the Ti—O bond distance is correspondingly long [2.154(6) Å versus Ti—C 1.99(1) Å] (Fig. 3). A similar structure is adopted by $Cp^*_2Zr(CH_2SiMe_3)(THF)^+$ (5) (54). The THF lies nearly in the plane between the Cp^* ligands (C—O—C/O—Zr—C dihedral angle 13.6°) and the Zr—O distance is long [2.243(3) Å versus Zr—C 2.238(6) Å]. Evidently, repulsive steric interactions between the THF and the Cp^* ligands in 4 and 5 preclude the electronically favored perpendicular THF orientation. The Zr—O π interaction in $Cp_2Zr(R)(THF)^+$ complexes strongly influences their reactivity as discussed below.

The structure of the cationic alkenyl complex $Cp'_2Zr[(Z)—C(Me)=C(^iPr)(Me)](THF)^+$, which is derived from insertion of 2-butyne into

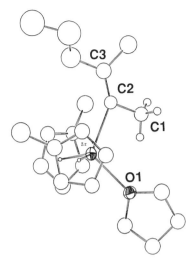

FIG. 4.   X-Ray structure of $(C_5H_4Me)_2Zr(C(Me){=}C({}^iPr)(Me))(THF)^+$ cation.

$Cp'_2Zr({}^iPr)(THF)^+$, has also been reported (Fig. 4) (64). The alkenyl ligand lies in the plane between the C_5H_4Me ligands, and thus Zr—C π bonding is precluded by symmetry. The alkenyl ligand is quite distorted [angles at C-2: Zr—C-2—C-3 145.2(8)°, Zr—C-2—C-1 92.4(6)°] as a result of steric interactions with the C_5H_4Me and THF ligands and a weak Zr—H agostic interaction involving one of the C-1 methyl hydrogens [Zr—H-1A 2.4 Å, Zr—C-1 2.72(1) Å]. The THF ligand is rotated 50° from the angle of maximum Zr—O π overlap (C—O—C/O—Zr—C dihedral angle 39.7°), and the Zr—O distance is relatively long [2.289(6) Å versus Zr—C 2.227(9) Å]. This difference probably results from the increased steric crowding in this complex and from a weaker Zr—O π interaction. The Me substituents on the C_5H_4Me rings, and possibly the agostic interaction, should decrease the Lewis acidity of the Zr center and disfavor Zr—O π bonding. Distorted alkenyl ligands of this type have been observed in neutral $Cp_2Zr(alkenyl)X$ complexes (67).

B. $Cp_2Zr(CH_2Ph)(L)^+$ Complexes

Replacement of the THF ligand of a $Cp_2Zr(R)(THF)^+$ species with a two-electron donor ligand opens up an empty orbital on Zr and invariably results in a compensating structural distortion. Spectroscopic data establish that the cationic benzyl complex $(C_5H_5)_2Zr(CH_2Ph)(THF)^+$ has a normal η^1-benzyl ligand. This species reacts irreversibly with CH_3CN to yield the

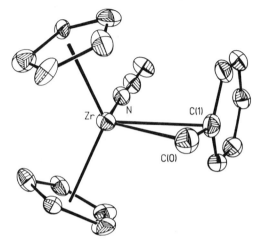

FIG. 5.   X-Ray structure of $(C_5H_5)_2Zr(\eta^2\text{-}CH_2Ph)(CH_3CN)^+$ cation.

η^2-benzyl complex $(C_5H_5)_2Zr(\eta^2\text{-}CH_2Ph)(CH_3CN)^+$, which is also formed directly by oxidation of $Cp_2Zr(CH_2Ph)_2$ in CH_3CN. The X-ray structure of this complex (Fig. 5) reveals that the benzyl ligand is highly distorted [Zr—C-0—C-1 84.9(4)°] to allow bonding through both the methylene carbon and the phenyl π system [Zr—C-0 2.344(8), Zr—C-1$_{ipso}$ 2.648(6) Å].

Similar structures have been observed for group 4 metal tetrabenzyls $M(CH_2Ph)_4$ and other d^0 and d^0f^n benzyl complexes (68,69). ^1H and ^{13}C NMR data establish that this structure is maintained in solution. Very similar results were obtained for the analogous $Cp'_2Zr(CH_2Ph)(L)^+$ and $(EBTHI)Zr(CH_2Ph)(L)^+$ systems; in both cases, the THF complexes have normal structures and the CH_3CN complexes have η^2-benzyl structures (35,64). The structure of $(EBTHI)Zr(CH_2Ph)(CH_3CN)^+$ is shown in Fig. 6. In these complexes the Zr—Ph interaction is favored over π donation from CH_3CN for electronic and steric reasons: the CN π orbitals are too stable to be effective donors, and the CH_3CN ligand is small so that steric constraints on the Zr—benzyl ligation mode are minimized (70). Interestingly, whereas these η^2-benzyl complexes undergo rapid associative exchange with free CH_3CN [Eq. (18)], stable bis-CH_3CN species with normal benzyl ligands are not formed.

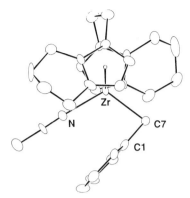

FIG. 6.   X-Ray structure of ethylenebis(tetrahydroindenyl)Zr(η^2-CH$_2$Ph)(CH$_3$CN)$^+$ cation.

C. Agostic Interactions in Cp'$_2$Zr(CH$_2$CH$_2$R)(L)$^+$ Complexes

Cationic alkyl complexes Cp'$_2$Zr(CH$_2$CH$_2$R)(THF)$^+$ contain normal undistorted alkyl groups as established by NMR and IR data. Reaction with PMe$_3$ yields free THF and the corresponding PMe$_3$ complexes Cp'$_2$Zr(CH$_2$CH$_2$R)(PMe$_3$)$^+$ [R = H (**6**), Et (**7**), Ph (**8**), CMe$_3$ (**9**)] which exhibit strong β-agostic interactions [Eq. (19) (*71,72*). X-ray analysis of the

$$\text{Cp'}_2\overset{+}{\text{Zr}}\diagdown_{\text{O}}\diagup^{R} \xrightarrow[\text{-THF}]{\text{PMe}_3} \text{Cp'}_2\overset{+}{\text{Zr}}\begin{smallmatrix}\text{H}\diagdown\diagup\text{R}\\ \diagup\diagdown\text{H}\\ \diagup\\ \diagdown_{\text{PMe}_3}\end{smallmatrix} \underset{\text{-PMe}_3}{\xrightarrow{\text{PMe}_3}} \left[\text{Cp'}_2\overset{+}{\text{Zr}}\begin{smallmatrix}\diagup^{\text{PMe}_3}\\ \diagdown\diagup^{R}\\ \diagdown_{\text{PMe}_3}\end{smallmatrix}\right] \quad (19)$$

6 - 9

ethyl complex **6** (Fig. 7) reveals an acute Zr—Cα—Cβ angle [84.7(5)°], a short Zr—Cβ contact [2.629(9) Å], and a reduced Cα—Cβ distance [1.47(2) Å]. The bridging H was located in the plane containing the Zr LUMO as required for a three-center, two-electron interaction (Zr—H$_{\text{bridge}}$ 2.16 Å). The IR spectrum of **6** exhibits bands at 2395 and 2312 cm^{-1} assigned to $\nu_{\text{C-H}}$ for the agostic hydrogen (confirmed by IR spectra of deuterated material). Thus, the Cβ—H bond is significantly weakened, and the ethyl ligand is distorted partway along the β-H elimination reaction coordinate.

NMR studies establish that similar β-H interactions are present in the higher alkyls **7–9**, that these structures are maintained in solution, and

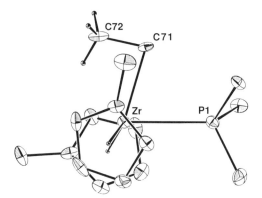

FIG. 7.   X-Ray structure of $(C_5H_4Me)_2Zr(CH_2CH_3)(PMe_3)^+$ cation.

that exchange of the agostic and terminal β hydrogens is rapid. Spectra of these complexes exhibit unusually high-field β-CH_2R ^1H and ^{13}C resonances and large values for $J_{C\alpha-H}$ (135–145 Hz), consistent with a Zr—βH interaction and distortion toward sp^2 hybridization at $C\alpha$ and/or incorporation of $C\alpha$ in a small ring. Low-temperature ^1H NMR spectra (to $-120°C$) exhibit some broadening but no splitting, indicating that exchange of bridging and terminal β-hydrogens (rotation about the $C\alpha$—$C\beta$ bond) is rapid. Substantial temperature-dependent isotope perturbation effects are observed in spectra of labeled compounds, which confirm that a single Zr—H—C bridging interaction is present. The $J_{C\beta-H}$ value for 6 (124 Hz) is the expected average of two large (terminal Hs) and one small (agostic H) J_{C-H} values. Similarly, the $J_{C\beta-H}$ value for 7–9 is approximately 110 Hz, the expected average of one large and one reduced value. The agostic structures of these complexes are similar to that proposed for isoelectronic $(C_5Me_5)_2ScEt$ on the basis of IR and NMR data (73). In the Sc series, however, the higher alkyls have normal structures owing to steric crowding associated with the bulky C_5Me_5 ligands.

Complex 6 is structurally similar to the agostic d^0 ethyl complex $(dmpe)Cl_3Ti(Et)$ (74) but is less distorted toward a hydride olefin species than late-metal agostic complexes. Two useful parameters for assessing the degree of distortion are the M—$C\alpha$—$C\beta$ angle θ and the difference Δ between the M—$C\alpha$ and M—$C\beta$ distances. Smaller values for these parameters indicate greater distortion toward an olefin hydride structure. The values for 6 ($\theta = 84.7°$, $\Delta = 0.34$ Å) are nearly equal to those for $(dmpe)Cl_3Ti(Et)$ (86.3°, 0.37 Å), indicating that the distortions are very similar. These values are significantly larger than those for d^6 $(C_5Me_5)(P$-p-tolyl$_3)Co(Et)^+$ (74.5°, 0.15 Å) (75) and d^8 $\{(^tBu)_2PCH_2CH_2P$-$(^tBu)_2\}Pt(norbornyl)^+$ (78.3°, 0.21 Å) (76), indicating greater distortion in the latter complexes. This difference is reflected in the barriers to β-H elimination. Labeling studies establish that $Cp'_2Zr(Et)(PMe_3)^+$ does not undergo reversible β-H elimination at room temperature (77). In contrast, the d^6 Co and d^8 Pt complexes undergo rapid reversible β-H elimination (76,78). These differences in structure and dynamics may be traced to the backbonding ability of the d^n ($n > 0$) metals, which would stabilize an olefin hydride complex.

Interestingly, although agostic complexes **6–9** undergo rapid associative exchange with free PMe_3, bis-PMe_3 complexes with normal alkyl ligands are not stable even in the presence of excess PMe_3 at low temperature [Eq. (19)]. To some extent this probably reflects the high steric crowding expected for a $Cp'_2Zr(CH_2CH_2R)(PMe_3)_2^+$ species. However, the corresponding Ph species $(C_5H_5)_2Zr(Ph)(PMe_3)_2^+$ (**10**), which is formed by reaction of $(C_5H_5)_2Zr(Ph)(THF)^+$ with excess PMe_3, *is* stable at low temperature; in this case, coordination of a second PMe_3 rather than an agostic aryl $C\beta$—H—Zr interaction is preferred [Eq. (20)] (*38*). Complex

$$\tag{20}$$

10

10 undergoes rapid exchange with free PMe_3 at room temperature, but the extent of PMe_3 dissociation is small. For comparison, the corresponding hydride $(C_5H_5)_2Zr(H)(PMe_3)_2^+$ is stable, whereas the methyl complex $(C_5H_5)_2Zr(CH_3)(PMe_3)_2^+$ is isolable but undergoes extensive PMe_3 dissociation at 25°C (*63*).

Two isomers are possible for $Cp'_2Zr\{CH_2CH(\mu\text{-}H)R\}(PMe_3)^+$ complexes **6–9**: an endo isomer with the agostic C—H "ligand" in the central site and an exo isomer with the agostic H in the outer site as observed in the X-ray structure of **6**. However, NMR spectra show only single sets of resonances, even at low temperature, indicating that one isomer is highly favored or that exchange is extremely rapid. It is likely that the PMe_3 exchange process in Eq. (19) catalyzes isomer exchange during the synthesis of **6–9**, so that the more stable isomer is isolated in each case.

D. Zr—βCH_2—Si Interactions

The reaction of $Cp'_2Zr(H(THF)^+$ with vinyltrimethylsilane yields the 1,2-insertion product $Cp'_2Zr(CH_2CH_2SiMe_3)(THF)^+$ [**11**, Eq. (21)]

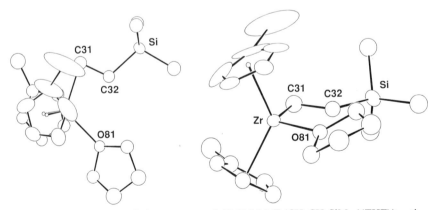

(21)

(79,83). The NMR spectra of **11** exhibit high-field shifts for the βCH_2 group (^1H δ -0.99, ^{13}C δ -3.25) and J_{C-H} values ($J_{C\alpha-H}$ 141 Hz, $J_{C\beta-H}$ 111 Hz) which are similar to those of the β-agostic Cp'_2Zr-$(CH_2CH_2R)(PMe_3)^+$ complexes, indicating that the alkyl group is distorted. The X-ray structure of **11** (Fig. 8) reveals an acute $Zr-C\alpha-C\beta$ angle ($83.0°$), a short $Zr-C\beta$ distance (2.58 Å), and a short $C\alpha-C\beta$ bond (1.49 Å), also characteristic of a β-agostic structure. However, the placement of the β-$SiMe_3$ group *in the plane* between the two Cp ligands implies that the β hydrogens lie above and below this plane and that neither $\beta C-H$ bond is positioned for optimal overlap with the Zr LUMO. The β hydrogens were not accurately located in this study.

As there is no obvious steric factor which favors this structure, the distortion is likely electronic in origin. $Cp/SiMe_3$ steric interactions are not expected to be severe enough to restrict the $SiMe_3$ group to the plane between the Cp ligands; for comparison, the bulky alkyls $Cp'_2Zr(CH_2CH_2R)(PMe_3)^+$ ($R = {}^tBu$, Ph), which are equally crowded, adopt singly β-H bridged structures in which the β substitutent points toward one of the Cp ligands [Eq. (19)]. There are no detectable low-fre-

FIG. 8. Two views of the structure of $(C_5H_4Me)_2Zr(CH_2CH_2SiMe_3)(THF)^+$ cation showing the distorted $CH_2CH_2SiMe_3$ ligand.

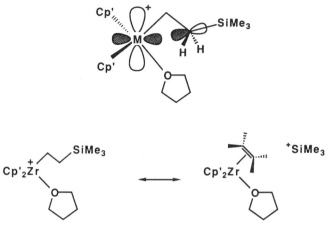

FIG. 9. Two representations of the Zr—$\beta CH_2 Si$ interaction in $Cp'_2Zr(CH_2CH_2$-$SiMe_3)(THF)^+$.

quency IR ν_{C-H} bands for 11, and NMR spectra of specifically labeled material do not exhibit isotope perturbation effects. These observations indicate that any Zr—β-H interactions are weak and symmetrical, and they suggest that the Zr interacts primarily with the β carbon, which is relatively electron-rich owing to the inductive electron-donating effect of the electropositive Si. The interaction of the Zr LUMO with the back lobe of the Cβ—Si bond is optimized by the in-plane placement of the $SiMe_3$ group. Thus, the bonding is best described as a Zr—βCH_2—Si interaction, as represented in Fig. 9. The Si—Cβ bond is slightly (0.05 Å) longer than the Si—CH_3 bonds, which is consistent with this interpretation.

The THF ligand of 11 is rotated 60° from the optimum orientation for O—Zr π bonding, and the Zr—O bond is long (2.29 Å versus Zr—C 2.27 Å), indicating that the THF acts as a two-electron σ donor in this complex. Thus, the Zr—βCH_2—Si interaction must be quite strong as it occurs in preference to π donation from THF. The interaction of the β carbon with both Zr and Si is related to, but much less symmetrical than, the bridging CH_3 in $(\mu$-$CH_3)[(C_5H_5)_2Zr(OCCHR)]_2AlMe_2$ and related complexes (80). Significant Zr—βC interactions may also be present in several neutral complexes which contain electropositive substituents on the β carbon (30,81,82).

Complex 11 undergoes ligand substitution reactions yielding PMe_3 and CH_3CN adducts 12 and 13 [Eq. (21)]. NMR and IR data establish that 12 and 13 adopt agostic Zr—βCH_2—Si structures analogous to that of 11. Generation of 13 at low temperature yields a mixture of endo and exo isomers [Eq. (22)] which is converted to the more stable endo isomer in a

$$
\text{endo} + CH_3CN \rightleftharpoons \left[Cp'_2Zr(...)SiMe_3 \right] \rightleftharpoons \text{exo} + CH_3CN \tag{22}
$$

CH_3CN-catalyzed process. This exchange presumably proceeds via an intermediate bis-CH_3CN complex in which the alkyl group has a normal structure. This is the only cation to date for which such endo/exo isomerism has been observed (79).

E. Relative Strengths of Nonclassical Bonding Interactions in Cp₂Zr(R)(L)⁺ Complexes

It is difficult to assess the relative strengths of the nonclassical bonding interactions described above for $Cp_2Zr(R)(L)^+$ complexes owing to differences in steric effects. However, some insight to general trends is provided by comparison of the relative ease with which ligands displace these interactions. The only interaction which is maintained when a THF ligand is present is the $Zr-\beta CH_2Si$ interaction in $Cp'_2Zr(CH_2CH_2SiMe_3)(THF)^+$. The hydrocarbyl groups in $Cp'_2Zr(R)(THF)^+$ ($R = CH_2CH_2R$, CH_2Ph, Ph) are all normal. This suggests that the $Zr-\beta CH_2Si$ interaction is stronger than the $Zr-O$ THF π donor interaction whereas $Zr-\beta H$ interactions are not. Similarly, the $Zr-\beta CH_2Si$ interaction in $Cp'_2Zr(CH_2CH_2SiMe_3)(CH_3CN)^+$ and the $Zr-Ph$ interactions $Cp_2Zr(\eta^2\text{-}CH_2Ph)(CH_3CN)^+$ complexes are not displaced by coordination of a second CH_3CN (although reversible exchange occurs), while the alkyls $Cp'_2Zr(CH_2CH_2R)(CH_3CN)^+$ form bis-CH_3CN complexes with undistorted alkyl groups in the presence of excess CH_3CN. This suggests that $Zr-\beta CH_2Si$ and $Zr-Ph$ interactions are stronger than $Zr-\beta H$ interactions. Also, both $Cp'_2Zr(CH_2CH_2SiMe_3)(PMe_3)^+$ and $Cp'_2Zr(CH_2CH_2R)(PMe_3)^+$ are stable in the presence of excess PMe_3 (although reversible exchange occurs) whereas $(C_5H_5)_2Zr(Ph)(PMe_3)^+$ coordinates a second PMe_3 ligand. Overall, these results suggest that the strengths of these interactions vary in the following qualitative order: $Zr-\beta CH_2Si >$ $Zr-\beta Ph$ in $Zr-\eta^2\text{-}CH_2Ph > Zr-\beta HC_{alkyl} > Zr-\beta HC_{aryl}$.

There is as yet no evidence for agostic $C\alpha-H$ interactions in these cationic systems, even in 16-electron methyl complexes where such interactions might be expected. No structural features or low-frequency IR bands attributable to an α-agostic interaction were noted for $Cp^*_2Zr(CH_3)(THF)^+$or $(C_5Me_4Et)_2Zr(CH_3)(C_2B_9H_{12})$ (47,58). Low-temperature NMR studies of the 16-electron pyridine complexes $(C_5H_4R)_2Zr(CH_3)(NC_5H_4R)^+$ (R = H, Me) suggest that the ortho C—H bond of the coordinated pyridine interacts weakly with the Zr center (83). α-Agostic interactions have been observed in several related neutral systems which lack β hydrogens (84).

F. Structures of Base-Free $Cp_2M(R)^+$ Systems

Three base-free $Cp_2M(R)^+$ complexes have been structurally characterized, and all exhibit nonclassical bonding interactions as a result of the high Lewis acidity of the metal centers. The Sinn–Kaminsky complex $[(C_5H_5)_2Zr\{CH_2CH(AlEt_2)_2\}][C_5H_5]$ [Eq. (4)] contains discrete cations and anions which are strongly ion-paired via C_5H_5/Al interactions (30). Two views of the key elements of the cation (based on the reported atomic coordinates) are shown in Fig. 10. The Cp_2Zr framework is normal (centroid–Zr–centroid 128.9°). However, the $CH_2CH(AlEt_2)_2$ ligand is highly distorted, as indicated by the acute Zr—Cα—Cβ angle [75.7(3)°] and short Zr—Cβ distance [2.393(4) Å], and lies in the plane between the two Cp ligands (centroid–Zr–centroid/Zr—Cα—Cβ dihedral angle 96.9°). This is the result of a strong Zr—Cβ interaction analogous to that in the $ZrCH_2CH_2SiMe_3$ complex 11. The β carbon is electron-rich owing to the two electropositive Al substituents, and one Al group (Al-2) lies

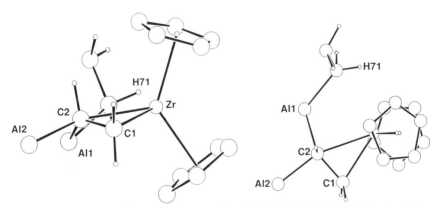

FIG. 10. Two views of the key elements of the $(C_5H_5)_2Zr\{CH_2CH(AlEt_2)_2\}^+$ cation. Two ethyl groups from Al-2 and one ethyl group from Al-1 have been removed for clarity. The Cp^- counterion is strongly ion-paired with Al-1 and Al-2 but is also not shown.

within 0.3 Å of the Zr—Cα—Cβ plane, allowing effective overlap between the back lobe of the Al-2—Cβ bond and one of the Zr LUMOs (Al-2—Cβ—Zr angle 158.6°). The distorted trigonal bipyramidal geometry at Cβ is very similar to that of the μ-CH$_3$ group in $(\mu$-CH$_3)$-[(C$_5$H$_5$)$_2$Zr(OCCHR)]$_2$AlMe$_2$ (80). There is a short Zr—Hβ contact (2.64 Å), but this interaction is weak as this distance is longer than the Zr—Cβ distance and the Cβ—Hβ bond is not in the equatorial plane containing the two Zr LUMOs. There is also a short contact between Zr and H-71 (2.64 Å) which lies in the equatorial plane. The chloride salt (C$_5$H$_5$)$_2$Zr{CH$_2$CH(AlEt$_2$)$_2$}Cl was also structurally characterized and contains a similarly distorted CH$_2$CH(AlEt$_2$)$_2$ ligand; the Cl$^-$ bridges Zr and Al-1 (30,81).

The Eisch cation (C$_5$H$_5$)$_2$Ti{C(SiMe$_3$)=C(Ph)(Me)}$^+$, which was isolated and structurally characterized as the AlCl$_4^-$ salt [Eq. (2)], adopts a normal bent-metallocene structure with the alkenyl ligand in the plane between the Cp ligands in an orientation which precludes C—Ti π bonding (24). The angles at Cα are distorted (Ti—Cα—Si 88.9°, Ti—Cα—Cβ 144.9°), and the SiMe$_3$ group is bent toward the metal center. This has been rationalized in terms of hyperconjugation involving the Cα—Si bond and one of the empty orbitals on Ti, that is, an agostic Ti—Si—C interaction as in Fig. 11 (85). The stability of this AlCl$_4^-$ salt and the absence of strong, site-specific ion pairing are somewhat surprising given Bochmann's finding that Cp$_2$Ti(CH$_3$)(L)$^+$ complexes are not stable as the PF$_6^-$ or BF$_4^-$ salts (39,46), and it probably reflects the steric crowding at the metal center.

One of the most interesting cations to be structurally characterized is Turner and Hlatky's zwitterionic compound (C$_5$Me$_5$)$_2$Zr{2-Et,5-[B(p-ethylphenyl)$_3$]phenyl}$^+$ in which the cation and anion are linked by a metallated Ph ring [see Eq. (12) for the BPh$_4^-$ analog] (47). The metallated ring is restricted to the plane between the two bulky C$_5$Me$_5$ ligands so there is no Zr—C π bonding. However, there is a strong Zr—(β-aryl-H) agostic interaction (Zr—H 2.14 Å) which results in an acute Zr—C-13—C-12 angle (87.2°). The agostic hydrogen appears at δ 4.77 in the ^{1}H NMR spectrum. Agostic interactions involving aryl C—H bonds have been observed in several related neutral systems (86,87).

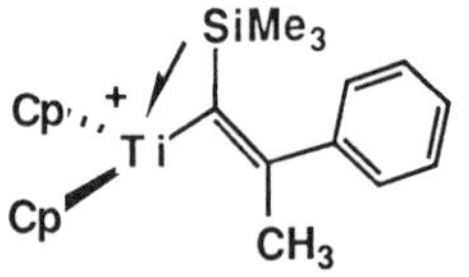

FIG. 11.   Agostic Zr—Si interaction in the (C$_5$H$_5$)$_2$Ti{C(SiMe$_3$)=C(Me)(Ph)}$^+$ cation.

V

REACTIVITY

A. Lewis Acid–Base Chemistry

The high Lewis acidity of $Cp_2M(R)(L)^+$ and $Cp_2M(R)^+$ species implied by the structural data above is manifested in their reactivity, as illustrated by several simple reactions of $Cp_2Zr(R)(THF)^+$ complexes (Scheme 2). The THF ligands in these complexes are labile and generally undergo rapid (NMR time scale) exchange with free THF by dissociative or associative mechanisms at ambient temperature. One exception is the alkenyl series $Cp'_2Zr\{C(Me)=C(R)(Me)\}(THF)^+$, for which THF exchange is slow on the NMR time scale but rapid on the laboratory time scale. The extent of THF dissociation in CH_2Cl_2 solution is generally small. Significant THF dissociation is observed for $(C_5H_5)_2Zr(CH_2Ph)(THF)^+$ however, where the THF dissociation product, $(C_5H_5)_2Zr(\eta^2\text{-}CH_2Ph)^+$ (or its CH_2Cl_2 solvate), is stabilized by an η^2-bonding interaction (33). Bis-THF complexes are also not observed. For example, the low-temperature ($-78°C$) NMR spectrum of $Cp'_2Zr(Et)(THF)^+$ in CD_2Cl_2 containing excess THF exhibits separate resonances for coordinated (1 equiv) and free THF; these are coalesced at $23°C$ owing to rapid exchange (64).

Cationic $Cp_2Zr(R)(THF)^+$ complexes undergo ligand substitution by

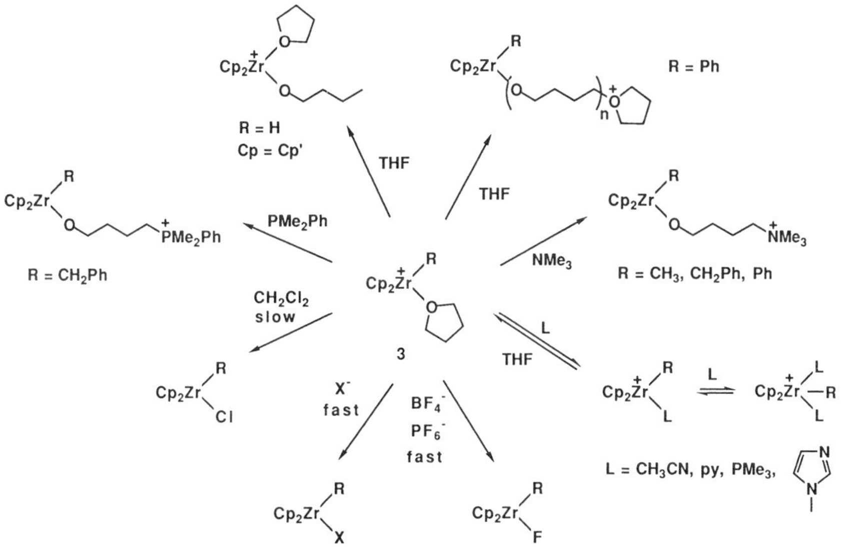

SCHEME 2

nitriles, small phosphines, and pyridine and other heterocycles. For small, two-electron donor ligands such as PMe_3, CH_3CN, and N-methylimidazole, 18-electron $Cp_2Zr(R)(L)_2^+$ complexes are sometimes formed. Representative examples include $(C_5H_4R)_2Zr(H)(PMe_3)_2^+$ ($R = H$, Me), $(C_5H_4Me)_2Zr(R)(CH_3CN)_2^+$ (R = alkyl), $(C_5H_5)_2Zr(R)(PMe_3)_2^+$ (R = CH_3, Ph), and $(C_5H_5)_2Zr(CH_3)(N$-methylimidazole)$_2^+$. In contrast, neutral 16-electron $Cp_2Zr(R)(X)$ complexes generally do not form stable 18-electron $Cp_2Zr(R)(X)(L)$ adducts, although such species are important intermediates or transition states in many reactions ($88,89$). This difference reflects the increased electrophilicity of the cationic complexes. The tetrahydrothiophene ligands of Teuben's $Cp^*_2M(CH_3)(THT)^+$ (M = group 4) complexes are even more labile because of the increased crowding associated with the bulky C_5Me_5 ligands and the lower basicity of THT, and they undergo rapid substitution by THF, 2-Me-THF, and PMe_3 (53).

Salts of $Cp_2Zr(R)(L)^+$ cations incorporating BF_4^-, PF_6^-, or related anions are almost invariably unstable with respect to F^- transfer, which initially yields neutral $Cp_2Zr(R)(F)$ (Scheme 2) ($31,36$). These reactions imply that $Cp_2Zr(R)^+$ is a stronger Lewis acid for F^- than BF_3 or PF_5; in other words, $Cp_2Zr(R)^+$ is a strong, hard Lewis acid. $Cp_2Zr(R)(L)^+$ complexes are reasonably stable as BPh_4^- salts, although base-free $Cp_2Zr(R)^+$ species strongly ion-pair with and can undergo electrophilic attack on BPh_4^- ($47,51$). The hard character of $Cp_2Zr(R)(L)^+$ species is underscored by comparison of the anion tolerance trend $BPh_4^- \gg PF_6^-$, BF_4^- with that for other organometallic cations. For example, the group 5 and 6 cations $(C_5H_5)_2Ta(CH_3)_2^+$ and $(C_5H_5)_2W(R)(L)^+$ are stable as BF_4^- or PF_6^- salts, and $(C_5H_5)W(NO)_2^+$ is stable as the BF_4^- salt but abstracts Ph^- from BPh_4^- ($29,62,90$).

Addition of halide salts to $Cp_2Zr(R)(L)^+$ complexes yields neutral $Cp_2Zr(R)X$ complexes as expected. In slower reactions, $Cp_2Zr(R)(THF)^+$ complexes decompose in CH_2Cl_2 via Cl^- abstraction, yielding $Cp_2Zr(R)Cl$ as the initial product. At 25°C, this reaction is slow for $Cp'_2Zr(Et)(THF)^+$ ($t_{1/2}$ 28 hours) and relatively rapid for $(C_5H_5)_2Zr(Ph)(THF)^+$ ($t_{1/2} < 12$ hours). The 18-electron complexes $(C_5H_5)_2Zr(\eta^2$-pyridyl)(L)$^+$ (L = THF, PMe_3, py) and the chelated species $(C_5H_5)_2Zr(\eta^2$-N,C-CH_2CH_2py)$^+$ are stable in CH_2Cl_2 (83). Teuben's THT complexes react very rapidly with CH_2Cl_2, C_6H_5Cl, and $C_6H_5CF_3$ to yield initially $Cp^*_2M(CH_3)X$, consistent with their increased lability (53).

Lewis acids promote nucleophilic attack at the α carbon of THF, leading to ring-opening reactions or ring-opening polymerization (91). However, despite the high Lewis acidity the $Cp_2Zr(R)^+$ fragment, this is not a general feature of $Cp_2Zr(R)(THF)^+$ chemistry, probably as a result of steric crowding and the tendency of nucleophiles to attack at Zr. Of the

$[Cp_2Zr(R)(THF)][BPh_4]$ complexes studied to date, only $(C_5H_5)_2$-$Zr(Ph)(THF)^+$ initiates THF polymerization at detectable rates (Scheme 2) (38). Nucleophilic ring opening leading to Zr alkoxide species is observed in the reactions of $(C_5H_5)_2Zr(R)(THF)^+$ (R = CH_3, CH_2Ph, Ph) with the small, nucleophilic amine NMe_3 and in the reaction of the benzyl derivative with PMe_2Ph (33). The hydride $Cp'_2Zr(H)(THF)^+$ undergoes rearrangement to $Cp'_2Zr(OCH_2CH_2CH_2CH_3)(THF)^+$ in THF at 40°C (92). The relative stability of $Cp_2Zr(R)(THF)^+$ complexes toward THF ring-opening reactions and the lability of the coordinated THF make these complexes extremely useful synthetic intermediates in early-metal cation chemistry.

Cationic Ti complexes $Cp_2Ti(CH_3)(L)^+$ exhibit properties similar to the Zr analogs except that 18-electron $Cp_2Ti(CH_3)(L)_2^+$ ions generally do not form, owing to the smaller Ti(IV) radius (39,46,93). The NH_3 complex $(C_5H_5)_2Ti(CH_3)(NH_3)^+$ is stable as the PF_6^- salt in THF at ambient temperature. However, attempted ligand exchange reactions with CH_3CN, pyridine, or aniline yield ultimately $(C_5H_5)_2TiF_2$ via F^- abstraction from the anion by more labile/electrophilic $(C_5H_5)_2Ti(CH_3)(L)^+$ species. The BPh_4^- salts are much more stable, and a variety of $(C_5H_5)_2Ti(CH_3)(L)^+$ and $(Ind)_2Ti(CH_3)(L)^+$ complexes (L = RCN, PR_3, py) have been characterized (Scheme 3). The nitrile ligands in $Cp_2Ti(CH_3)(RCN)^+$ complexes are labile and are rapidly replaced by pyridine or PR_3 but not by THF (Scheme 3).

Interestingly, Bochmann et al. have characterized several Ti(IV) RNC adducts which are rare examples of d^0 metal isocyanide complexes (39). The reaction of $(C_5H_5)_2Ti(CH_3)(CH_3CN)^+$ with RNC (R = ^tBu, $SiMe_3$) yields the iminoacyl complexes $(C_5H_5)_2Ti\{\eta^2\text{-}C(Me)\!=\!NR\}(CNR)^+$ via insertion of and trapping by RNC. The ^tBuNC complex $Cp^*_2Ti(CH_3)(CN^tBu)^+$ is observed as an intermediate in the analogous

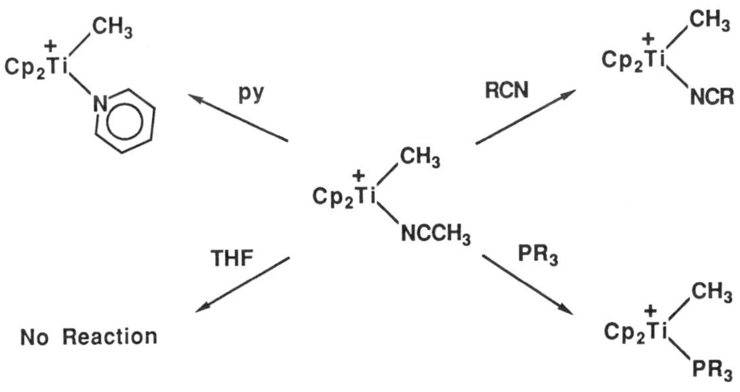

SCHEME 3

insertion reaction of ^{t}BuNC and Cp*$_2$Ti(CH$_3$)(THF)$^+$ (58). The strong binding of soft RNC and PR$_3$ ligands by the hard Cp$_2$M(R)$^+$ Lewis acid fragments is surprising and reflects the high Lewis acid strength of these species, which leads to robust adducts despite the hard/soft mismatch.

B. Insertion Chemistry

The reactions of Cp$_2$Zr(H)(THF)$^+$ hydride complexes with olefins and other substrates were noted in Section III,D. Cationic alkyl complexes also exhibit a rich insertion chemistry. In this section simple *single* insertion reactions are reviewed; reactions with olefins are discussed in Section V,F.

The high insertion reactivity of Cp$_2$M(R)(L)$^+$ complexes was noted in the earliest reports of these compounds (31,46). Representative insertion chemistry of the simplest Zr alkyl cation, (C$_5$H$_5$)$_2$Zr(CH$_3$)(THF)$^+$ (3), is summarized in Scheme 4. Cation 3 reacts irreversibly with CO to yield an η^2-acyl species and also inserts ketones, nitriles, and acetylenes under mild conditions (minutes, 23°C, CH$_2$Cl$_2$). Cation 3 also catalyzes ethylene polymerization and propylene oligomerization. The reaction of 3 with 2-butyne in CH$_2$Cl$_2$ solution is strongly inhibited by added THF which, given that the bis-THF complex (C$_5$H$_5$)$_2$Zr(CH$_3$)(THF)$_2$$^+$ does not form, implies that loss of THF and substrate coordination precede insertion. Intermediate ketone complexes (C$_5$H$_5$)$_2$Zr(CH$_3$)(OCR$_2$)$^+$ are observed in ketone insertions, and the nitrile complexes Cp$_2$Zr(R)(CH$_3$CN)n^+ can be isolated as noted above.

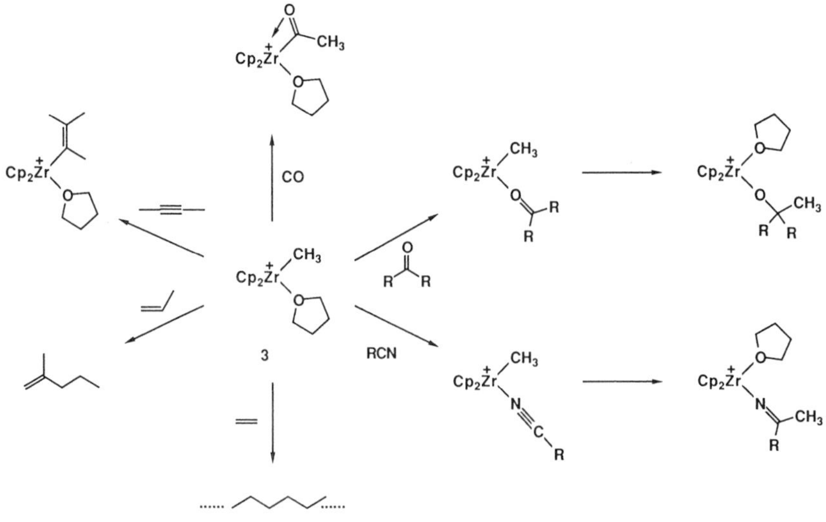

SCHEME 4

The higher alkyls $Cp'_2Zr(CH_2CH_2R)(THF)^+$ also readily insert 2-butyne, but the cationic phenyl complex $(C_5H_5)_2Zr(Ph)(THF)^+$ (14) and alkenyl complexes $Cp'_2Zr\{C(Me)=C(Me)(R)\}(THF)^+$ (15) are unreactive with this substrate. This is probably because 2-butyne does not displace THF in these cases. The observations of initiation of THF polymerization by 14 (38) and slow (NMR time scale) THF exchange for 15 (64) suggest that the THF ligands in these cases are more strongly bound and less easily displaced than in the analogous alkyls. Consistent with this interpretation, both 14 and 15 undergo rapid ligand exchange with the more nucleophilic CH_3CN and subsequent rapid CH_3CN insertion [Eq. (23)] (94). For com-

parison, neutral $Cp_2Zr(IV)$ alkyls insert CO reversibly and do not insert nitriles, ketones, or acetylenes under mild conditions (5,6,95,96). The increased reactivity of the cationic systems is due to the combination of (i) the lability of the THF ligand and (ii) the charge at Zr, which promotes coordination and activation of substrates. This concept is, of course, well developed for other metal systems (97).

The scope and mechanisms of nitrile insertions of $(C_5H_5)_2Zr(R)$-$(CH_3CN)_n^+$ and $Cp'_2Zr(R)(CH_3CN)_n^+$ complexes [which yield $Cp_2Zr\{N=C(R)CH_3\}(CH_3CN)^+$; see Scheme 4 and Eq. (23)] have been investigated recently (94). The qualitative trend in migratory aptitude is H, Ph (seconds at 23°C) ≫ CH_3 (hours at 23°C) > η^2-CH_2Ph (no reaction at 60°C). The rapid CH_3CN insertions of the hydrides are consistent with the high insertion reactivity observed for early-metal hydrides in general, which has been ascribed to (i) effective bonding of the nondirectional H $1s$ orbital in the bridged insertion transition state and (ii) operation of only minimal steric effects (98). The rapid insertion of the phenyl complexes was also ascribed to enhanced bonding in the bridged transition state, owing to participation of the phenyl π system. NMR studies show that the phenyl group in $(C_5H_5)_2Zr(Ph)(THF)^+$ either rotates rapidly or lies perpendicular to the plane between the Cp ligands. Rotation of the Ph group in the reactive $(C_5H_5)_2Zr(Ph)(CH_3CN)_n^+$ intermediate to an orientation which allows overlap of the ipso carbon p orbital with the LUMO of the coordinated CH_3CN should thus be facile. The lack of insertion for $(C_5H_5)_2Zr(\eta^2\text{-}CH_2Ph)(CH_3CN)^+$ was ascribed to the $Zr\text{—}Ph$ interaction, which prevents the migrating $Zr\text{—}CH_2$ bond from being cis to the coordinated CH_3CN.

The kinetics of CH_3CN insertions of $(C_5H_4R)_2Zr(CH_3)(CH_3CN)_n^+$ (R = H, Me) in CH_2Cl_2 solution have been studied in detail (94). NMR studies establish that the bis-CH_3CN complexes ($n = 2$, two isomers) are strongly favored over the mono-CH_3CN complexes in the presence of excess CH_3CN. Kinetic results are in accord with a mechanism involving rate-limiting insertion of the bis-CH_3CN adducts [Eq. (24)]. This contrasts

$$(24)$$

directly with the Ti system where bis-CH_3CN adducts are unimportant as shown earlier by Bochmann et al. (vide infra) (99). It could not be determined whether the symmetric bis-CH_3CN adduct **A** or the unsymmetric adduct **B** is more reactive. The CH_3CN insertion rate is faster by a factor of three for the C_5H_4Me system versus the C_5H_5 system. This was ascribed to more effective stabilization of the developing electron deficiency at the metal in the transition state leading from the five-coordinate reactant to the four-coordinate product by the stronger C_5H_4Me donor (100).

Bochmann et al. have developed the insertion chemistry of the analogous Ti systems (Scheme 5) (99). The methyl complex $(C_5H_5)_2Ti(CH_3)(CH_3CN)^+$ inserts CO and isocyanides to yield dihapto acyl and iminoacyl complexes. As noted above, the intermediate isocyanide adduct is observed in the reaction of $(C_5Me_5)_2Ti(CH_3)(THF)^+$ with 'BuNC. The nitrile complexes $(C_5H_5)_2Ti(CH_3)(RCN)^+$ and $(Ind)_2Ti(CH_3)(RCN)^+$ undergo RCN insertion under mild conditions to generate azaalkenylidene complexes. Kinetic studies are consistent with a mechanism involving rate-limiting migratory insertion to yield a three-coordinate azaalkenyli-

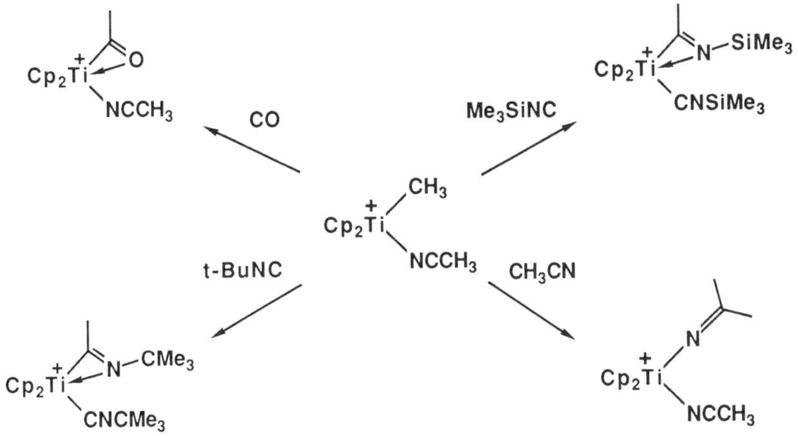

SCHEME 5

dene intermediate that is rapidly trapped by a second equivalent of nitrile as shown in Eq. (25). In general, the $(Ind)_2Ti$ complexes insert nitriles

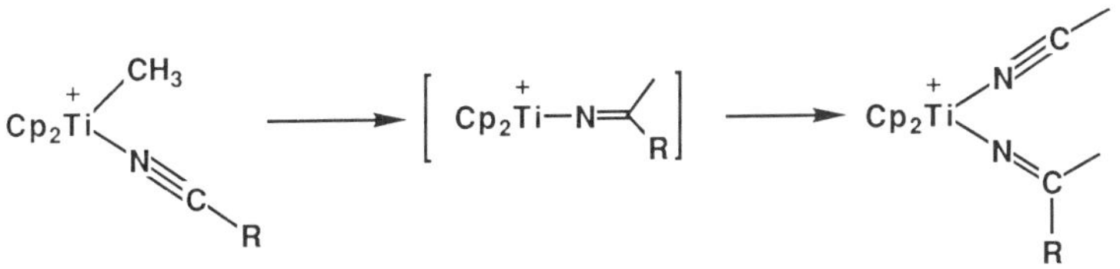

$$(25)$$

faster than the corresponding $(C_5H_5)_2Ti$ analogs. This is consistent with the faster CH_3CN insertions observed for $(C_5H_4Me)_2Zr$ versus $(C_5H_5)_2Zr$ complexes. Recent studies by Gassman establish that idenyl is a stronger donor than Cp in ruthenocene systems, and this is probably true for the group 4 systems as well (101). For comparison, neutral Cp_2TiR_2 complexes do not react with nitriles, and the d^1 Cp_2TiR complexes form nitrile adducts but do not insert (28).

The CH_3CN chemistry of $Cp_2Hf(CH_3)(L)_n^+$ complexes is quite different. The bis-CH_3CN adduct $Cp_2Hf(CH_3)(CH_3CN)_2^+$ does not undergo CH_3CN insertion at ambient temperature, but rather partially disproportionates to $Cp_2Hf(CH_3)_2$ and the dication $Cp_2Hf(CH_3CN)_3^{2+}$ as shown in Eq. (26), for which K_{eq} was determined to be 0.19 at 23°C (56). Analogous equilibria were observed previously for the cationic halide complexes $Cp_2Zr(X)(CH_3CN)_n^+$ [Eq. (27)] (43). The dications are prepared by reaction of Cp_2MI_2 with $Ag[BPh_4]$ in CH_3CN [Eq. (28)].

$$Cp_2Hf(CH_3)(CH_3CN)_2{}^+ \; \rightleftharpoons \; Cp_2Hf\overset{CH_3}{\underset{CH_3}{\diagup}} \; + \; \overset{2+}{Cp_2Hf}\overset{NCCH_3}{\underset{NCCH_3}{\diagup}}\!\!-\!NCCH_3$$

(26)

$$Cp_2\overset{+}{Hf}\overset{\diagup N \diagup \overset{CH_3}{\underset{}{}}\!\!-\!CH_3}{\underset{NCCH_3}{}}$$

$$2\; Cp_2\overset{+}{Zr}(X)(CH_3CN)_n \; \rightleftharpoons \; Cp_2Zr\overset{X}{\underset{X}{\diagup}} \; + \; \overset{2+}{Cp_2Zr}\overset{NCCH_3}{\underset{NCCH_3}{\diagup}}\!\!-\!NCCH_3$$ (27)

X = F, Cl, Br, I

$$Cp_2MI_2 \; + \; Ag^+BPh_4{}^- \; \xrightarrow{CH_3CN} \; \overset{2+}{Cp_2M}\overset{NCCH_3}{\underset{NCCH_3}{\diagup}}\!\!-\!NCCH_3$$ (28)

M = Zr, Hf

C. β-H Elimination Reactions

Schwartz's observation that $Cp_2Zr(H)(Cl)$ reacts with internal olefins to yield terminal Zr alkyls implies that secondary $Cp_2Zr(R)(Cl)$ alkyls undergo rapid β-H elimination (and reinsertion) (3,102). However, primary $(C_5H_5)_2Zr(R)Cl$ alkyls are more resistant to β-H elimination. For example, $(C_5H_5)_2Zr(R)Cl$ complexes do not react with ethylene to yield free olefin and $(C_5H_5)_2Zr(Et)Cl$, the expected products of β-H elimination and formation of $(C_5H_5)_2Zr(H)Cl$, even at elevated temperature (3,103). Neutral Sc alkyls $(C_5Me_5)_2Sc(CH_2CH_2R)$ do undergo β-H elimination at ambient temperatures (98). Structure–reactivity studies by Bercaw establish a mechanism involving rate-limiting hydride transfer with some buildup of positive charge on the β carbon in a four-center migration transition state. The β-H elimination chemistry of $Cp'_2Zr(CH_2CH_2R)(L)^+$ complexes has been studied in detail and follows the patterns established for these neutral systems (77,79,83).

The THF complexes $Cp'_2Zr(CH_2CH_2R)(THF)^+$ are rather resistant to β-H elimination. For example, $Cp'_2Zr(^nBu)(THF)^+$ reacts with vinyltrimethylsilane in THF ($t_{1/2} > 24$ hours, 23°C) to yield the stable

$Cp'_2Zr(CH_2CH_2SiMe_3)(THF)^+$ (**11**) and 1-butene, via slow β-H elimination followed by trapping by vinyltrimethylsilane [Eq. (29)]. Additional substituents on the β carbon promote β-elimination; for example, the isobutyl complex $Cp'_2Zr(CH_2CHMe_2)(THF)^+$ is unstable in the absence of isobutene above 0°C [Eq. (30)] (65). In contrast, the corresponding

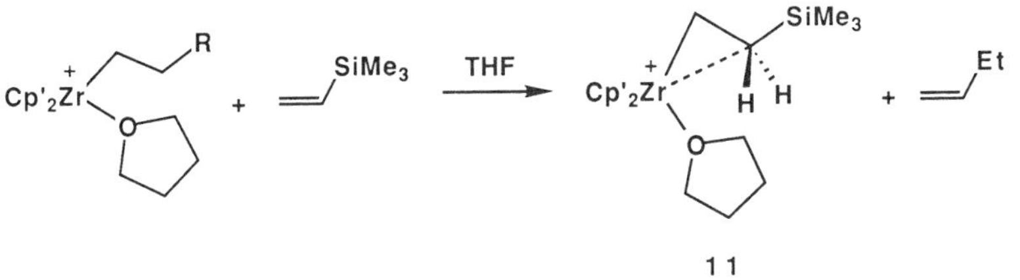

$$(29)$$

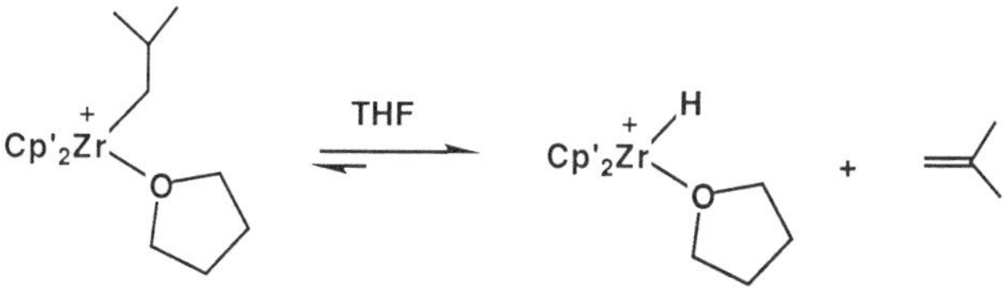

11

$$(30)$$

PMe$_3$ complexes $Cp'_2Zr(CH_2CH_2R)(PMe_3)^+$ (**6–9**) undergo much faster β-H elimination. The butyl complex $Cp'_2Zr(Bu)(PMe_3)^+$ (**7**) reacts rapidly ($t_{1/2}$ 55 min) at 25°C in the presence of vinyl TMS, yielding the β-H elimination-resistant complex $Cp'_2Zr(CH_2CH_2TMS)(PMe_3)^+$ (**12**) and 1-butene [Eq. (31)]. A similar reaction occurs in the presence of PMe$_3$, yielding $Cp'_2Zr(H)(PMe_3)_2^+$ (**16**) and 1-butene [Eq. (31)].

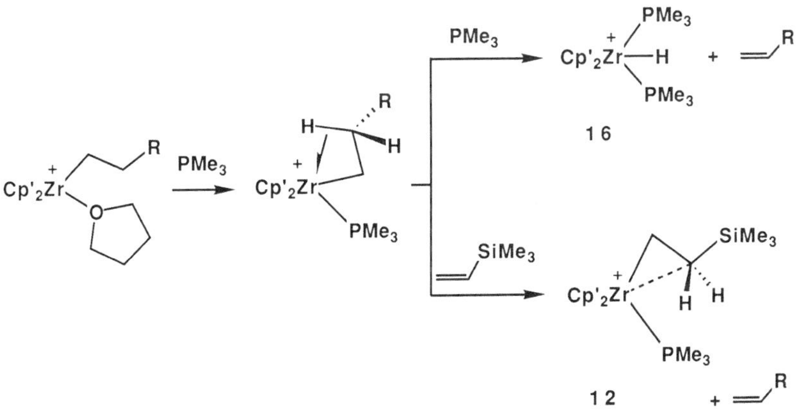

16

12

$$(31)$$

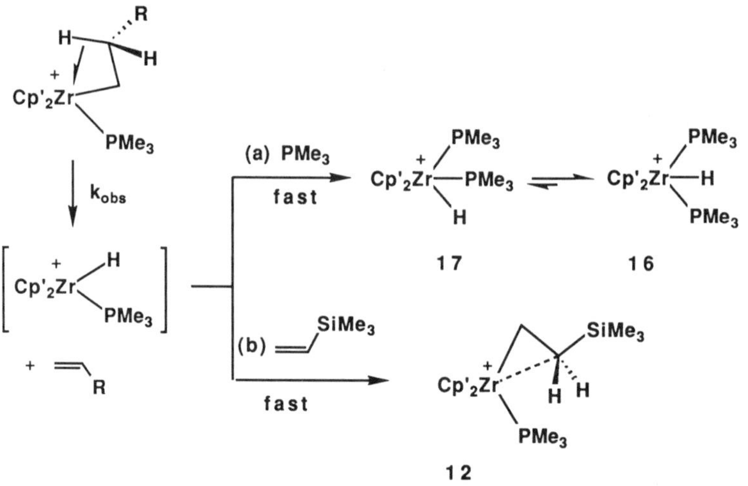

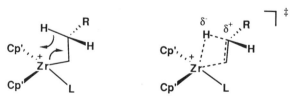

$$(32)$$

Kinetic studies show that both reactions involve rate-limiting β-H elimination followed by rapid trapping [Eq. (32)]. Key observations for **7** are as follows: (i) the rates of both reactions are identical [k_{obs} 2.1(1) $\times$ 10^{-4} second^{-1} at 25.4°C], first order in [Zr], and independent of [PMe$_3$] or [vinyl-TMS], (ii) k_{obs} of the PMe$_3$ reaction is independent of [added THF] and [added butene], (iii) a deuterium kinetic isotope effect [k_{obs}-$(Zr-C_4H_9)/k_{obs}(Zr-C_4D_9)$ 2.1(1) at 25.4°C] is observed (PMe$_3$ reaction), (iv) the activation entropy is small (+5.2 eu), and (v) no H/D scrambling in unreacted Zr(CH$_2$CHDCH$_2$CH$_3$)(PMe$_3$)$^+$ is observed. The phenethyl complex Cp$'_2$Zr(CH$_2$CH$_2$Ph)(PMe$_3$)$^+$ (**8**) undergoes an analogous but faster reaction with PMe$_3$ ($t_{1/2}$ 25 minutes, 0°C); in this case the unsymmetrical hydride Cp$'_2$Zr(H)(PMe$_3$)$_2$$^+$ (**17**) is observed as the initial product and isomerizes via PMe$_3$ dissociation and readdition to the symmetrical hydride **16** [Eq. (32)]. This same intermediate is also observed in the reaction of **7** at high [PMe$_3$] where isomerization is slow. The ethyl complex Cp$'_2$Zr(Et)(PMe$_3$)$^+$ (**6**) undergoes much slower β-H elimination ($t_{1/2}$ > 24 hours, 25°C). The observed rate trend for β-H elimination

FIG. 12.   Transition state for β-H elimination from Cp$'_2$Zr(CH$_2$CH$_2$R)(PMe$_3$)$^+$.

$(ZrCH_2CH_2Ph > ZrCH_2CH_2R > ZrCH_2CH_3)$, the kinetic isotope effect, and the activation entropy are all consistent with a somewhat polar, cyclic transition state for H^- migration of the type proposed by Bercaw for the Sc systems (Fig. 12). A near zero activation entropy is observed because this transition state differs only slightly from the agostic starting complex.

The reaction of $Cp'_2Zr(CH_2CH_2R)(THF)^+$ complexes with CH_3CN yields the labile bis-CH_3CN complexes $Cp'_2Zr(CH_2CH_2R)(CH_3CN)_2{}^+$, which undergo competitive insertion and/or β-H elimination as shown in Eq. (33) (104). The latter pathway yields the hydride

(33)

$Cp'_2Zr(H)(CH_3CN)^+$ which is rapidly and irreversibly trapped by insertion. This is confirmed by the reaction of $Cp'_2Zr(H)(THF)^+$ with CH_3CN, which proceeds by rapid ligand substitution and insertion (64). For the ethyl complex $Cp'_2Zr(CH_2CH_3)(CH_3CN)^+$, the insertion pathway dominates the β-H elimination pathway (product ratio 85/15). As above, alkyl or aryl substituents on the β carbon accelerate the β-H elimination, and for $Cp'_2Zr(CH_2CH_2R)(CH_3CN)_2{}^+$ (R = alkyl, aryl) complexes only β-H elimination products are observed. Kinetic studies are in accord with the mechanism in Eq. (33) involving insertion and/or β-H elimination from the mono-CH_3CN species, which is in (very unfavorable) equilibrium with the bis-CH_3CN complexes. Key results are the following: (i) for the ethyl case, the product ratio is independent of [CH$_3$CN], and (ii) the rate is inhibited by [CH$_3$CN]. NMR studies suggest that the mono-CH$_3$CN complexes, which can be prepared at low temperature, adopt β-agostic structures analogous to those established for Cp'$_2$Zr(CH$_2$CH$_2$R)(PMe$_3$)$^+$ complexes 6–9.

Little is known to date about ligand effects on these β-H elimination reactions. For the Cp'$_2$Zr(CH$_2$CH$_3$)(L)$^+$ series, the rate trend is L = CH$_3$CN > PMe$_3$ ≫ THF. The slow β-H elimination of the THF complexes results from O—Zr π donation which utilizes the empty Zr orbital required for H$^-$ migration (Fig. 3). This orbital is clearly available to accept the β-H$^-$ in the PMe$_3$ and CH$_3$CN complexes as these have β-agostic structures.

D. Zr—R Bond Hydrogenolysis

A wide variety of neutral, unsaturated, d^0 and $d^0 f^n$ metal alkyls undergo σ-bond metathesis reactions in which H$_2$ or substrate C—H bonds are activated (22,23,73,87,105–107). These reactions are believed to proceed via four-center transition states that are accessed by initial coordination of the H—H or C—H bonds to the electrophilic metal center [Eq. (34)]. The

$$R'H \; + \; L_nMR \; \rightleftharpoons \; \left[L_nM \begin{matrix} \diagup R' \\ \diagdown H \\ R \end{matrix} \right]^{\ddagger} \; \rightleftharpoons \; RH \; + \; L_nMR' \qquad (34)$$

requirement that a low-lying metal-centered LUMO be available for interaction with the incoming H—H or C—H bond was demonstrated by Schwartz and co-workers in studies of Cp$_2$Zr(R)(X) hydrogenolysis reactions. Hydrogenolysis of the Zr—R bond was found to be slow when X is a good π-donor ligand such as OR or Cl (5). Cationic Cp$_2$M(R)(L)$^+$ and Cp$_2$M(R)$^+$ species have been found to be highly reactive in both H—H and ligand C—H activation reactions.

The H$_2$ reactions of (C$_5$H$_5$)$_2$Zr(CH$_3$)(L)$_n$$^+$ complexes have been studied in detail (63,64). The rate depends critically on the number and donor properties of the ligands L which influence the energy and availability of a Zr-centered LUMO for H$_2$ activation.

The cationic methyl complex (C$_5$H$_5$)$_2$Zr(CH$_3$)(THF)$^+$ (3) reacts slowly with H$_2$ in THF solution ($t_{1/2}$ 21 hours, 1 atm, 23°C) to yield the insoluble hydride (C$_5$H$_5$)$_2$Zr(H)(THF)$^+$ and CH$_4$ [Eq. (35)]. Reaction with D$_2$ yields the analogous Zr—D complex and CH$_3$D. The hydride was assigned a monomeric, terminal hydride structure on the basis of the ν_{Zr-H} value of

$$Cp_2\overset{+}{Zr}\underset{O}{\overset{CH_3}{\diagup}} \xrightarrow{PMe_3} \left[Cp_2\overset{+}{Zr}\underset{PMe_3}{\overset{CH_3}{\diagup}}\right] \xrightarrow[fast]{H_2} \left[Cp_2\overset{+}{Zr}\underset{PMe_3}{\overset{H}{\diagup}}\right]$$

3

$$slow \downarrow H_2 \qquad\qquad \| PMe_3 \qquad\qquad fast \downarrow PMe_3 \qquad (35)$$

$$Cp_2\overset{+}{Zr}\underset{O}{\overset{H}{\diagup}} \qquad Cp_2\overset{+}{Zr}\underset{PMe_3}{\overset{PMe_3}{\diagdown}}-CH_3 \qquad Cp_2\overset{+}{Zr}\underset{PMe_3}{\overset{PMe_3}{\diagdown}}-H$$

18

1450 cm^{-1}, which is closer to the range exhibited by terminal (1500–1560 cm^{-1}) hydrides than that of bridging hydrides (1300–1390 cm^{-1}) in neutral Cp$_2$Zr(IV) systems (108). The related C$_5$H$_4$Me complexes Cp'$_2$Zr(R)(THF)$^+$ (R = CH$_3$, CH$_2$Ph) react at similar rate (days at $23\,^\circ$C, 1 atm, THF) to yield the soluble hydride Cp'$_2$Zr(H)(THF)$^+$, which also was assigned a terminal hydride structure on the basis of the ^{1}H NMR Zr—H chemical shift of δ 5.88. These reactions are about four times faster than the reaction of (C$_5$H$_5$)$_2$Zr(CH$_3$)$_2$ with H$_2$ under the same conditions, which yields the insoluble dimer [(C$_5$H$_5$)Zr(μ-H)(CH$_3$)]$_2$. The hydride Cp'$_2$Zr(H)(THF)$^+$ undergoes slow exchange with D$_2$ ($t_{1/2}$ 7 hours, 1 atm, $23\,^\circ$C, THF).

Reaction of **3** with excess PMe$_3$ in THF yields the isolable bis-PMe$_3$ adduct (C$_5$H$_5$)$_2$Zr(CH$_3$)(PMe$_3$)$_2$$^+$ [**18**, Eq. (35)]. The PMe$_3$ ligands of this species are quite labile, and substantial dissociation and formation of the mono-PMe$_3$ adduct (C$_5$H$_5$)$_2$Zr(CH$_3$)(PMe$_3$)$^+$ occurs in THF and CH$_2$Cl$_2$ solution. Hydrogenolysis of **18** in THF or CH$_2$Cl$_2$ is extremely rapid ($t_{1/2} < 2$ minutes, 1 atm, $23\,^\circ$C) and yields the bis-PMe$_3$ hydride (C$_5$H$_5$)$_2$Zr(H)(PMe$_3$)$_2$$^+$, which was characterized by X-ray diffraction. The related bisphosphine complex (C$_5$H$_5$)$_2$Zr(CH$_3$)(dmpe)$^+$, which contains a chelated, nonlabile dmpe ligand, does not react with H$_2$. On this basis it was concluded that the active species in the reaction of (C$_5$H$_5$)$_2$Zr(CH$_3$)(PMe$_3$)$_2$$^+$ with H$_2$ is the monoadduct, and substitution of the THF in **3** by PMe$_3$ thus increases the hydrogenolysis rate by more than a factor of 600. PMe$_2$Ph also displaces THF from **3** and accelerates hydrogenolysis in the same manner as PMe$_3$. In this case, (C$_5$H$_5$)$_2$Zr(H)(PMe$_2$Ph)$_2$$^+$ is formed. The bulkier phosphines PMePh$_2$ and PPh$_3$ do not displace THF and do not accelerate hydrogenolysis.

The trend in hydrogenolysis reactivity is shown in Fig. 13. The extremely

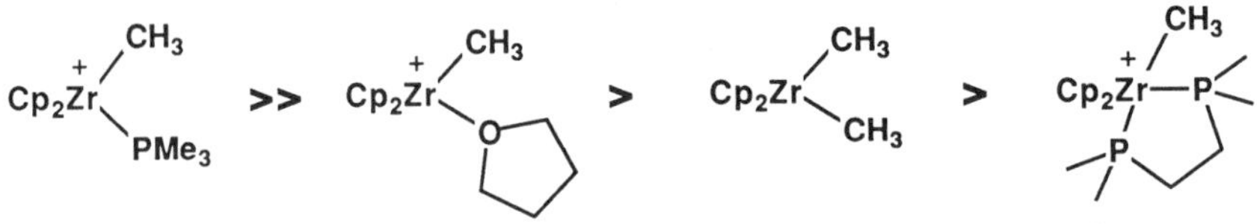

FIG. 13.   Trend in H_2 reactivity of cationic and neutral Cp_2Zr methyl complexes.

rapid hydrogenolysis of $(C_5H_5)_2Zr(CH_3)(PR_3)^+$ reflects the high electrophilicity of the cationic 16-electron Zr center which contains a low-lying Zr LUMO for H_2 activation. This LUMO is utilized for $O-Zr$ π bonding in the THF complex, and thus hydrogenolysis is strongly inhibited. Formal replacement of THF by CH_3^- in $(C_5H_5)_2Zr(CH_3)_2$ slows hydrogenolysis further owing to the substantial decrease in metal acidity accompanying conversion of a cationic species to a neutral species. Finally, the 18-electron complex $(C_5H_5)_2Zr(CH_3)(dmpe)^+$ lacks a low-lying Zr LUMO and does not react with H_2. This trend provides a general guide to the hydrogenolysis chemistry of other cationic complexes. For example, the bis-CH_3CN complex $Cp_2Zr(CH_3)(CH_3CN)_2^+$ does not react with H_2 whereas the 16-electron chelated C,N complexes $(C_5H_5)_2Zr\{\eta^2\text{-}CH_2CH(R)py\}^+$ react very rapidly. The latter reaction has been exploited in a catalytic $C-H$ activation/$C-C$ coupling reaction (109).

E. Ligand C—H Activation

Neutral d^0 $Cp_2M(R)$ complexes (M = group 3, lanthanide) undergo $C-H$ bond activation reactions with a variety of hydrocarbons (22,23,73,87,106). The first example of a $C-H$ activation reaction by a cationic $Cp_2M(R)^+$ complex was the report by Hlatky et al. of metallation of BPh_4^- by the base-free species $(C_5Me_5)_2Zr(CH_3)^+$ generated by protonolysis [Eq. (12)] (47). More systematic studies with $Cp_2M(R)(L)^+$ complexes have focused on ligand $C-H$ activations because of the anticipated difficulties in effecting reactions of hydrocarbon substrates in the presence of more reactive ligand and counterion $C-H$ bonds.

The cationic methyl complex $(C_5H_5)_2Zr(CH_3)(THF)^+$ (3) reacts with α-picoline via initial ligand substitution followed by rapid $C-H$ activation/CH_4 elimination ($t_{1/2}$ 6 minutes, 23°C, CH_2Cl_2) to yield the η^2-pyridyl complex $(C_5H_5)_2Zr(\eta^2\text{-}N,C\text{-picolyl})(THF)^+$ (19), which is isolated as a mixture of two isomers [Eq. (36)] (83, 109). Low-temperature NMR studies of the intermediate picoline adduct and the $(C_5H_4Me)_2Zr$ analog reveal a high-field ^1H resonance and a slightly reduced J_{C-H} value for the ortho $C-H$ of the coordinated picoline and suggest that a weak β-aryl-H agostic

$$
\text{(36)}
$$

interaction is present in these compounds. Complex **19** reacts with PMe_3 to yield the corresponding PMe_3 adduct; a mixture of two isomers is observed initially and is slowly converted to a single isomer. The X-ray structure of the thermodynamic isomer is shown in Fig. 14 and confirms the η^2-pyridyl bonding mode. This C—H activation reaction is similar to pyridine orthometallations by $(C_5Me_5)_2MR$ complexes (M = Lu, Sc, Y) (*23,73,87*) and $(C_5H_5)_2TiR$ (*110*) as well as to Cα—H abstractions leading to benzyne complexes (*7,111*).

Methyl cation **3** undergoes a much slower reaction (hours, 50°C, CH_2Cl_2) with pyridine itself to yield a mixture of pyridyl complexes containing coordinated THF or pyridine [Eq. (37)]. The acceleration of C—H activation by the α-methyl group of picoline is probably due to steric crowding which enhances the overlap of the ortho C—H bond with the Zr LUMO (*112*). A similar trend is observed in the $(C_5H_5)_2Ti(R)$ reactions

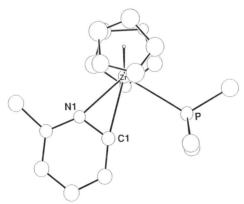

FIG. 14.   X-Ray structure of $(C_5H_5)_2Zr(\eta^2$-pyridyl)$(PMe_3)^+$ cation.

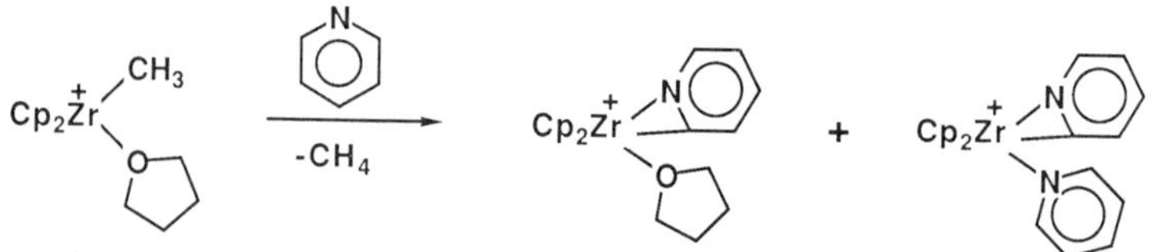

(37)

(110). The isolation of a mixture of THF and pyridine complexes in this case reflects the stronger Lewis basicity of pyridine versus α-picoline.

More extensive studies of the reactions of $(C_5H_5)_2Zr(CH_3)(THF)^+$ (**3**) with other heterocycles have provided insight into the requirements for and selectivity of ligand C—H activation *(113)*. Complex **3** reacts with a variety of pyridines and quinolines in CH_2Cl_2 solution by ortho C—H activation as shown by the representative examples in Scheme 6. These reactions all proceed via initial displacement of THF by substrate followed by C—H activation, as in Eq. (36). Exceptionally crowded or weakly basic

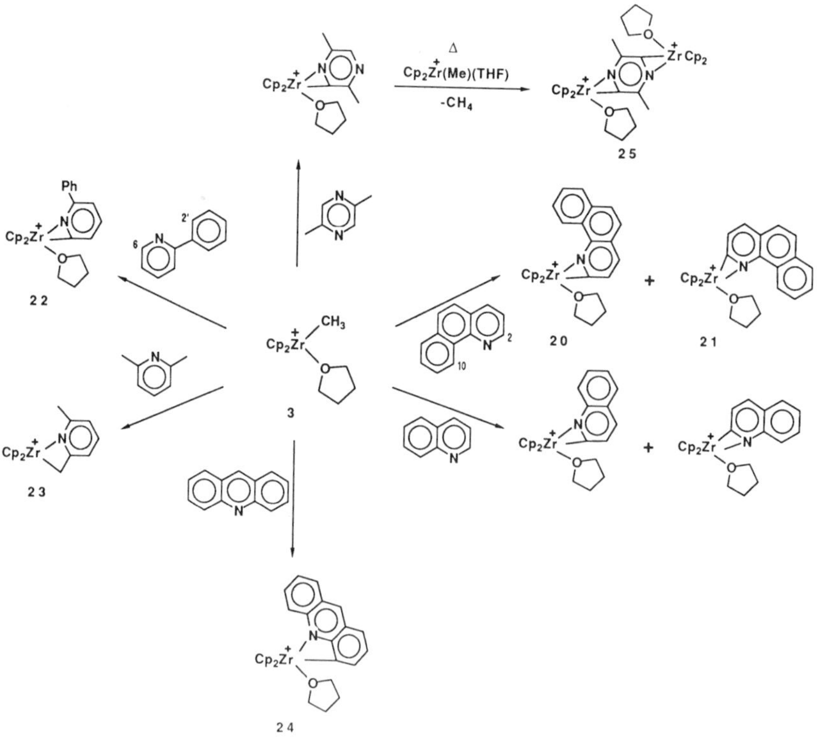

SCHEME 6

substrates such as 2-phenylquinoline or 2-methylthiophene do not displace THF from **3** and thus do not undergo C—H activation. The five-membered heterocycle N-methylimidazole and the bidentate ligand 4,4'-dimethylbipyridine form unreactive 18-electron complexes with $(C_5H_5)_2Zr(CH_3)^+$ [Eq. (38)]. In these cases the lack of a low-energy Zr LUMO for activation of a C—H bond precludes orthometallation.

$$(38)$$

Scheme 6 also illustrates a general trend of activation of ortho C—H bonds and formation of three-membered metallacycles in preference to activation of more remote alkyl or aryl C—H bonds and formation of four- or five-membered metallacycles. This selectivity contrasts directly with the general trend of middle and late transition metal systems to activate more remote C—H bonds of these ligands (114). For example, **3** reacts with 7,8-benzoquinoline to yield a mixture of isomers (**20** and **21**) resulting from C—H activation at C-2. In contrast, $Ir(PPh_3)_2$-$(CO)(CH_3CN)^+$ reacts with 7,8-benzoquinoline via C—H activation at C-10 leading to a five-membered metallacycle product (115,116). Similarly, **3** reacts with the ortho C—H bond (C-6) of 2-phenylpyridine to yield three-membered metallacycle **22**, whereas $(CH_3)Mn(CO)_5$ reacts with this substrate via activation at C-2' to yield a five-membered metallacycle (117). The selectivity for ortho C—H activation in the reactions of $Cp_2Zr(R)(L)^+$ probably reflects the shape of the metal LUMO (lobes 45° from Zr—L bond, Fig. 1) and steric factors which promote interaction of the ortho C—H bond with Zr. The activation of ortho C—H bonds of picoline [Eq. (36)] and dimethylpyrazine (Scheme 6) rather than the more acidic methyl C—H bonds also suggests that the orientation of the C—H bonds in the $Cp_2Zr(CH_3)(L)^+$ complex determines the regioselectivity of C—H activation.

When both ortho sites on the pyridine substrate are blocked, activation of more remote C—H bonds is possible in the coordination sphere of $(C_5H_5)_2Zr(CH_3)^+$ (Scheme 6). The reaction of **3** with 2,6-lutidine results in activation of the methyl C—H bonds, leading to the four-membered metallacycle $(C_5H_5)_2Zr(\eta^2$-C,N-$CH_2(6$-Me-pyridine)]^+$ (**23**), which is isolated free of coordinated THF. Similarly, the reaction of **3** with 2,3-benzoquinoline results in aryl C—H activation, producing a four-membered metallacycle **24**. Dimethylpyrazine undergoes an interesting double metallation ultimately yielding **25**.

F. Olefin Reactions

The first example of an olefin insertion by an isolable cationic $Cp_2M(R)^+$ complex was the 1976 report of Kaminsky *et al.* that the ion pair $[Cp_2ZrCH_2CH(AlEt_2)_2][Cp]$ inserts α-olefins and dienes (*30,81*). However, no information about the rates or the products of these reactions was provided. The issues of whether or not the distorted alkyl structure, the Al substituents, and/or the ion pair structure were important for reactivity were not addressed. It is interesting in a historical context to note that two closely related complexes containing coordinating Cl^- or $ClAlEt_3^-$ counterions, $[Cp_2Zr(ClAlEt_3)]_2(\mu\text{-}CH_2CH_2)$ and $Cp_2Zr\{CH_2CH(AlEt_2)_2\}Cl$, were reported to be active α-olefin polymerization catalysts, but only on addition of Al alkyls. In retrospect, this is a strong hint that cationic $Cp_2M(R)^+$ species are likely to be highly reactive with olefins.

In 1986 it was reported that CH_2Cl_2 solutions of the simple methyl complex $[Cp_2ZrCH_3)(THF)][BPh_4]$ (**3**) catalyze ethylene polymerization in the absence of Al cocatalysts or supports under mild conditions [1–4 atm, 25°C, Eq. (39)] (*32*). The activity of **3** is very low [~ 12 g polyethylene/(mmol Zr·atm·hour) at 23°C] compared to Cp_2MX_2/alumoxane systems owing to the presence of the THF ligand, which inhibits coordination and insertion of ethylene. Additionally, catalyst decomposition via reaction with solvent, which yields unreactive $Cp_2Zr(R)Cl$ species, contributes to the low activity. Addition of THF strongly inhibits polymerization, and no activity is observed in CH_3CN or THF solution at room temperature. As **3** does not coordinate a second equivalent of THF, this implies that loss of THF precedes insertion; this is presumed to occur by the dissociative process in Eq. (39). Further studies showed that the analogous benzyl complex $Cp_2Zr(CH_2Ph)(THF)^+$ also polymerizes ethylene under mild conditions. In this case, the base-free complex $Cp_2Zr(\eta^2\text{-}CH_2Ph)^+$ could be observed by low-temperature ^1H NMR, although the reaction of this species with ethylene could not be directly probed because of rapid exchange with the THF complex under polymerization conditions (*33*).

The polymer produced by **3** under the conditions noted above is essentially linear polyethylene with a molecular weight in the range of 30,000, with an M_w/M_n ratio of 2.5. End group analysis reveals approximately 25% vinyl end groups, implying that chain transfer/termination occurs by both β-H elimination and *net* Zr—R bond protonolysis processes. The latter

$$\text{Cp}_2\overset{+}{\text{Zr}} \underset{\text{O}}{\overset{\text{Me}}{\diagdown}} \qquad \text{Cp}_2\overset{+}{\text{Zr}} \underset{\text{O}}{\diagdown}^{\text{}}{}_n \longrightarrow \quad \cdots \diagup\diagdown\diagup\diagdown \cdots \tag{39}$$

$$\text{Cp}_2\overset{+}{\text{Zr}} \overset{\text{Me}}{\diagdown} \quad \xrightarrow{\;n\;=\;} \quad \text{Cp}_2\overset{+}{\text{Zr}} \diagup\diagdown{}_n \quad + \quad \text{O}$$

possibly involve adventitious H_2O, acidic species derived from the CH_3Cl^+ (or equivalent) which is produced by the reaction of $Cp_2Zr(R)^+$ with solvent, or C—H activation reactions. The insertion process in Eq. (39) is consistent with (i) the model insertion chemistry established for isolated $Cp_2Zr(R)(L)^+$ complexes, (ii) the linear polymer product, and (iii) results obtained for neutral $Cp*_2M(R)$ and Ti/Al catalyst systems (22,118).

The analogous Cp′ alkyls $Cp'_2Zr(R)(THF)^+$ catalyze ethylene polymerization in CH_2Cl_2 and slow ethylene oligomerization to 1-butene, 1-hexene, and 1-octene at 23–50°C in THF solution (64). This reaction proceeds by the standard insertion/β-elimination process shown in Scheme 7 (119). Similar activity (~0.3 turnover/hour, 50°C, 1 atm) is observed for $Cp'_2Zr(R)(THF)^+$ (R = Et, Bu) as expected. The ethyl cation does not react with 1-butene at ambient temperature, implying that α-olefin insertion is not important for chain growth. ^{1}H NMR monitoring of active solutions reveals the presence of $Cp_2Zr(R)(THF)^+$ (R = Et, Bu, He cannot be distinguished after significant α-olefin production owing to overlapping resonances) but no hydride $Cp'_2Zr(H)(THF)^+$. Thus, the catalyst resting state(s) is the alkyl stage, not the hydride stage. This is consistent with the observed, very rapid olefin insertion rates for cationic Zr hydrides and with thermodynamic considerations (120). The lower activity and product molecular weight (oligomer versus polymer) of this system in THF as compared to CH_2Cl_2 are due to the inhibition of THF dissociation and the resulting inhibition of olefin insertion.

The results discussed above for $Cp_2Zr(R)(THF)^+$ complexes are significant in that they represent the first cases in which simple group 4 metallocene alkyl complexes react directly with olefins. Much higher olefin reactiv-

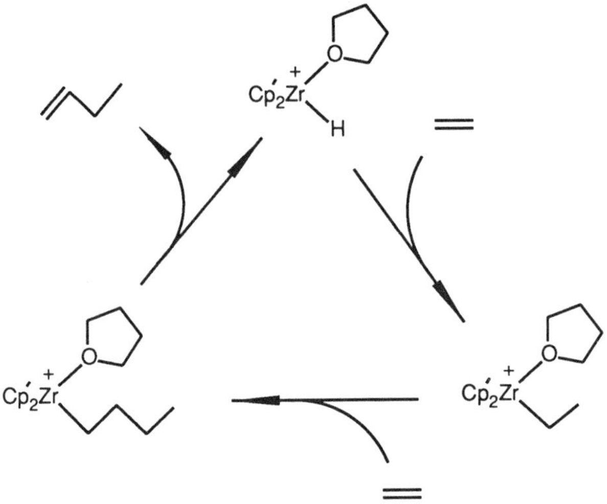

SCHEME 7

ity is observed for base-free cations and for complexes with ligands which are more labile than THF. Turner and Hlatky have studied the ethylene polymerization chemistry of the zwitterion $Cp^*_2Zr[2\text{-Me},5\text{-}B(C_5H_4Me)_3Ph]$ and the ion-paired system $Cp^*_2Zr(Me)(C_2B_9H_{12})$ (47). Both produce linear polyethylene with high activity in toluene solution. Reported activities are 375 g polyethylene/(mmol $Zr \cdot atm \cdot hour$) at 90 psig, 90°C, and 265 g polyethylene/(mmol $Zr \cdot atm \cdot hour$) at 240 psig, 40°C, respectively. Teuben's tetrahydrothiophene complexes $[Cp^*_2M\text{-}(CH_3)(THT)][BPh_4]$ (M = Zr, Hf) are also quite active for ethylene polymerization. In PhOMe or $PhNMe_2$ solvents, both complexes exhibit activities of approximately 170 g polyethylene/(mmol $M \cdot atm \cdot hour$) at 25°C, 1 atm, and yield linear polyethylene. No activity is observed in THT, and these catalysts are not stable in chlorinated solvents. The activities of the Turner–Hlatky complexes and the THT complexes are competitive with those reported for $Cp^*_2ZrX_2$/alumoxane catalysts (121,122).

The ethylene chemistry of isolable $Cp_2Ti(CH_3)(L)^+$ complexes is similar to that of the Zr systems. Complexes with relatively labile ligands (L = THF, Et_2O, EtOPh) exhibit low polymerization activity in CH_2Cl_2 solution. For the THF complex, Taube reported an activity of about 3 g polyethylene/(mmol $Ti \cdot atm \cdot hour$) (CH_2Cl_2 solution, 25°C, 1 atm) (52). The corresponding pyridine, RCN, and PR_3 complexes, which are less labile, are inactive under these conditions (39). Surprisingly, the pentamethylcyclopentadienyl systems $Cp^*_2Ti(CH_3)(L)^+$ (L = THF, THT) do not polymerize ethylene despite the fact that the THF and THT ligands are quite labile (53,58). This is most likely due to the increased steric crowding and decreased Lewis acidity at the metal, which inhibit monomer coordination.

Bochmann *et al.* have studied the ethylene polymerization reactions of *in situ* generated base-free $Cp_2M(CH_3)^+$ species (M = Ti, Zr) by ^1H NMR (51). The indenyl complex $(Ind)_2Ti(CH_3)^+$ polymerizes ethylene at $-40°C$ in CH_2Cl_2 solution. The polymer produced under these conditions has a narrow molecular weight distribution (M_w/M_n 2.0), characteristic of a single-sited catalyst. The activity varies in the order $(Ind)_2Ti(CH_3)^+ > (C_5H_5)_2Ti(CH_3)^+$, which is the same order observed for the CH_3CN insertions of $Cp_2Ti(CH_3)(CH_3CN)^+$ complexes.

Turner and co-workers and Bochmann *et al.* have reported that propylene is polymerized by *in situ* generated base-free cations $Cp_2M(R)^+$ (49,51). In contrast, $Cp*_2M(CH_3)(THT)^+$ (M = Zr, Hf) complexes catalyze propylene oligomerization under mild conditions ($PhNMe_2$, 25°C, 1 atm) by an insertion/β-CH_3 elimination process (Scheme 8) (53). The conventional β-H elimination chain transfer process is disfavored by severe steric interactions between the $C\beta$-substituents and the bulky Cp* ligands in the transition state (Scheme 8). The Hf system produces only C_6 and C_9 products whereas the Zr system produces oligomers up to C_{24} under these conditions.

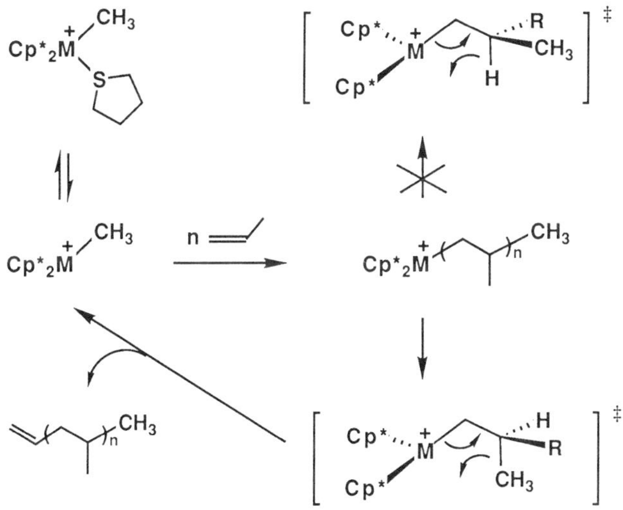

SCHEME 8

VI

APPLICATIONS

A. Cationic Alkyl Complexes in Ziegler–Natta Catalysis

Early studies of soluble Cp_2TiX_2/AlR_nX_{3-n} Ziegler–Natta catalysts were noted in Section II. The 1980s have witnessed a resurgence of interest in Cp_2MX_2-based catalysts as a result of the development by Kaminsky *et al.* of high activity Cp_2MX_2/alumoxane catalysts (*20,121,123*). These systems consist of a Cp_2MX_2 complex and a (typically) large excess of methylalumoxane, which is a mixture of oligomeric $[Al(CH_3)(\mu\text{-}O)]_n$ species prepared by controlled hydrolysis of $AlMe_3$. Notable features of these catalysts include high activity for ethylene and α-olefin polymerizations, high activity for Zr and Hf systems which are relatively stable to reduction, stereoselective α-olefin polymerization to isotactic or syndiotactic polymers with chiral or otherwise structurally tailored Cp_2MX_2 complexes (*124–126*), good H_2 response (i.e., significant molecular weight reduction on addition of H_2) (*127*), conversion of a high percentage of Cp_2MX_2 components to active species (*128*), and high activity in other reactions, including olefin hydrooligomerization and hydrogenation and diene cyclization/polymerization (*129–131*). A more detailed discussion of these catalyst systems is beyond the scope of this review.

As mentioned in Section II, one of the principal reasons for studying the chemistry of $Cp_2M(R)^+$ and $Cp_2M(R)(L)^+$ complexes has been to provide a reactivity data base for assessing the role that these ions may play in Cp_2MX_2/alumoxane catalysts and other Cp_2MX_2-based Ziegler–Natta catalysts. There is now strong circumstantial and some direct evidence that cationic alkyl complexes $Cp_2M(R)^+$ are in fact the active species in these catalyst systems.

The organometallic chemistry discussed in Sections III, IV, and V establishes that three-coordinate, d^0 $Cp_2M(R)^+$ cations, and sufficiently labile $Cp_2M(R)(L)^+$ complexes, undergo all of the key reactions involved in olefin polymerization, including olefin insertion, β-H and β-CH_3 elimination, Zr—R bond hydrogenolysis, and certain types of ligand C—H activation reactions. The observation of facile olefin and acetylene insertions for $Cp_2M(R)^+$ species [formed by direct synthesis or ligand dissociation from $Cp_2M(R)(L)^+$] and polymerization activities of base-free $Cp^*_2Zr(R)^+$ species approaching those of alumoxane catalysts are particularly notable. The chemistry of these cationic species parallels that of neutral group 3 and lanthanide $Cp^*_2M(R)$ complexes. In contrast, d^0 $Cp_2M(R)X$, d^1 $Cp_2M(R)$, and d^2 $Cp_2M(R)(olefin)$ complexes are all unreactive toward olefin insertion ($3,28,29$). This strongly implies that a d^0, three-coordinate species is required for facile olefin insertion chemistry in Cp_2M systems. That the Cp_2M fragment retains its integrity in complex Cp_2MX_2-based catalysts is implied by extensive studies of stereocontrolled α-olefin polymerizations by (EBTHI)MX_2/alumoxane catalysts and related metallocene catalysts ($124,125,132$). The unique reactivity of d^0 $Cp_2M(R)^+$ and $Cp_2M(R)$ species can be traced to the high Lewis acidity and coordinative unsaturation of the three-coordinate, 14-electron metal center which promotes coordination and activation of olefins. Four-coordinate neutral $Cp_2M(R)(X)$ species are insufficiently electrophilic to coordinate/activate olefins, whereas backbonding stabilizes d^2 $Cp_2M(R)$(olefin) adducts and inhibits insertion.

There are three lines of more direct evidence which support the proposed role of $Cp_2M(R)^+$ ions as active species. Ewen has compared the α-olefin polymerization chemistry of (EBTHI)MX_2/alumoxane, $Me_2C(Cp)$(fluorenyl)MX_2/alumoxane, and related catalysts containing modified Cp_2MX_2 metallocene components with that of cationic $Cp_2M(R)^+$ species (containing the same Cp_2M frameworks) generated *in situ* from $Cp_2M(CH_3)_2$ using Turner's $[HNR_3][B(C_6F_5)_4]$ reagent ($49,133$). The alumoxane systems and the cation systems exhibit comparable propylene polymerization activities and nearly identical polymer microstructures (type and frequency of regiochemical/stereochemical errors) for both isotactic and syndiotactic polypropylenes. As the microstructures are extremely sensitive to the active site structure, these observations strongly imply that the active species in both systems are the same.

Gassman and Callstrom have studied Cp_2ZrX_2/alumoxane catalysts by X-ray photoelectron spectroscopy (XPS) (134). XPS $Zr(3d_{5/2})$ core binding energies (CBEs) provide a direct and sensitive probe of the electronic environment at Zr, with higher binding energies indicating more electron-deficient species. Reaction of Cp_2ZrCl_2 (CBE 181.7 eV) with methylalumoxane (MAO) yields a new species with higher binding energy (CBE 182.4 eV, ΔCBE = 0.7 eV). Reaction of $Cp_2Zr(CH_3)Cl$ or $Cp_2Zr(CH_3)_2$ with MAO yields a product with the same CBE, suggesting that the product is the same for all three cases. Studies of model compounds establish that $Cp_2Zr(R)(OR)$, $Cp_2Zr(R)(H)$, $Cp_2Zr\{CHR(AlR_2)\}Cl$, and related neutral species which might be expected to form on reaction of Cp_2ZrX_2 with MAO should all exhibit *lower* CBE values (< 181.7 eV). On the other hand, comparison of $Ti(2p_{3/2})$ CBEs of Cp_2TiCl_2 (457.1 eV) and the Eisch cation $Cp_2Ti\{C(SiMe_3) = C(Me)(Ph)\}^+$ (457.6) shows that conversion of Cp_2MX_2 to a three-coordinate $Cp_2M(R)^+$ cation results in substantial *increase* in metal CBE. On this basis it was proposed that the species with increased CBE formed from Cp_2ZrX_2 and MAO is $Cp_2Zr(CH_3)^+$ (Fig. 15).

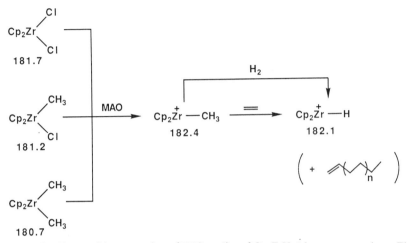

FIG. 15. Proposed interpretation of XPS studies of Cp_2ZrX_2/alumoxane catalysts. The numbers are measured core binding energies (CBEs), and the cationic methyl and hydride complexes are proposed species.

Addition of ethylene to this first-formed species results in polymerization as expected and, interestingly, leads to a decrease in Zr CBE to 182.1 eV. On the basis of CBE values for neutral Zr hydrides, and the observation that reaction of the first-formed species with H_2 also gave a product with a CBE value of 182.1 eV, it was suggested that the second species is the hydride cation $Cp_2Zr(H)^+$, formed by β-H elimination of $Cp_2Zr\{-(CH_2CH_2)_n-CH_3\}^+$ alkyl species formed by ethylene insertion. It is also possible that the lowering of CBE values upon ethylene addition to "$Cp_2Zr(CH_3)^+$" is due to formation of higher alkyl species $Cp_2Zr\{-(CH_2CH_2)_n-CH_3\}^+$ with agostic structures (83). The combination of XPS studies of actual catalysts *and* model compounds promises to be very useful approach to active site/species identification for complex catalysts.

Marks has studied supported catalysts of the type $Cp_2M(CH_3)_2$/dehydroxylated alumina and $Cp_2M(CH_3)_2$/$MgCl_2$ (M = actinide, Zr) by solid-state CP-MAS ^{13}C NMR and has found strong evidence for CH_3^- transfer to the support, concomitant formation of surface $Cp_2M(CH_3)^+$ species, and subsequent olefin insertion of these species (27,135,136). Reaction of $Cp^*_2Th(CH_3)_2$ with anhydrous $MgCl_2$ yields a supported ethylene polymerization catalyst. Solid-state ^{13}C NMR spectra of this material exhibit a high-field resonance at $\delta -8.0$, assigned to a $Mg-CH_3$ group derived from CH_3^- transfer to the support, and a $Th-CH_3$ resonance at $\delta 69.0$ of approximately equal intensity. When the catalyst is exposed to ethylene, the Th—CH_3 resonance decreases and resonances for polyethylene grow in, but the $\underline{C}_5Me_5$ and Mg—CH_3 resonances remain unchanged, indicating that ethylene insertion into the Th—CH_3 bond occurs and that approximately 50% of the surface Th species are active. The Th—CH_3 species cannot be conclusively identified on the basis of the Th—CH_3 shift as the observed value is in the range for both cationic compounds $[Cp^*_2Th(CH_3)(THF)_2][BPh_4]$ and $[Cp^*_2Th(CH_3)][BPh]$ and neutral $Cp^*_2Th(CH_3)_2$ and $Cp^*_2Th(CH_3)Cl$. However, as the cationic complexes are active ethylene polymerization catalysts whereas the neutral complexes are unreactive, it was concluded that the surface species is a $Cp^*_2Th(CH_3)^+$ complex which interacts with the surface by coulombic forces or via weak μ-Cl or μ-CH_3 bridges. Similar conclusions were reached for alumina-supported $Cp_2M(CH_3)_2$ catalysts, although a much lower fraction of the surface species are active.

These studies of alumoxane-activated and $MgCl_2$- and alumina-supported Cp_2MX_2 catalysts show that the MAO cocatalyst and the supports are capable of alkylating Cp_2MX_2 and/or abstracting and sequestering an anionic (halide or alkyl) ligand to generate reactive $Cp_2M(R)^+$ species. Presumably, bulky alumoxane species and alkylated oxide/halide surfaces function as non- or weakly coordinating anions for these cations. Thus, the simplified mechanism in Eq. (40) summarizes the key features of these Cp_2MX_2-based catalyst systems.

$$Cp_2ZrX_2 + \text{Al cocatalyst or support} \longrightarrow Cp_2\overset{+}{Z}rR \overset{=}{\longrightarrow} Cp_2\overset{+}{Z}r \; \diagup\!\!\!\diagdown\!\!\!\diagup\{\text{---}\}_n R \quad (40)$$

further chain
growth by insertion

chain transfer
by β-H elimination,
Zr-C bond hydrogenolysis,
C-H activation,
Zr/Al transmetallation,
etc...

The charge of the $Cp_2M(R)^+$ species is probably not the critical feature for olefin polymerization activity. Rather, the high polymerization activities observed for neutral group 3 and lanthanide $Cp^*_2M(R)$ complexes suggest that it is the high coordinative unsaturation of the 14-electron three-coordinate metal center which is important. Electron-counting constraints simply dictate that for a group 4 metal, a d^0 $Cp_2M(R)$ fragment must be cationic. Consistent with this view, recent work shows that neutral, mixed-ring metallocene alkyls Cp*(dicarbollide)M(R) (M = Zr, Hf), which are related to $Cp_2M(R)^+$ by formal replacement of a uninegative Cp^- by the isolobal, dinegative $(\eta^5\text{-}C_2B_9H_{11})^{2-}$ (dicarbollide) ligand, are also very active ethylene polymerization catalysts (*137*).

B. *Carbon–Carbon Coupling Chemistry of Cationic Metallacycles*

Buchwald and others have developed a rich insertion/C—C coupling chemistry for $Cp_2Zr(\text{benzyne})(PMe_3)$, $Cp_2Zr\{\eta^2\text{-}(N,C)N(SiMe_3)=CH(R)\}(THF)$, and related neutral three-membered zirconocene metallacycles (*7,138,139*). For example, $Cp_2Zr(\text{benzyne})$, generated *in situ* from the corresponding PMe_3 complex, inserts acetylenes, nitriles, and ketones to yield expanded metallacycles. The cationic azazirconacycle $Cp_2Zr(\eta^2\text{-}\text{picolyl})(THF)^+$ (**19**) produced in Eq. (36), and related complexes formed by ligand C—H activation in Eq. (37) and Scheme 6 are isoelectronic with $Cp_2Zr(\text{benzyne})(L)$ and exhibit a rich parallel insertion/C—C coupling chemistry (*83,109,140*).

Representative insertion/C—C coupling reactions of **19** are shown in Scheme 9. This metallacycle reacts with olefins and acetylenes to give stable five-membered metallacycles. These reactions are strongly inhibited by THF, indicating that THF dissociation precedes insertion as shown in Scheme 9. The corresponding PMe_3 complex, $Cp_2Zr(\eta^2\text{-}\text{picolyl})(PMe_3)^+$, does not react with these substrates because the PMe_3 ligand is not labile. The 1,2-regiochemistry of the α-olefin insertions is analogous to that observed for $Cp'_2Zr(H)(THF)^+$ and is rationalized on electronic and steric grounds. A 1,2-insertion yields a primary rather than a less stable secondary Zr alkyl, and Cp/olefin substituent steric interactions are minimized. Complex **19** undergoes a single insertion of ethylene yielding **26** rather than multiple insertions (i.e., polymerization) as observed for simple $Cp_2Zr(R)(THF)^+$ alkyls (Section V,F). The lack of further ethylene insertion by **26** reflects the stability of the five-membered chelate ring which does not open to allow ethylene coordination. The X-ray structure of **26** reveals that the ring is unstrained, and NMR experiments establish that the ring is not opened by THF. Evidently, the C—N—Zr angle in **26** (77.3°) is large enough to prevent direct coordination/insertion of ethylene without ring opening. Interestingly, **19** reacts with styrene by 2,1-insertion to yield the α-substituted metallacycle **27**. In this case, the electronic stabili-

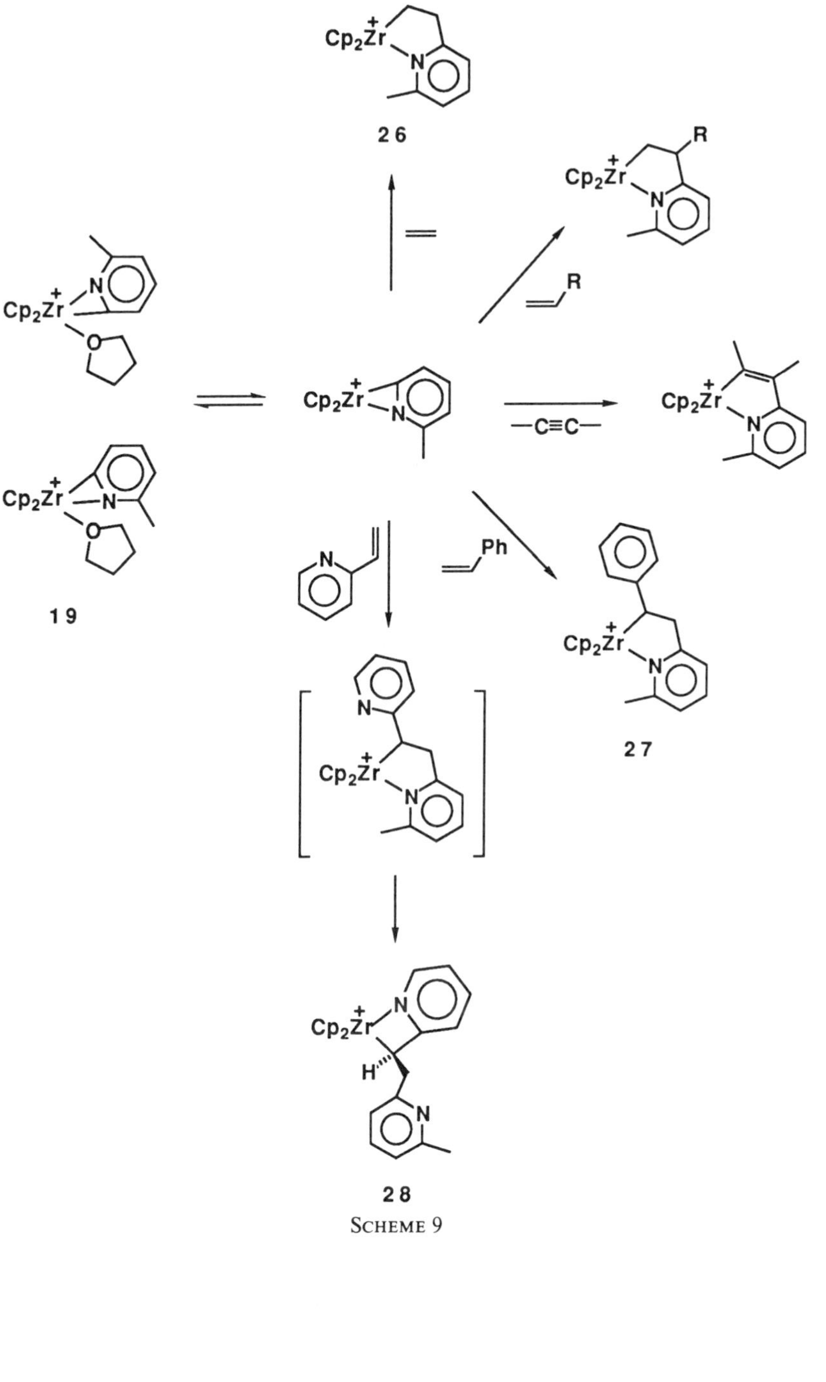

SCHEME 9

zation of the electron-rich Zr—C α carbon by the aryl substituent dominates the unfavorable Cp/aryl steric interactions. There are several cases of
similar styrene reactions with neutral Cp_2Zr hydrides (141). Vinylpyridine
reacts in a similar fashion with 19, but in this case the initial product
isomerizes rapidly to the four-membered metallacycle 28. The driving
force for this intramolecular ligand exchange is the greater binding strength
of the less crowded monosubstituted versus the disubstituted pyridine.

To probe the role of steric and electronic effects in more detail, the
reactions of $Cp_2Zr(\eta^2\text{-}6\text{-phenylpyridyl})(THF)^+$ (22, Scheme 6) with sily-
lated olefins were studied (142). As shown in Eq. (41), 22 reacts with

(41)

allyltrimethylsilane via normal 1,2-insertion yielding the β-substituted metallacycle 29. In contrast, 22 reacts with vinyltrimethylsilane via 2,1-inser-
tion to yield the sterically disfavored secondary Zr alkyl 30. As the alternate regioisomers were not observed when these reactions were monitored
by ^{1}H NMR, and no isomerization occurs under conditions where these
reactions are reversible (65 and 110°C, respectively), it was suggested that
these products are both kinetically and thermodynamically favored. Complex 30 is thermodynamically more stable than its regioisomer with β-
$SiMe_3$ substitution owing to stabilization of the Zr^+ center and the electron-rich Zr-alkyl carbon by Si, which is β and α to these centers,
respectively (143). Additionally, the regioselectivity in Eq. (41) can be
rationalized on the basis of polar insertion transition states (Fig. 16) which
are analogous to those proposed by Bercaw and co-workers for olefin
insertions into early-metal M—H and M—R bonds (98). These reactions,

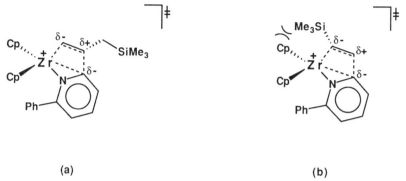

(a)   (b)

FIG. 16. Transition states for insertion/C—C coupling of allyltrimethylsilane (a) and vinyltrimethylsilane (b) with $Cp_2Zr(\eta^2$-6-phenylpyridyl)$^+$.

and the reactions of styrene and vinylpyridine in Scheme 9, show that electronic effects in these insertions are quite strong.

In contrast to the rich insertion chemistry developed for neutral three-membered Zr metallacycles, the insertion chemistry of four-membered metallacycles such as $Cp_2Zr\{(CH_2)_2SiMe_2\}$ and $Cp_2Zr\{(CH_2)_2CH_2\}$ is limited to polar substrates (144–147). However, the cationic four-membered azazirconacycle $Cp_2Zr(\eta^2$-$CH_2(6$-methylpyridyl))$^+$ (23 in Scheme 6), derived from C—H activation of a lutidine CH_3 group by $Cp_2Zr(CH_3)(THF)^+$, is highly reactive (148). Representative chemistry of 23, generated in situ from the corresponding THF complex, is shown in Scheme 10. Ethylene and α-olefins insert at 55°C in $ClCH_2CH_2Cl$, whereas more reactive substrates including acetylenes and carbonyl compounds react at room temperature. The six-membered metallacycles are stable toward further insertions. The higher reactivity of 23 compared to $Cp_2Zr\{(CH_2)_2CH_2\}$ and related neutral four-membered metallacycles may be traced to the cationic charge which promotes substrate coordination and activation. Additionally, in some cases ring opening via cleavage of the dative Zr—N bond may be important.

The alkylpyridine ligands of these insertion products can be removed from Zr by normal hydrolysis or other electrophilic Zr—R bond cleavage reactions developed for $Cp_2Zr(R)Cl$ compounds (3). The five-membered metallacycles are quite resistant to β-H elimination. The β hydrogens cannot attain the correct orientation for transfer to Zr owing to the chelated structures, and, as for other Zr alkyls, β-H elimination is likely to be endothermic in any case (120). However, β-H elimination can be induced by ligands which open the chelate rings and can trap the cationic Zr

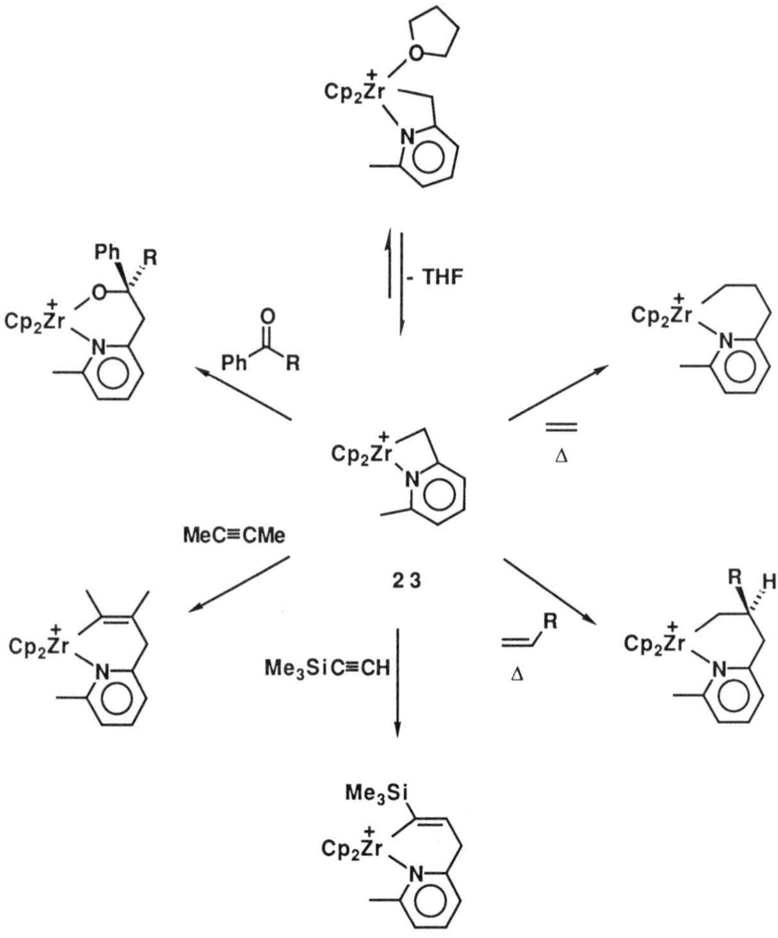

SCHEME 10

hydride product. For example, CH_3CN and PMe_3 induce β-H elimination from $Cp_2Zr\{\eta^2\text{-}CH_2CH(R)py\}^+$ via the mechanisms shown in Eq. (42), which are analogous to reactions of simple cationic alkyls discussed in Section V,C. Overall, the reaction schemes in Eq. (42) represent high-yield Zr-mediated alkenylations of α-picoline by olefins.

C. Catalytic, Productive Ligand C—H Activation

The C—H activation/C—C coupling chemistry described above is stoichiometric because the hydrolysis and β-H elimination reactions used to

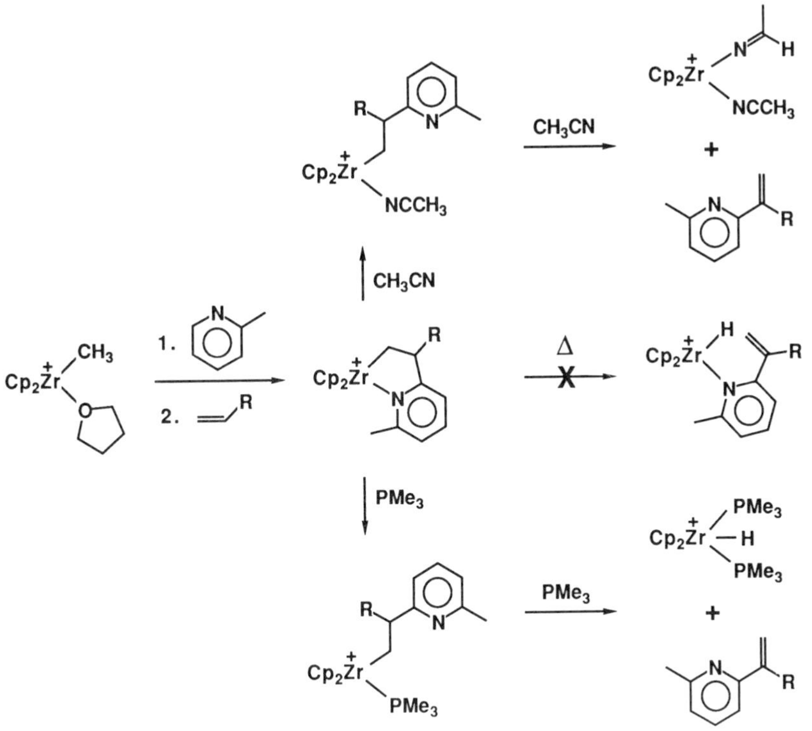

(42)

cleave the elaborated pyridine from Zr produce unreactive Zr species. For example, the bis-PMe$_3$ hydride product in Eq. (42) is unreactive with olefins because the PMe$_3$ ligands are not labile, and the azaalkenylidene product of CH$_3$CN-induced β-H elimination in Eq. (42) is unreactive owing to strong Zr—N π bonding. However, Zr—R bond hydrogenolysis provides a mild way to remove the elaborated pyridines from these systems which can be incorporated into catalytic processes (109,149).

Olefins and ortho-substituted pyridines are catalytically coupled by Cp$_2$Zr(R)$^+$ in the presence of H$_2$ according to Eq (43). This remarkable

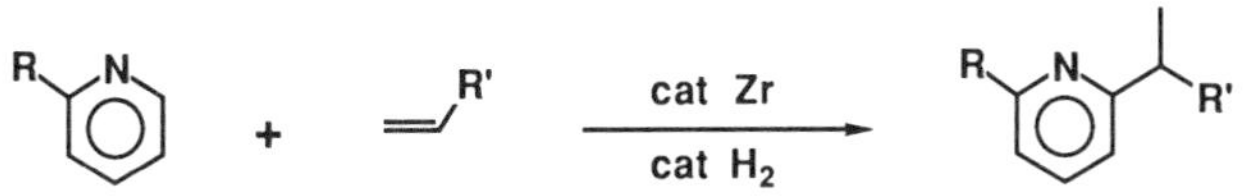

(43)

reaction is highly selective and proceeds at a moderate rate under very mild conditions. For example, propene and α-picoline are catalytically coupled by Cp$_2$Zr{η^2-CH$_2$CH(Me)(6-methylpyridyl)}$^+$ (31) to 2-methyl-6-isopropylpyridine at a rate of 2 turnovers/hour at 23°C, 2 atm H_2; at 50°C in $ClCH_2CH_2Cl$, a rate of 15 turnovers/hour is observed. As many as 50 total turnovers have been observed at 23°C. Studies of stoichiometric model reactions, NMR monitoring experiments, and activity versus substrate concentration/pressure studies have established that this reaction proceeds by the mechanism in Scheme 11 (*109*). In Scheme 11, hydride pyridine complex **33** undergoes orthometallation/H_2 elimination to yield η^2-pyridyl species **37**, which exists in equilibrium with 18-electron complex **34**. Olefin insertion of **37** yields metallacycle **31**, which, as expected for a 16-electron cationic alkyl, reacts with H_2, yielding hydride **32**. This species contains a very bulky, disubstituted pyridine ligand which is rapidly displaced by the monosubstituted pyridine substrate, yielding product and hydride **33**, and completing the cycle. The reaction is catalytic in H_2, and activity is indeed observed at low H_2 pressures.

For the $(C_5H_5)_2Zr$-catalyzed coupling of propene and α-picoline, the turnover-limiting steps are the insertion (**34**→**31**), and the hydrogenolysis steps (**31**→**32**), and only **34** and **31** are observed in active solutions by ^1H NMR. Catalysis is inhibited by added picoline, owing to inhibition of picoline dissociation from **34**, but is promoted by added propene and H_2. Thus, highest activities are observed at low picoline concentrations and higher propene and H_2 pressures. Catalysis can be initiated by either **34** or

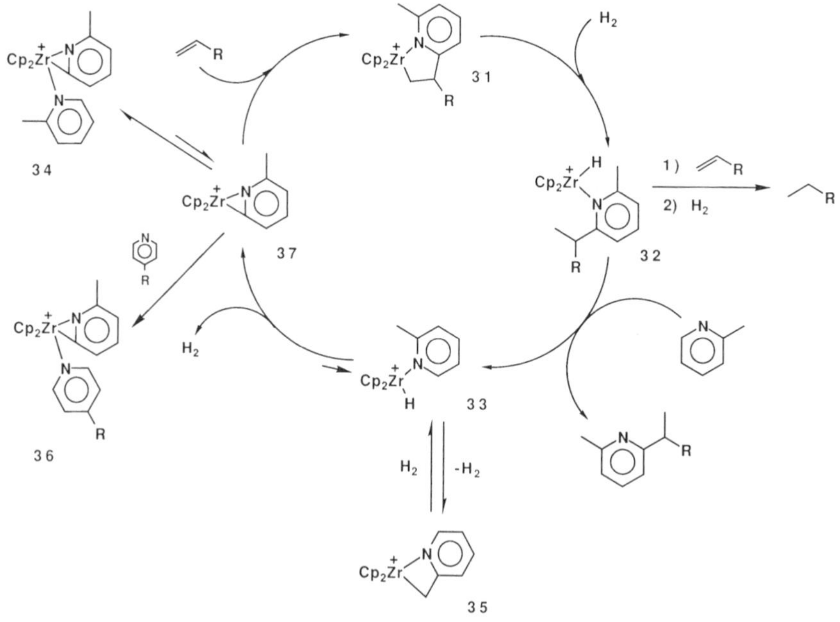

SCHEME 11

31, which are isolable compounds. Alternatively, **34** may be generated *in situ* by reaction of $Cp_2Zr(CH_3)_2$ with $[HNBu_3][BPh_4]$ in the presence of picoline.

Several side reactions are also included in Scheme 11. Olefin hydrogenation is observed as a minor side reaction and most likely occurs by olefin insertion reactions of hydrides **32** or **33**, followed by Zr—C bond hydrogenolysis. In the α-picoline/propene reaction, 10 mol % propane/2-methyl-6-isopropylpyridine is observed. On the other hand, styrene is completely hydrogenated to ethylbenzene, and no coupling with α-picoline is observed. Labeling experiments establish that intermediate **33** undergoes reversible methyl group C—H activation, although products derived from insertion reactions of the resulting four-membered metallacycle **35** are not observed.

The steric properties of the pyridine substrate are critical. Pyridines which lack ortho substituents form nonlabile, unreactive 18-electron pyridyl pyridine complexes **36** [see also Eq. (37)]. In fact, one of the principal catalyst deactivation processes in α-picoline coupling reactions is catalyst poisoning via formation of **36** by trace amounts of 3- and 4-methylpyridine impurities in the α-picoline feed.

The combination of the ligand C—H activation chemistry in Section V,E and the insertion/C—C coupling chemistry in Section VI,B provides a powerful approach to productive, C—H activation chemistry. Three features of this chemistry are notable. First, the products of the C—H activation step are highly reactive in further chemistry, so that productive reaction schemes are possible. This is a general feature of d^0 systems which activate C—H bonds by σ-bond metathesis mechanisms. In contrast, 18-electron complexes $L_nM(R)(H)$ produced by C—H activation of 16-electron L_nM species (generated thermally or photochemically) are typically resistant to insertion and other reactions (*150–153*). Second, the extent of olefin insertion is controlled (i.e., single rather than multiple insertion/polymerization occurs) by manipulation of the ring size of the metallacycles: three- and four-membered azametallacycles are highly reactive but four- and five-membered metallacycles are not. Finally, catalytic chemistry is possible when Zr—R bond hydrogenolysis is used for product removal from the metal.

VII

SUMMARY

Cationic, d^0 $Cp_2M(R)(L)^+$ and base-free $Cp_2M(R)^+$ complexes are easily prepared from readily available $Cp_2M(R)_2$ compounds. These electrophilic alkyl complexes exhibit a rich ligand exchange, insertion, β-H elimination, and σ-bond metathesis chemistry. In general, $Cp_2M(R)(L)^+$ complexes are more reactive than neutral $Cp_2M(R)(X)$ or $Cp_2M(R)_2$ compounds as a result of their increased unsaturation and their charge, and their chemistry resembles that of d^0 $Cp^*_2M(R)$ (M = group 3, lanthanide) complexes. There is strong evidence that $Cp_2M(R)^+$ ions are the active species in Cp_2MX_2-based Ziegler–Natta olefin polymerization catalysts. Isolable $Cp_2M(R)(L)^+$ complexes catalyze olefin polymerization and also undergo other, more controlled, C—C coupling reactions which are of interest for organic synthesis.

ACKNOWLEDGMENTS

The author is grateful to the co-workers listed in the references for their creative and dedicated efforts, and to Sam Borkowsky, Dr. Tom Gardner, Dr. Anil Guram, Dr. Garry Hinch, Karen Jordan, and Dr. Yun Wang for assistance in the preparation of this article. Our work in this area was funded by National Science Foundation Grant CHE-8816445 and U.S. Department of Energy Grant DE-FG02-88ER13935, and was assisted by a Sloan Fellowship and a Union Carbide Research Innovation Award to R.F.I.

REFERENCES

1. D. J. Cardin, M. F. Lappert, and C. L. Raston, "Chemistry of Organo-Zirconium and -Hafnium Compounds." Ellis Horwood, Ltd., West Sussex, England, 1986.
2. P. T. Wolczanski and J. E. Bercaw, *Acc. Chem. Res.* **13**, 121 (1980).
3. Leading references: J. Schwartz, and J. A. Labinger, *Angew. Chem., Int. Ed. Engl.* **15**, 333 (1976); E. Negishi, *Aldrichimica Acta* **18**, 31 (1985).
4. G. Erker, *Acc. Chem. Res.* **17**, 103 (1984).
5. K. I. Gell, J. Schwartz, and G. M. Williams, *J. Am. Chem. Soc.* **104**, 1846 (1982); J. A. Marsella, K. G. Moloy, and K. G. Caulton, *J. Organomet. Chem.* **201**, 389 (1980).
6. D. B. Carr and J. Schwartz, *J. Am. Chem. Soc.* **101**, 3521 (1979).
7. S. L. Buchwald and R. B. Nielsen, *Chem. Rev.* **88**, 1047 (1988); H. Yasuda and A. Nakamura, *Angew. Chem., Int. Ed. Engl.* **26**, 723 (1987); S. L. Buchwald, R. T. Lum, and J. C. Dewan, *J. Am. Chem. Soc.* **108**, 7441 (1986); S. L. Buchwald, R. T. Lum, R. A. Fisher, and W. M. Davis, *J. Am. Chem. Soc.* **111**, 9113 (1989); D. R. Swanson, C. J. Rousset, E. Negishi, T. Takahashi, T. Seki, M. Saburi, and Y. Uchida, *J. Org. Chem.* **54**, 3521 (1989).
8. K. A. Wensley-Brown, S. L. Buchwald, L. Cannizzo, L. Clawson, S. Ho, D. Meinhardt, J. R. Stille, D. Straus, and R. H. Grubbs, *Pure Appl. Chem.* **55**, 1733 (1983).
9. R. H. Grubbs and W. Tumas, *Science* **243**, 907 (1989).
10. For previous reviews, see R. F. Jordan, *J. Chem. Ed.* **65**, 285 (1988); R. F. Jordan, P. K. Bradley, R. E. LaPointe, and D. F. Taylor, *New J. Chem.* **14**, 505 (1990).
11. For theoretical studies, see J. C. Green, M. L. H. Green, and C. K. Prout, *J. Chem. Soc., Chem. Commun.*, 421 (1972); J. L. Petersen, D. L. Lichtenberger, R. F. Fenske, and L. F. Dahl, *J. Am. Chem. Soc.* **97**, 6433 (1975); J. W. Lauher and R. Hoffmann, *J. Am. Chem. Soc.* **98**, 1729 (1976); P. Hofmann, P. Stauffert, and N. E. Schore, *Chem. Ber.* **115**, 2153 (1982); L. Zen and N. M. Kostic, *J. Organomet. Chem.* **335**, 395 (1987).
12. C. S. Christ, Jr., J. R. Eyler, and D. E. Richardson, *J. Am. Chem. Soc.* **112**, 596, 4778 (1990).

13. U. Thewalt and W. Lasser, *Z. Naturforsch., B: Anorg. Chem., Org. Chem.* **38**, 1501 (1983); B. Honold and U. Thewalt, *J. Organomet. Chem.* **316**, 291 (1986).
14. B. Longato, B. D. Martin, J. R. Norton, and O. P. Anderson, *Inorg. Chem.* **24**, 1389 (1985).
15. K. H. Theopold, *Acc. Chem. Res.* **23**, 263 (1990).
16. A. K. Zefirova and A. E. Shilov, *Dokl. Acad. Nauk SSSR* **136**, 599 (1961); F. S. Dyachkovskii, A. K. Shilova, and A. E. Shilov, *J. Polym. Sci., Part C: Polym. Symp.* **16**, 2333 (1967).
17. W. P. Long and D. S. Breslow, *J. Am. Chem. Soc.* **82**, 1953 (1960); D. S. Breslow and N. R. Newburg, *J. Am. Chem. Soc.* **81**, 81 (1959).
18. P. Cossee, *J. Catal.* **3**, 80 (1964); E. J. Arlman and P. Cossee, *J. Catal.* **3**, 99 (1964).
19. W. P. Long and D. S. Breslow, *Liebigs Ann. Chem.,* 463 (1975).
20. H. Sinn and W. Kaminsky, *Adv. Organomet. Chem.* **18**, 99 (1980).
21. J. Boor "Ziegler–Natta Catalysts and Polymerizations." Academic Press, New York, 1979.
22. P. L. Watson and G. W. Parshall, *Acc. Chem. Res.* **18**, 51 (1985); M. E. Thompson and J. E. Bercaw, *Pure Appl. Chem.* **56**, 1 (1985); G. Jeske, H. Lauke, H. Mauermann, P. N. Sweptson, H. Schumann, and T. J. Marks, *J. Am. Chem. Soc.* **107**, 8091 (1985).
23. P. L. Watson, *J. Am. Chem. Soc.* **104**, 337 (1982); P. L. Watson and D. C. Roe, *J. Am. Chem. Soc.* **104**, 6471 (1982); P. L. Watson, *J. Chem. Soc., Chem. Commun.,* 276 (1983).
24. J. J. Eisch, A. M. Piotrowski, S. K. Brownstein, E. J. Gabe, and F. L. Lee, *J. Am. Chem. Soc.* **107**, 7219 (1985).
25. J. A. Ewen, *J. Am. Chem. Soc.* **106**, 6355 (1984).
26. E. Giannetti, G. Martino Nicoletti, and R. Mazzocchi, *J. Polym. Sci., Polym. Chem. Ed.* **23**, 2117 (1985).
27. M.-Y. He, G. Xiong, P. J. Toscano, R. L. Burwell, and T. J. Marks, *J. Am. Chem. Soc.* **107**, 641 (1985); P. J. Toscano and T. J. Marks, *J. Am. Chem. Soc.* **107**, 653 (1985).
28. J. H. Teuben, in "Fundamental and Technological Aspects of Organo-f-Element Chemistry" (T. J. Marks and J. L. Fragala, eds.), p. 195. Reidel, Dordrecht, The Netherlands, 1985.
29. F. N. Tebbe and G. W. Parshall, *J. Am. Chem. Soc.* **93**, 3793 (1971); L. J. Guggenberger, P. Meakin, and F. N. Tebbe, *J. Am. Chem. Soc.* **96**, 5420 (1974); M. L. H. Green, and R. Mahtab, *J. Chem. Soc., Dalton Trans.,* 262 (1979); F. W. S. Benfield, N. J. Cooper, and M. L. H. Green, *J. Organomet. Chem.* **76**, 49 (1974).
30. W. Kaminsky, J. Kopf, H. Sinn, and H.-J. Vollmer, *Angew. Chem., Int. Ed. Engl.* **15**, 629 (1976); J. Kopf, H.-J. Vollmer, and W. Kaminsky, *Cryst. Struct. Commun.* **9**, 271 (1980).
31. R. F. Jordan, W. E. Dasher, and S. F. Echols, *J. Am. Chem. Soc.* **108**, 1718 (1986).
32. R. F. Jordan, C. S. Bajgur, R. Willett, and B. Scott, *J. Am. Chem. Soc.* **108**, 7410 (1986).
33. R. F. Jordan, R. E. LaPointe, C. S. Bajgur, and R. Willett, *J. Am. Chem. Soc.* **109**, 4111 (1987).
34. R. F. Jordan, C. Kenyon, S. L. Borkowsky, and R. E. LaPointe, unpublished results.
35. R. F. Jordan, R. E. LaPointe, N. Baenziger, and G. D. Hinch, *Organometallics* **9**, 1539 (1990).
36. R. F. Jordan, *J. Organomet. Chem.* **294**, 321 (1985).
37. D. M. Roddick, R. H. Heyn, and T. D. Tilley, *Organometallics* **8**, 324 (1989); B. D. Martin, S. A. Matchett, J. R. Norton, and O. P. Anderson, *J. Am. Chem. Soc.* **107**, 7952 (1985); A. R. Siedle, R. A. Newmark, W. B. Gleason, and W. M. Lamanna, *Organometallics* **9**, 1290 (1990).

38. S. L. Borkowsky, R. F. Jordan, and G. D. Hinch, *Organometallics* **10**, 1268 (1991).
39a. M. Bochmann, L. M. Wilson, M. B. Hursthouse, and R. L. Short, *Organometallics* **6**, 2556 (1987).
39b. K. Mashima, K. Jyodoi, A. Ohyoshi, and H. Takaya, *Organometallics* **6**, 885 (1987).
40. R. F. Jordan and S. L. Borkowsky, manuscript in preparation.
41. M. J. Burk, W. Tumas, M. D. Ward, and D. R. Wheeler, *J. Am. Chem. Soc.* **112**, 6133 (1990).
42. A. L. J. Beckwith and R. J. Leydon, *Aust. J. Chem.*, 3055 (1966); A. L. J. Beckwith and R. J. Leydon, *Tetrahedron* **20**, 791 (1964).
43. R. F. Jordan and S. F. Echols, *Inorg. Chem.* **26**, 383 (1987).
44. D. J. Crowther, R. F. Jordan, N. C. Baenziger, and A. Verma, *Organometallics* **9**, 2574 (1990).
45. G. D. Hinch, R. F. Jordan, and A. Verma, unpublished results.
46. M. Bochmann and L. M. Wilson, *J. Chem. Soc., Chem. Commun.*, 1610 (1986).
47. G. G. Hlatky, H. W. Turner, and R. R. Eckman, *J. Am. Chem. Soc.* **111**, 2728 (1989).
48. D. A. T. Young, G. R. Willey, M. F. Hawthorne, A. H. Reis, Jr., and M. R. Churchill, *J. Am. Chem. Soc.* **92**, 6663 (1970); D. M. Schubert, M. A. Bandman, W. S. Rees, Jr., C. B. Knobler, P. Lu, W. Nam, and M. F. Hawthorne, *Organometallics* **9**, 2046 (1990).
49. H. W. Turner and G. G. Hlatky, Eur. Patent Appl. 0,277,003 (1988); H. W. Turner, Eur. Patent Appl. 0,277,004 (1988).
50. Z. Lin, J. F. LeMarechal, M. Sabat, and T. J. Marks, *J. Am. Chem. Soc.* **109**, 4127 (1987).
51. M. Bochmann, A. J. Jagger, and J. C. Nicholls, *Angew. Chem., Int. Ed. Engl.* **29**, 780 (1990).
52. R. Taube and L. Krukowka, *J. Organomet. Chem.* **347**, C9 (1988).
53. J. J. W. Eshuis, Y. Y. Tan, and J. H. Teuben, *J. Molec. Catal.* **62**, 277 (1990).
54. J. L. Petersen, personal communication.
55. E. Negishi, *in* "Comprehensive Organometallic Chemistry" (G. Wilkinson, F. G. A. Stone, and E. W. Abel, eds.) Vol. 7, p. 326. Pergamon, New York, 1982; F. R. Jensen B. Rickborn, "Electrophilic Substitution of Organomercurials," McGraw-Hill, New York, 1968.
56. G. D. Hinch, R. F. Jordan, and N. C. Baenziger, unpublished results.
57. M. Bochmann, G. Karger, and A. J. Jagger, *J. Chem. Soc., Chem. Commun.*, 1038 (1990).
58. M. Bochmann, A. J. Jagger, L. M. Wilson, M. B. Hursthouse, and M. Motevalli, *Polyhedron* **8**, 1838 (1989).
59. T. Cuenca and P. Royo, *J. Organomet. Chem.* **294**, 321 (1985).
60. J. W. Park, L. M. Henling, W. P. Schaefer, and R. H. Grubbs, *Organometallics* **9**, 1650 (1990).
61. D. A. Straus, C. Zhang, and T. D. Tilley, *J. Organomet. Chem.* **369**, C13 (1989).
62. R. R. Schrock and P. R. Sharp, *J. Am. Chem. Soc.* **100**, 2389 (1978).
63. R. F. Jordan, C. S. Bajgur, W. E. Dasher, and A. L. Rheingold, *Organometallics* **6**, 1041 (1987).
64. R. F. Jordan, R. E. LaPointe, P. K. Bradley, and N. Baenziger, *Organometallics* **8**, 2892 (1989).
65. R. F. Jordan, P. K. Bradley, and Z. Guo, unpublished results.
66. W. J. Evans, R. Dominguez, and T. P. Hanusa, *Organometallics* **5**, 263 (1986).
67. I. Hyla-Krypsin, R. Gleiter, C. Kruger, R. Zwettler, and G. Erker, *Organometallics* **9**, 517 (1990); G. Erker, R. Zwettler, C. Kruger, I. Hyla-Krypsin, and R. Gleiter, *Organometallics* **9**, 524 (1990).

68. G. R. Davies, J. A. J. Jarvis, B. T. Kilbourn, and A. J. P. Pioli, *J. Chem. Soc., Chem. Commun.*, 677 (1971); G. R. Davies, J. A. J. Jarvis, and B. T. Kilbourn, *J. Chem. Soc., Chem. Commun.*, 1511 (1971); I. W. Bassi, G. Allegra, R. Scordamaglia, and G. J. Chioccola, *J. Am. Chem. Soc.* **93**, 3787 (1971).
69. E. A. Mintz, K. G. Moloy, T. J. Marks, and V. W. Day, *J. Am. Chem. Soc.* **104**, 4692 (1982); G. S. Girolami, G. Wilkinson, M. Thornton-Pett, and M. B. Hursthouse, *J. Chem. Soc., Dalton Trans.*, 2789 (1984); P. G. Edwards, R. A. Andersen, and A. Zalkin, *Organometallics* **3**, 293 (1984); S. L. Latesky, A. K. McMullen, G. P. Niccolai, I. P. Rothwell, and J. C. Huffman, *Organometallics* **4**, 902 (1985); J. Scholz, M. Schlegel, and K.-H. Theile, *Chem. Ber.* **120**, 1369 (1987).
70. The PES-derived ionization potentials are as follows: CH_3CN CN-π orbital 12.1. eV; THF O-centered b_1 π-donor orbital 9.6 eV. See D. C. Frost, F. G. Herring, C. A. McDowell, and I. A. Stenhouse, *Chem. Phys. Lett.* **4**, 533 (1970); S. Pignataro and G. Distefano, *Chem. Phys. Lett.* **26**, 356 (1974); H. Schmidt and A. Schweig, *Chem. Ber.* **107**, 725 (1974).
71. R. F. Jordan, P. K. Bradley, N. C. Baenziger, and R. E. LaPointe, *J. Am. Chem. Soc.* **112**, 1289 (1990).
72. M. Brookhart, M. L. H. Green, and L. Wong, *Prog. Inorg. Chem.* **36**, 1 (1988); R. H. Crabtree and D. G. Hamilton, *Adv. Organomet. Chem.* **28**, 299 (1988).
73. M. E. Thompson, S. M. Baxter, A. R. Bulls, B. J. Burger, M. C. Nolan, B. D. Santarsiero, W. P. Schaefer, and J. E. Bercaw, *J. Am. Chem. Soc.* **109**, 203 (1987).
74. Z. Dawoodi, M. L. H. Green, V. S. B. Mtetwa, K. Prout, A. J. Schultz, J. M. Williams, and T. F. Koetzle, *J. Chem. Soc., Dalton Trans.*, 1629 (1986).
75. R. B. Cracknell, A. G. Orpen, and J. L. Spencer, *J. Chem. Soc., Chem. Commun.*, 326 (1984).
76. N. Carr, B. J. Dunne, A. G. Orpen, and J. L. Spencer, *J. Chem. Soc., Chem. Commun.*, 926 (1988).
77. P. K. Bradley, Y. Wang, and R. F. Jordan, manuscript in preparation.
78. M. Brookhart, D. M. Lincoln, M. A. Bennett, and S. Pelling, *J. Am. Chem. Soc.* **112**, 2691 (1990).
79. Y. Wang, R. F. Jordan, P. K. Bradley, and N. C. Baenziger, *Polymer Preprints* in press (1991); Y. Wang, R. F. Jordan, P. K. Bradley and N. C. Baenziger, **32**, 457 (1991).
80. R. M. Waymouth, B. D. Santarsiero, R. J. Coots, M. J. Bronikowski, and R. H. Grubbs, *J. Am. Chem. Soc.* **108**, 1427 (1986), and references therein.
81. J. Kopf, W. Kaminsky, and H.-J. Vollmer, *Cryst. Struct. Commun.* **9**, 197 (1980).
82. R. M. Bullock, F. R. Lemke, and D. J. Szalda, *J. Am. Chem. Soc.* **112**, 3244 (1990).
83. R. F. Jordan, D. F. Taylor, and N. C. Baenziger, *Organometallics* **9**, 1546 (1990).
84. K. H. den Haan, J. L. de Boer, J. H. Teuben, A. L. Spek, B. Kojic-Prodic, G. R. Hays, and R. Huis, *Organometallics* **5**, 1726 (1986); J. W. Bruno, G. M. Smith, T. J. Marks, C. K. Fair, A. J. Schultz, and J. M. Williams, *J. Am. Chem. Soc.* **108**, 40 (1986).
85. N. Koga, and K. Morokuma, *J. Am. Chem. Soc.* **110**, 108 (1988).
86. K. H. den Haan, Y. Wielstra, and J. H. Teuben, *Organometallics* **6**, 2053 (1987).
87. W. J. Evans, D. K. Drummond, S. G. Bott, and J. L. Atwood, *Organometallics* **5**, 2389 (1986).
88. J. A. Marsella, C. J. Curtis, J. E. Bercaw, and K. G. Caulton, *J. Am. Chem. Soc.* **102**, 7244 (1980).
89. K. I. Gell and J. Schwartz, *J. Am. Chem. Soc.* **103**, 2687 (1981); J. Jeffery, M. F. Lappert, N. T. Luong-Thi, M. Webb, J. L. Atwood, and W. E. Hunter, *J. Chem. Soc., Dalton Trans.*, 1593 (1981).
90. P. Legzdins and D. T. Martin, *Organometallics* **2**, 1785 (1983).

91. P. Dreyfuss, "Poly(tetrahydrofuran)." Gordon and Breach; New York, 1981; J. S. Hrkach and K. Matyjaszewski, *Macromolecules* **23**, 4042 (1990).
92. Z. Guo, P. K. Bradley, and R. F. Jordan, unpublished results.
93. Ti(IV) 0.88, Zr(IV) 0.98 Å in eight-coordinate geometries; R. D. Shannon, *Acta Crystallogr., Sect. A* **32**, 751 (1976).
94. Y. Wang, R. F. Jordan, S. F. Echols, S. L. Borkowsky, and P. K. Bradley, *Organometallics* **10**, 1406 (1991).
95. G. Fachinetti, G. Fochi, and C. Floriani, *J. Chem. Soc., Dalton Trans.,* 1946 (1977).
96. T. Yoshida and E. Negishi, *J. Am. Chem. Soc.* **103**, 4985 (1981).
97. T. C. Flood and J. A. Statler, *Organometallics* **3**, 1795 (1984); H. C. Clark, C. R. Jablonski, and C. S. Wong, *Inorg. Chem.* **14**, 1332 (1975); J. Kress and J. A. Osborn, *J. Am. Chem. Soc.* **105**, 6346 (1983); J. H. Wengrovious and R. R. Schrock, *Organometallics* **1**, 148 (1982).
98. B. J. Burger, M. E. Thompson, D. Cotter, and J. E. Bercaw, *J. Am. Chem. Soc.* **112**, 1566 (1990).
99. M. Bochmann, L. M. Wilson, M. B. Hursthouse, and M. Motevalli, *Organometallics* **7**, 1148 (1988).
100. P. G. Gassman, D. W. Macomber, and J. W. Hershberger, *Organometallics* **2**, 1470 (1983); P. G. Gassman, W. H. Campbell, and D. W. Macomber, *Organometallics* **3**, 385 (1984).
101. P. G. Gassman and C. H. Winter, *J. Am. Chem. Soc.* **110**, 6130 (1988).
102. D. W. Hart and J. Schwartz, *J. Am. Chem. Soc.* **96**, 8115 (1974).
103. See U. Annby, J. Alvhall, S. Gronowitz, and A. Hallberg, *J. Organomet. Chem.* **377**, 75 (1989).
104. Y. Wang, R. F. Jordan, P. K. Bradley, and R. E. LaPointe, manuscript in preparation.
105. Z. Lin and T. J. Marks, *J. Am. Chem. Soc.* **109**, 7979 (1987); S. P. Nolan and T. J. Marks, *J. Am. Chem. Soc.* **111**, 8538 (1989); S. L. Latesky, A. K. McMullen, I. P. Rothwell, and J. C. Huffman, *J. Am. Chem. Soc.* **107**, 5981 (1985); I. P. Rothwell, *Polyhedron* **4**, 177 (1985).
106. W. J. Evans, L. R. Chamberlain, T. A. Ulibarri, and J. W. Ziller, *J. Am. Chem. Soc.* **110**, 6423 (1988).
107. For theoretical studies see: H. H. Brintzinger, *J. Organomet. Chem.* **171**, 337 (1979); M. L. Steigerwald and W. A. Goddard III *J. Am. Chem. Soc.* **106**, 308 (1984); H. Rabaa, J.-Y. Saillard, and R. Hoffmann, *J. Am. Chem. Soc.* **108**, 4327 (1986).
108. S. B. Jones and J. L. Petersen, *Inorg. Chem.* **20**, 2889 (1981).
109. R. F. Jordan and D. F. Taylor, *J. Am. Chem. Soc.* **111**, 778 (1989).
110. E. Klei and J. H. Teuben, *J. Organomet. Chem.* **214**, 53 (1981).
111. G. Erker, *J. Organomet. Chem.* **134**, 189 (1977); M. D. Rausch and E. A. Mintz, *J. Organomet. Chem.* **190**, 65 (1980); L. E. Schock, C. P. Brock, and T. J. Marks, *Organometallics* **6**, 232 (1987).
112. A. J. Cheney, B. E. Mann, B. L. Shaw, and R. M. Slade, *J. Chem. Soc. D,* 1176 (1970); S. L. Buchwald, R. T. Lum, R. A. Fisher, and W. M. Davis, *J. Am. Chem. Soc.* **111**, 9113 (1989).
113. R. F. Jordan and A. S. Guram, *Organometallics* **9**, 2116 (1990).
114. S. Sprouse, K. A. King, P. J. Spellane, and R. J. Watts, *J. Am. Chem. Soc.* **106**, 6647 (1984); G. R. Newkome, W. E. Puckett, V. K. Gupta, and G. E. Kiefer, *Chem. Rev.* **86**, 451 (1988).
115. F. Neve, M. Ghedini, A. Tiripicchio, and F. Ugozzoli, *Inorg. Chem.* **28**, 3084 (1989).
116. M. Lavin, E. M. Holt, and R. H. Crabtree, *Organometallics* **8**, 99 (1989).
117. M. I. Bruce, B. L. Goodall, and I. Matsuda, *Aust. J. Chem.* **28**, 1259 (1975).

118. J. Soto, M. L. Steigerwald, and R. H. Grubbs, *J. Am. Chem. Soc.* **104**, 4479 (1982); L. Clawson, J. Soto, S. L. Buchwald, M. L. Steigerwald, and R. H. Grubbs, *J. Am. Chem. Soc.* **107**, 3377 (1985).
119. R. Cramer, *Acc. Chem. Res.* **1**, 186 (1968).
120. L. Schock and T. J. Marks, *J. Am. Chem. Soc.* **110**, 7701 (1988).
121. W. Kaminsky, K. Kulper, and S. Niedoba, *Makromol. Chem., Macromol. Symp.* **3**, 377 (1986).
122. J. A. Ewen, *in* "Catalytic Polymerization of Olefins" (T. Keii and K. Soga, eds.) p. 217. Elsevier, New York, 1986.
123. A. Ahlers and W. Kaminsky, *Makromol. Chem., Rapid Commun.* **9**, 457 (1988).
124. J. A. Ewen, R. L. Jones, A. Razavi, and J. D. Ferrara, *J. Am. Chem. Soc.* **110**, 6255 (1988); J. A. Ewen, L. Haspeslagh, J. L. Atwood, and H. Zhang, *J. Am. Chem. Soc.* **109**, 6544 (1987).
125. W. Kaminsky, K. Kulper, H. H. Brintzinger, and F. R. P. Wild, *Angew. Chem., Int. Ed. Engl.* **24**, 507 (1985); K. Soga, T. Shiono, S. Takemura, and W. Kaminsky, *Makromol. Chem., Rapid Commun.* **8**, 305 (1987).
126. D. T. Mallin, M. D. Rausch, Y. Lin, S. Dong, and J. C. W. Chien, *J. Am. Chem. Soc.* **112**, 4953 (1990).
127. W. Kaminsky and H. Luker, *Makromol. Chem., Rapid Commun.* **5**, 228 (1984).
128. D. T. Mallin, M. D. Rausch, and J. C. W. Chien, *Polymer Bull.* **20**, 421 (1988); J. C. W. Chien and B.-P. Wang, *J. Polymer Sci. A., Polymer Chem.* **27**, 1539 (1989).
129. P. Pino, P. Coini, and J. Wei, *J. Am. Chem. Soc.* **109**, 6189 (1987).
130. R. M. Waymouth and P. Pino, *J. Am. Chem. Soc.* **112**, 4911 (1990).
131. L. Resconi and R. M. Waymouth, *J. Am. Chem. Soc.* **112**, 4953 (1990).
132. A. Zambelli, P. Longo, and A. Grassi, *Macromolecules* **22**, 2186 (1989).
133. J. A. Ewen, M. J. Elder, A. Razavi, and H. N. Cheng, Abstracts of the 198th National ACS Meeting, 1989, PETR 0017.
134. P. G. Gassman and M. R. Callstrom, *J. Am. Chem. Soc.* **109**, 7875 (1987).
135. D. Heddon and T. J. Marks, *J. Am. Chem. Soc.* **110**, 1647 (1988).
136. P. J. Toscano and T. J. Marks, *Langmuir* **2**, 820 (1986).
137. D. J. Crowther, N. C. Baenziger, and R. F. Jordan, *J. Am. Chem. Soc.* **113**, 1455 (1991).
138. S. L. Buchwald, B. T. Watson, M. W. Wannamaker, and J. C. Dewan, *J. Am. Chem. Soc.* **111**, 4486 (1989).
139. G. Erker and K. Kropp, *J. Am. Chem. Soc.* **101**, 3659 (1979).
140. A. S. Guram and R. F. Jordan, submitted for publication.
141. J. E. Nelson, J. E. Bercaw and J. A. Labinger, *Organometallics* **8**, 2484 (1989), and references therein.
142. A. S. Guram and R. F. Jordan, *Organometallics* **9**, 2190 (1990).
143. J. B. Lambert, *Tetrahedron* **46**, 2677 (1990).
144. G. Erker, P. Czisch, C. Kruger, and J. M. Wallis, *Organometallics* **4**, 2059 (1985).
145. J. W. F. L. Seetz, G. Schat, O. S. Akkerman, and F. Bickelhaupt, *Angew. Chem., Int. Ed. Engl.* **22**, 248 (1983).
146. K. M. Doxsee and J. B. Farahi, *J. Am. Chem. Soc.* **110**, 7239 (1988).
147. F. J. Berg and J. L. Petersen, *Organometallics* **8**, 2461 (1989).
148. A. S. Guram, R. F. Jordan, and D. F. Taylor, *J. Am. Chem. Soc.* **113**, 1833 (1991).
149. R. F. Jordan and D. F. Taylor, *Pet. Div. Preprints* **34**, 583 (1989).
150. C. K. Ghosh and W. A. G. Graham, *J. Am. Chem. Soc.* **111**, 375 (1989).
151. W. D. McGhee and R. G. Bergman, *J. Am. Chem. Soc.* **110**, 4246 (1988).
152. M. J. Burk and R. H. Crabtree, *J. Am. Chem. Soc.* **109**, 8025 (1987).
153. W. D. Jones and W. P. Kosar, *J. Am. Chem. Soc.* **108**, 5640 (1986).

Index

Cumulative List of Contributors

Abel, E. W., **5**, 1; **8**, 117
Aguilo, A., **5**, 321
Akkerman, O. S., **32**, 147
Albano, V. G., **14**, 285
Alper, H., **19**, 183
Anderson, G. K., **20**, 39
Angelici, R. J., **27**, 51
Aradi, A. A., **30**, 189
Armitage, D. A., **5**, 1
Armor, J. N., **19**, 1
Ash, C. E., **27**, 1
Ashe III, A. J., **30**, 77
Atwell, W. H., **4**, 1
Baines, K. M., **25**, 1
Barone, R., **26**, 165
Bassner, S. L., **28**, 1
Behrens, H., **18**, 1
Bennett, M. A., **4**, 353
Bickelhaupt, F., **32**, 147
Birmingham, J., **2**, 365
Blinka, T. A., **23**, 193
Bogdanović, B., **17**, 105
Bottomley, F., **28**, 339
Bradley, J. S., **22**, 1
Brinckman, F. E., **20**, 313
Brook, A. G., **7**, 95; **25**, 1
Brown, H. C., **11**, 1
Brown, T. L., **3**, 365
Bruce, M. I., **6**, 273, **10**, 273; **11**, 447, **12**,
    379; **22**, 59
Brunner, H., **18**, 151
Buhro, W. E., **27**, 311
Cais, M., **8**, 211
Calderon, N., **17**, 449
Callahan, K. P., **14**, 145
Cartledge, F. K., **4**, 1
Chalk, A. J., **6**, 119
Chanon, M., **26**, 165
Chatt, J., **12**, 1
Chini, P., **14**, 285
Chisholm, M. H., **26**, 97; **27**, 311
Chiusoli, G. P., **17**, 195
Chojnowski, J., **30**, 243
Churchill, M. R., **5**, 93
Coates, G. E., **9**, 195
Collman, J. P., **7**, 53

Compton, N. A., **31**, 91
Connelly, N. G., **23**, 1; **24**, 87
Connolly, J. W., **19**, 123
Corey, J. Y., **13**, 139
Corriu, R. J. P., **20**, 265
Courtney, A., **16**, 241
Coutts, R. S. P., **9**, 135
Coyle, T. D., **10**, 237
Crabtree, R. H., **28**, 299
Craig, P. J., **11**, 331
Csuk, R., **28**, 85
Cullen, W. R., **4**, 145
Cundy, C. S., **11**, 253
Curtis, M. D., **19**, 213
Darensbourg, D. J., **21**, 113, **22**, 129
Darensbourg, M. Y., **27**, 1
Davies, S. G., **30**, 1
Deacon, G. B., **25**, 237
de Boer, E., **2**, 115
Deeming, A. J., **26**, 1
Dessy, R. E., **4**, 267
Dickson, R. S., **12**, 323
Dixneuf, P. H., **29**, 163
Eisch, J. J., **16**, 67
Ellis, J. E., **31**, 1
Emerson, G. F., **1**, 1
Epstein, P. S., **19**, 213
Erker, G., **24**, 1
Ernst, C. R., **10**, 79
Errington, R. J., **31**, 91
Evans, J., **16**, 319
Evans, W. J., **24**, 131
Faller, J. W., **16**, 211
Farrugia, L. J., **31**, 301
Faulks, S. J., **25**, 237
Fehlner, T. P., **21**, 57; **30**, 189
Fessenden, J. S., **18**, 275
Fessenden, R. J., **18**, 275
Fischer, E. O., **14**, 1
Ford, P. C., **28**, 139
Forniés, J., **28**, 219
Forster, D., **17**, 255
Fraser, P. J., **12**, 323
Fritz, H. P., **1**, 239
Fürstner, A., **28**, 85
Furukawa, J., **12**, 83

Mrowca, J. J., **7**, 157
Müller, G., **24**, 1
Mynott, R., **19**, 257
Nagy, P. L. I., **2**, 325
Nakamura, A., **14**, 245
Nesmeyanov, A. N., **10**, 1
Neumann, W. P., **7**, 241
Norman, N. C.,, **31**, 91
Ofstead, E. A., **17**, 449
Ohst, H., **25**, 199
Okawara, R., **5**, 137; **14**, 187
Oliver, J. P., **8**, 167; **15**, 235; **16**, 111
Onak, T., **3**, 263
Oosthuizen, H. E., **22**, 209
Otsuka, S., **14**, 245
Pain, G. N., **25**, 237
Parshall, G. W., **7**, 157
Paul, I., **10**, 199
Peres, Y., **32**, 121
Petrosyan, W. S., **14**, 63
Pettit, R., **1**, 1
Pez, G. P., **19**, 1
Poland, J. S., **9**, 397
Poliakoff, M., **25**, 277
Popa, V., **15**, 113
Pourreau, D. B., **24**, 249
Powell, P., **26**, 125
Pratt, J. M., **11**, 331
Prokai, B., **5**, 225
Pruett, R. L., **17**, 1
Rao, G. S., **27**, 113
Raubenheimer, H. G., **32**, 1
Rausch, M. D., **21**, 1; **25**, 317
Reetz, M. T., **16**, 33
Reutov, O. A., **14**, 63
Rijkens, F., **3**, 397
Ritter, J. J., **10**, 237
Rochow, E. G., **9**, 1
Rokicki, A., **28**, 139
Roper, W. R., **7**, 53; **25**, 121
Roundhill, D. M., **13**, 273
Rubezhov, A. Z., **10**, 347
Salerno, G., **17**, 195
Salter, I. D., **29**, 249
Satgé, J., **21**, 241
Schade, C., **27**, 169
Schmidbaur, H., **9**, 259; **14**, 205
Schrauzer, G. N., **2**, 1
Schubert, U., **30**, 151

Schulz, D. N., **18**, 55
Schwebke, G. L., **1**, 89
Setzer, W. N., **24**, 353
Seyferth, D., **14**, 97
Shen, Yanchang (Shen, Y. C.), **20**, 115
Shriver, D. F., **23**, 219
Siebert, W., **18**, 301
Sikora, D. J., **25**, 317
Silverthorn, W. E., **13**, 47
Singleton, E., **22**, 209
Sinn, H., **18**, 99
Skinner, H. A., **2**, 49
Slocum, D. W., **10**, 79
Smallridge, A. J., **30**, 1
Smeets, W. J. J., **32**, 147
Smith, J. D., **13**, 453
Speier, J. L., **17**, 407
Spek, A. L., **32**, 147
Stafford, S. L., **3**, 1
Stańczyk, W., **30**, 243
Stone, F. G. A., **1**, 143; **31**, 53
Su, A. C. L., **17**, 269
Suslick, K. M., **25**, 73
Sutin, L., **28**, 339
Swincer, A. G., **22**, 59
Tamao, K., **6**, 19
Tate, D. P., **18**, 55
Taylor, E. C., **11**, 147
Templeton, J. L., **29**, 1
Thayer, J. S., **5**, 169; **13**, 1; **20**, 313
Theodosiou, I., **26**, 165
Timms, P. L., **15**, 53
Todd, L. J., **8**, 87
Touchard, D., **29**, 163
Treichel, P. M., **1**, 143; **11**, 21
Tsuji, J., **17**, 141
Tsutsui, M., **9**, 361; **16**, 241
Turney, T. W., **15**, 53
Tyfield, S. P., **8**, 117
Usón, R., **28**, 219
Vahrenkamp, H., **22**, 169
van der Kerk, G. J. M., **3**, 397
van Koten, G., **21**, 151
Veith, M., **31**, 269
Vezey, P. N., **15**, 189
von Ragué Schleyer, P., **24**, 353; **27**, 169
Vreize, K., **21**, 151
Wada, M., **5**, 137
Walton, D. R. M., **13**, 453